Probability

Probability

Davar Khoshnevisan

Graduate Studies
in Mathematics
Volume 80

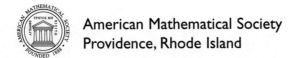

American Mathematical Society
Providence, Rhode Island

Editorial Board

David Cox
Walter Craig
Nikolai Ivanov
Steven G. Krantz
David Saltman (Chair)

2000 *Mathematics Subject Classification.* Primary 60–01;
Secondary 60–03, 28–01, 28–03.

For additional information and updates on this book, visit
www.ams.org/bookpages/gsm-80

Library of Congress Cataloging-in-Publication Data

Khoshnevisan, Davar.
    Probability / Davar Khoshnevisan.
       p. cm. — (Graduate studies in mathematics, ISSN 1065-7339 ; v. 80)
    Includes bibliographical references and index.
    ISBN-13: 978-0-8218-4215-7 (alk. paper)
    ISBN-10: 0-8218-4215-3 (alk. paper)
    1. Probabilities. I. Title.

QA273.K488   2007
519.2—dc22
                                                                                       2006052603

Copying and reprinting. Individual readers of this publication, and nonprofit libraries acting for them, are permitted to make fair use of the material, such as to copy a chapter for use in teaching or research. Permission is granted to quote brief passages from this publication in reviews, provided the customary acknowledgment of the source is given.

Republication, systematic copying, or multiple reproduction of any material in this publication is permitted only under license from the American Mathematical Society. Requests for such permission should be addressed to the Acquisitions Department, American Mathematical Society, 201 Charles Street, Providence, Rhode Island 02904-2294, USA. Requests can also be made by e-mail to reprint-permission@ams.org.

© 2007 by the American Mathematical Society. All rights reserved.
The American Mathematical Society retains all rights
except those granted to the United States Government.
Printed in the United States of America.

∞ The paper used in this book is acid-free and falls within the guidelines
established to ensure permanence and durability.
Visit the AMS home page at http://www.ams.org/

10 9 8 7 6 5 4 3 2 1   12 11 10 09 08 07

To my family

Contents

Preface	xi
General Notation	xv
Chapter 1. Classical Probability	1
§1. Discrete Probability	1
§2. Conditional Probability	4
§3. Independence	6
§4. Discrete Distributions	6
§5. Absolutely Continuous Distributions	10
§6. Expectation and Variance	12
Problems	13
Notes	15
Chapter 2. Bernoulli Trials	17
§1. The Classical Theorems	18
Problems	21
Notes	22
Chapter 3. Measure Theory	23
§1. Measure Spaces	23
§2. Lebesgue Measure	25
§3. Completion	28
§4. Proof of Carathéodory's Theorem	30
Problems	33

	Notes	34
Chapter 4.	Integration	35
§1.	Measurable Functions	35
§2.	The Abstract Integral	37
§3.	L^p-Spaces	39
§4.	Modes of Convergence	43
§5.	Limit Theorems	45
§6.	The Radon–Nikodým Theorem	47
	Problems	49
	Notes	52
Chapter 5.	Product Spaces	53
§1.	Finite Products	53
§2.	Infinite Products	58
§3.	Complement: Proof of Kolmogorov's Extension Theorem . . .	60
	Problems	62
	Notes	64
Chapter 6.	Independence	65
§1.	Random Variables and Distributions	65
§2.	Independent Random Variables	67
§3.	An Instructive Example	71
§4.	Khintchine's Weak Law of Large Numbers	71
§5.	Kolmogorov's Strong Law of Large Numbers	73
§6.	Applications	77
	Problems	84
	Notes	89
Chapter 7.	The Central Limit Theorem	91
§1.	Weak Convergence	91
§2.	Weak Convergence and Compact-Support Functions	94
§3.	Harmonic Analysis in Dimension One	96
§4.	The Plancherel Theorem	97
§5.	The 1-D Central Limit Theorem	100
§6.	Complements to the CLT	101
	Problems	111
	Notes	117

Contents

Chapter 8. Martingales 119
  §1. Conditional Expectations 119
  §2. Filtrations and Semi-Martingales 126
  §3. Stopping Times and Optional Stopping 129
  §4. Applications to Random Walks 131
  §5. Inequalities and Convergence 134
  §6. Further Applications 136
  Problems 151
  Notes 157

Chapter 9. Brownian Motion 159
  §1. Gaussian Processes 160
  §2. Wiener's Construction: Brownian Motion on $[0,1)$ 165
  §3. Nowhere-Differentiability 168
  §4. The Brownian Filtration and Stopping Times 170
  §5. The Strong Markov Property 173
  §6. The Reflection Principle 175
  Problems 176
  Notes 180

Chapter 10. Terminus: Stochastic Integration 181
  §1. The Indefinite Itô Integral 181
  §2. Continuous Martingales in $L^2(\mathrm{P})$ 187
  §3. The Definite Itô Integral 189
  §4. Quadratic Variation 192
  §5. Itô's Formula and Two Applications 193
  Problems 199
  Notes 201

Appendix 203
  §1. Hilbert Spaces 203
  §2. Fourier Series 205

Bibliography 209

Index 217

Preface

Say what you know, do what you must, come what may.

–Sofya Kovalevskaya

To us probability is the very guide of life.

–Bishop Joseph Butler

A few years ago the University of Utah switched from the quarter system to the semester system. This change gave the faculty a chance to re-evaluate their course offerings. As part of this re-evaluation process we decided to replace the usual year-long graduate course in probability theory with one that was a semester long. There was good reason to do so. The role of probability in mathematics, science, and engineering was, and still is, on the rise. There is increasing demand for a graduate course in probability. And yet the typical graduate student is not able to tackle a large number of year-long courses outside his or her own research area. Thus, we were presented with a non-trivial challenge: Can we offer a course that addresses the needs of our own, as well as other, graduate students, all within the temporal confines of one semester? I believe that the answer to the preceding question is "yes."

This book presents a cohesive graduate course in measure-theoretic probability that specifically has the one-semester student in mind. There is, in fact, ample material to cover an ordinary year-long course at a more leisurely pace. See, for example, the many sections that are entitled complements, and those that are called applications. However, the primary goals of this book are to maintain brevity and conciseness, and to introduce probability quickly and at a modestly deep level. I have used as my model a standard one-semester *undergraduate* course in probability. In that setting, the instructional issues are well understood, and most experts agree on what should be taught.

Giving a one-semester introduction to graduate probability necessarily involves making concessions. Mine form the contents of this book: No mention is made of Kolmogorov's theory of random series; Lévy's continuity theorem of characteristic functions is sadly omitted; Markov chains are not treated at all; and the construction of Brownian motion is Fourier-analytic rather than "probabilistic."

That is not to say that there is little coverage of the theory of stochastic processes. For example, included you will find an introduction to Itô's stochastic calculus and its connections to elliptic partial differential equations. This topic may seem ambitious, and it probably is for some readers. However, my experience in teaching this material has been that the reader who knows some measure theory can cover the book up to and including the last chapter in a single semester. Those who wish to learn measure theory from this book would probably aim to cover less stochastic processes.

Teaching Recommendations. In my own lectures I often begin with Chapter 2 and prove the De Moivre–Laplace central limit theorem in detail. Then, I spend two or three weeks going over basic results in analysis [Chapters 3 through 5]. Only a handful of the said results are actually proved. Without exception, one of them is Carathéodory's monotone class theorem (p. 30). The fundamental notion of independence is introduced, and a number of important examples are worked out. Among them are the weak and the strong laws of large numbers [Chapter 6], respectively due to A. Ya. Khintchine and A. N. Kolmogorov. Next follow elements of harmonic analysis and the central limit theorem [Chapter 7]. A majority of the subsequent lectures concern J. L. Doob's theory of martingales (1940) and its various applications [Chapter 8]. After martingales, there may be enough time left to introduce Brownian motion [Chapter 9], construct stochastic integrals, and deduce a striking computation, due to Chung (1947), of the distribution of the exit time from $[-1,1]$ of Brownian motion (p. 197). If at all possible, the latter topic should not be missed.

My personal teaching philosophy is to showcase the big ideas of probability by deriving very few, but central, theorems. François Marie Arouet [Voltaire] once wrote that "the art of being a bore is to tell everything." Viewed in this light, a chief aim of this book is to not bore.

I would like to leave the reader with one piece of advice on how to best use this book. Read it thoughtfully, and with pen and paper.

Acknowledgements: This book is based on the combined contents of several of my previous graduate courses in probability theory. Many of these were given at the University of Utah during the past decade or so. Also, I have used parts of some lectures that I gave during the formative stages of my career at MIT and the University of Washington. I wish to thank all three institutions for their hospitality and support, and the National Science Foundation, the National Security Agency, and the North-Atlantic Treaty Organization for their financial support of my research over the years.

All scholars know about the merits of library research. Nevertheless, the role of this lore is underplayed in some academic texts. I, for one, found the following to be enlightening: Billingsley (1995), Breiman (1992), Chow and Teicher (1997), Chung (1974), Cramér (1936), Dudley (2002), Durrett (1996), Fristedt and Gray (1997), Gnedenko (1967), Karlin and Taylor (1975, 1981), Kolmogorov (1933, 1950), Krickeberg (1963, 1965), Lange (2003), Pollard (2002), Resnick (1999), Stroock (1993), Varadhan (2001), Williams (1991), and Woodroofe (1975). Without doubt, there are other excellent references. The student is encouraged to consult other resources in addition to the present text. He or she would do well to remember that it may be nice to know facts, but it is vitally important to have a perspective.

I am grateful to the following for their various contributions to the development of this book: Nelson Beebe, Robert Brooks, Pieter Bowman, Rex Butler, Edward Dunne, Stewart Ethier, Victor Gabrenas, Frank Gao, Ana Meda Guardiola, Jan Hannig, Henryk Hecht, Lajos Horváth, Zsuzsanna Horváth, Adam Keenan, Karim Khader, Remigijus Leipus, An Le, David Levin, Michael Purcell, Pejman Mahboubi, Pedro Méndez, Jim Pitman, Natalya Pluzhnikov, Matthew Reimherr, Shang-Yuan Shiu, Josef Steinebach, James Turner, John Walsh, Jun Zhang, and Liang Zhang. Many of these people have helped find typographical errors, and even a few serious mistakes. All errors that remain are of course mine.

My family has been a stalwart pillar of patience. Their kindness and love were indispensable in completing this project. I thank them deeply.

And last but certainly not the least, my eternal gratitude is extended to my teachers, past and present, for introducing me to the joys of mathematics. I hope only that some of their ingenuity and spirit persists throughout these pages.

Davar Khoshnevisan
Salt Lake City, January 2007

General Notation

Here we set forth some of the general notation that is consistently used in the entire book. This is standard mathematical notation, and we may refer to it without further mention.

Logic and Set Theory. Throughout, $\cup$ and $\cap$ respectively denote union and intersection, and $\subseteq$ the subset relation. Occasionally we may write "$a := b$," where a and b could be sets, numbers, logical expressions, functions, etc. Depending on the context, this may mean either "define a to be b," or "define b to be a." We will not make a distinction between the two.

If $A, B \subseteq \mathbf{X}$, then A^c denotes the complement of A [in $\mathbf{X}$]. The dependence on $\mathbf{X}$ is usually suppressed as it is clear from the context. Let $A \setminus B := A \cap B^c$, and $A \triangle B := (A \setminus B) \cup (B \setminus A)$. The latter is called the *set difference* of A and B.

A set is *denumerable* if it is either countable or finite.

We frequently write "iff" as short-hand for "if and only if."

Finally, "$\forall$" and "$\exists$" respectively stand for "for all" and "there exists."

Euclidean Spaces. Throughout, $\mathbf{R} = (-\infty, \infty)$ denotes the real line, $\mathbf{Z} = \{0, \pm 1, \pm 2, \ldots\}$ the integers, $\mathbf{N} = \{1, 2, \ldots\}$ the natural numbers, and $\mathbf{Q}$ denotes the rationals. If $\mathbf{X}$ designates any one of these, then $\mathbf{X}_+$ denotes the non-negative elements of $\mathbf{X}$, and $\mathbf{X}_-$ denotes the non-positive elements. If $k \in \mathbf{N}$ then $\mathbf{X}^k$ denotes the collection of all k-tuples $(x_1, \ldots, x_k)$ such that $x_1, \ldots, x_k$ are in $\mathbf{X}$. For instance, $\mathbf{R}_+^k$ denotes the collection of all k-vectors that are non-negative coordinatewise. The complex plane is denoted by $\mathbf{C}$.

If $x, y \in \mathbf{R}$, then we write $x \wedge y = \min(x, y)$ for the minimum and $x \vee y = \max(x, y)$ for the maximum. Similarly, sup and inf respectively refer to supremum and infimum operations.

Functions. If $\mathbf{X}$ and $\mathbf{Y}$ are two sets, then "$f : \mathbf{X} \to \mathbf{Y}$" stands for "$f$ maps $\mathbf{X}$ into $\mathbf{Y}$," and "$x \mapsto f(x)$" refers to the map from x to $f(x)$. If $f : \mathbf{X} \to \mathbf{Y}$ and $A \subseteq \mathbf{Y}$, then $f^{-1}(A) = \{x \in \mathbf{X} : f(x) \in A\}$. This is the *inverse image* of A.

The Big-O/Little-o Notation. Suppose $a_1, a_2, \ldots, b_1, b_2, \ldots \in \mathbf{R}$. We say that $a_n \sim b_n$ [as $n \to \infty$] when $\lim_{n \to \infty}(a_n/b_n) = 1$. When the b_i's are also non-negative, "$\limsup_{n \to \infty} |a_n|/b_n < \infty$" is often written as "$a_n = O(b_n)$," and "$\lim_{n \to \infty} |a_n|/b_n = 0$" as "$a_n = o(b_n)$." Note that $a_n = O(b_n)$ iff there exists a constant C such that $|a_n| \leq C b_n$ for all $n \geq 1$.

The big-O/little-o notation is also applicable to functions: "$f(x) \sim g(x)$ as $x \to a$" means "$\lim_{x \to a}(f(x)/g(x)) = 1$"; and when $g \geq 0$ we may write "$f(x) = O(g(x))$ as $x \to a$" for "$\limsup_{x \to a} |f(x)| = O(g(x))$," and "$f(x) = o(g(x))$ as $x \to a$" in place of "$\lim_{x \to a}(f(x)/g(x)) = 0$."

Chapter 1

Classical Probability

Probability does not exist.

–Bruno de Finetti

How dare we speak of the laws of chance? Is not chance the antithesis of all law?

–Bertrand Russell

The original development of probability theory took place during the seventeenth through nineteenth centuries. In those times the subject was mainly concerned with games of chance. Since then an increasing number of scientific applications, many in mathematics itself, have spurred the development of probability in other directions. Nonetheless, classical probability remains a most natural place to start the subject. In this way we can work non-axiomatically and loosely in order to grasp a number of useful ideas without having to develop abstract machinery. That will be covered in later chapters.

1. Discrete Probability

Consider a game that can lead to a fixed denumerable (i.e., at most countable) set $\{\omega_i\}_{i=1}^{\infty}$ of possible outcomes. Suppose in addition that the outcome ω_j occurs with probability p_j $(j = 1, 2, \ldots)$, where we agree once and for all that probabilities are real numbers in $[0, 1]$, and that the ω_j's really are the only possible outcomes; i.e., that $\sum_{i=1}^{\infty} p_j = 1$.

Definition 1.1. We call $\Omega := \{\omega_j\}_{j=1}^\infty$ the *sample space* of our experiment. Subsets of Ω are called *events*, and the *probability* of an event A is

$$\text{(1.1)} \qquad \text{P}(A) := \sum_{j \geq 1:\ \omega_j \in A} p_j.$$

One can check directly that P has the following properties:

Proposition 1.2. $\text{P}(\Omega) = 1$, $\text{P}(\varnothing) = 0$, *and*

(1) $\text{P}(A_1 \cup A_2) = \text{P}(A_1) + \text{P}(A_2) - \text{P}(A_1 \cap A_2)$.
(2) $\text{P}(A^c) = 1 - \text{P}(A)$.
(3) *If the A_i's are pairwise disjoint, then* $\text{P}(\cup_{i=1}^\infty A_i) = \sum_{i=1}^\infty \text{P}(A_i)$.
(4) *If $A_1 \subseteq A_2$, then* $\text{P}(A_1) \leq \text{P}(A_2)$.
(5) (Boole's Inequality) $\text{P}(\cup_{i=1}^\infty A_i) \leq \sum_{i=1}^\infty \text{P}(A_i)$.
(6) (Inclusion-Exclusion Formula)

$$\text{P}\left(\bigcup_{i=1}^n A_i\right) = \sum_{i=1}^n \text{P}(A_i) - \sum\sum_{1 \leq i_1 < i_2 \leq n} \text{P}\left(A_{i_1} \cap A_{i_2}\right)$$
$$+ \sum\sum\sum_{1 \leq i_1 < i_2 < i_3 \leq n} \text{P}\left(A_{i_1} \cap A_{i_2} \cap A_{i_3}\right) - \cdots$$
$$+ \cdots + (-1)^{n-1} \text{P}\left(A_1 \cap \cdots \cap A_n\right).$$

The case where $\Omega = \{\omega_i\}_{i=1}^N$ and $p_1 = \cdots = p_N = 1/N$ deserves special mention. In this case, the ω_j's are *equally likely* to occur, and

$$\text{(1.2)} \qquad \text{P}(A) = \frac{|A|}{|\Omega|} = \frac{|A|}{N},$$

for all events A, where $|\cdots|$ denotes cardinality.

Example 1.3. If we are tossing a coin (the experiment), then we can think of Ω as $\{H, T\}$, where H stands for "heads" and T for "tails." If the coin is fair, then p_1 and p_2 are both equal to $1/2$, and the preceding definition states that: (a) $\text{P}(\varnothing) = 0$; (b) $\text{P}(\{H\}) = 1/2$; (c) $\text{P}(\{T\}) = 1/2$; and (d) $\text{P}(\Omega) = \text{P}(\{H, T\}) = (1/2) + (1/2) = 1$.

Example 1.4. Consider the set $\Omega := \{(1,1), (1,2), \ldots, (6,6)\}$; this could denote the sample space of a roll of a pair of fair dice. For example, (i, j) may correspond to i dots on the first die and j on the second. Then, $\omega_1 = (1,1), \omega_2 = (1,2), \ldots$ are equally likely, and each has probability $1/36$ because $|\Omega| = 36$. The collection $A := \{(1,1), (2,2), (3,3), (4,4), (5,5), (6,6)\}$ is the event that we roll doubles; it has probability $\text{P}(A) = 1/6$.

Example 1.5 (The Boltzmann Statistic). Suppose we wanted to put n distinct balls in N urns, where $N > n$. Let us assume that the balls are thrown at random into the N urns, so that all N^n possible outcomes of this game are equally likely.

To understand this model better, consider the case $N = 4$ and $n = 3$. We then have three balls x, y, and z, and four urns. Let $A_1 := \{xy|\ |z|\ \}$. This notation designates the event that balls x and y end up in the first urn, in this order, and ball z lands in the third urn; the three vertical lines represent the "dividers" that separate the different urns. Note that A_1 represents a different arrangement from $A_2 := \{yx|\ |z|\ \}$, but the two events have the same chance of occurring.

Going back to the general case, let A_* denote the event that the first n urns end up with exactly one ball in each (and, therefore, the remaining $N - n$ end up with no balls in them). Then the cardinality of A_* is the same as the number of different ways we can order the n balls; i.e., $n!$. The latter count is the so-called *Boltzmann statistic*, and we have found that

$$(1.3) \qquad \mathrm{P}(A_*) = \frac{n!}{N^n}.$$

Example 1.6 (The Bose–Einstein Statistic). Here, the setting is very much the same as it was in Example 1.5, but now the balls are indistinguishable from one another. As a result, the events A_1 and A_2 of Example 1.5 are now one and the same. Thus, $|A_*| = 1$ and $\mathrm{P}(A_*) = 1/|\Omega|$, and it remains to compute $|\Omega|$.

Observe that $|\Omega|$ denotes the number of ways that we can mix up the balls together with the urn-dividers and then lay them down all in a straight line. Because there are $(N - 1)$ dividers and n balls, $|\Omega| = \binom{N+n-1}{n}$. This number is called the *Bose–Einstein* statistic, and we have found that

$$(1.4) \qquad \mathrm{P}(A_*) = \frac{1}{\binom{N+n-1}{n}}.$$

Example 1.7 (The Fermi–Dirac Statistic). We continue to work within the general framework of Examples 1.5 and 1.6. But now, we assume that the balls are all indistinguishable and each urn contains at most one ball. Evidently, $|\Omega|$ is equal to $\binom{N}{n}$, the *Fermi–Dirac* statistic, and

$$(1.5) \qquad \mathrm{P}(A_*) = \frac{1}{\binom{N}{n}}.$$

Example 1.8 (Random Matchings). Consider n people each of whom drives a different car. If we assign a car to each person at random, then what is the probability that at least one person gets his or her own car?

Equivalently, consider all permutations $\pi : \{1, \ldots, n\} \to \{1, \ldots, n\}$; there are $n!$ of them. If we select a permutation at random, all being equally likely, then what is the probability that π has a fixed point? That is, what are the chances that there exists an i such that $\pi(i) = i$?

If $E_i := \{\pi(i) = i\}$ denotes the event that i is a fixed point, then

$$(1.6) \qquad \mathrm{P}(E_{i_1} \cap \cdots \cap E_{i_k}) = \frac{(n-k)!}{n!},$$

whenever $1 \leq i_1 < \cdots < i_k \leq n$. Therefore,

$$(1.7) \qquad \sum_{i_1 < \cdots < i_k} \mathrm{P}(E_{i_1} \cap \cdots \cap E_{i_k}) = \binom{n}{k}\frac{(n-k)!}{n!} = \frac{1}{k!}.$$

Thanks to the inclusion-exclusion principle,

$$(1.8) \qquad \mathrm{P}\left(\bigcup_{i=1}^{n} E_i\right) = 1 - \frac{1}{2!} + \frac{1}{3!} + \cdots + (-1)^{n+1}\frac{1}{n!}.$$

As $n \to \infty$, this converges to $\sum_{k=1}^{\infty}(-1)^{k+1}/k! = 1 - e^{-1}$. We can therefore conclude that, in a large random permutation, chances are nearly $1/e$ that no fixed points arise.

2. Conditional Probability

A certain population is comprised of a hundred adults, twenty of whom are women. Fifteen of the women and twenty of the men are employed. A statistician has selected a person at random from this population, and tells us that the person so sampled is employed. Given this information, what are the chances that a woman was sampled?

If we were not privy to the information given by the statistician, then the sample space would have $N = 100$ equally likely elements; one for each adult in the population. Therefore, $\mathrm{P}(W) = 20/100 = 0.2$, where W denotes the event that a woman is sampled.

On the other hand, once the statistician tells us that the sampled person is employed, this knowledge changes our original sample space to a new one that contains the employed people only. The new sample space is comprised of 35 people (15 women and 20 men), all of whom are employed and equally likely to be chosen. Given the knowledge imparted to us by the statistician, the chances are $15/35 \approx 0.4285$ that the sampled person is a woman.

In general, it is easier to not change the sample space as new information surfaces, but instead use the following conditional probability formula on the original sample space:

Definition 1.9. For any two events A and B such that $\mathrm{P}(B) > 0$, the *conditional probability of A given B* is defined as

$$(1.9) \qquad \mathrm{P}(A \mid B) := \frac{\mathrm{P}(A \cap B)}{\mathrm{P}(B)}.$$

2. Conditional Probability

If we apply (1.9) with $A = \{\text{woman}\}$ and $B = \{\text{employed}\}$, then we find that $P(A \mid B) = 15/35 \approx 0.4285$. This agrees with our earlier bare-hands calculation.

In general, $A \mapsto P(A \mid B)$ is a probability, and has the properties of Proposition 1.2. In other words, we have the following.

Proposition 1.10. *For any event B that has positive probability, $P(B \mid B) = 1$ and $P(\varnothing \mid B) = 0$. In addition, for all events $A, A_1, A_2, \ldots$:*

(1) $P(A_1 \cup A_2 \mid B) = P(A_1 \mid B) + P(A_2 \mid B) - P(A_1 \cap A_2 \mid B)$.
(2) $P(A^c \mid B) = 1 - P(A \mid B)$.
(3) *If the $A_i \cap B$'s are disjoint, then* $P(\cup_{i=1}^\infty A_i \mid B) = \sum_{i=1}^\infty P(A_i \mid B)$.
(4) *If $A_1 \subseteq A_2$, then* $P(A_1 \mid B) \leq P(A_2 \mid B)$.
(5) *(Boole's Inequality)* $P(\cup_{i=1}^\infty A_i \mid B) \leq \sum_{i=1}^\infty P(A_i \mid B)$.
(6) *(Inclusion-Exclusion Formula)*

$$P\left(\bigcup_{i=1}^n A_i \,\Big|\, B\right) = \sum_{i=1}^n P(A_i \mid B) - \sum\sum_{1 \leq i_1 < i_2 \leq n} P(A_{i_1} \cap A_{i_2} \mid B)$$
$$+ \sum\sum\sum_{1 \leq i_1 < i_2 < i_3 \leq n} P(A_{i_1} \cap A_{i_2} \cap A_{i_3} \mid B) - \cdots$$
$$+ \cdots + (-1)^{n-1} P(A_1 \cap \cdots \cap A_n \mid B).$$

The following is a particularly useful corollary.

The Law of Total Probability. *Consider events $B, A_1, A_2, \ldots, A_n$ where the A_i's are disjoint, $\cup_{i=1}^n A_i = \Omega$, and $P(A_j) > 0$ for all $j = 1, \ldots, n$. Then,*

$$(1.10) \qquad P(B) = \sum_{j=1}^n P(B \mid A_j) P(A_j).$$

The law of total probability provides us with a method for computing weighted averages, viz.,

Example 1.11. *Dichromatism* is a form of color blindness. It is a genetic disorder that is caused by a defect in the X-chromosome. Roughly, about 5% of all men and 1% of all women suffer from dichromatism. Suppose that 60% of a certain population are women. If we sample an individual at random from this population, then what is the probability that this person has dichromatism?

Define the events $D = \{\text{dichromatism}\}$ and $F = \{\text{female}\}$ in the obvious way. Because $P(D \mid F^c) \approx 0.05$, $P(D \mid F) \approx 0.01$, and $P(F) \approx 0.6$,

$$(1.11) \qquad P(D) = P(D \mid F) P(F) + P(D \mid F^c) P(F^c) \approx 0.026.$$

That is, about 2.6% of this population suffers from dichromatism.

3. Independence

Events A and B are independent if our assessment of the likelihood of A is not affected by whether or not we know B has occurred. This can be formalized as follows.

Definition 1.12. Two events A and B are *independent* if $P(A \mid B) = P(A)$. This is equivalent to the so-called *product rule* of probabilities: $P(A \cap B) = P(A)P(B)$. More generally, events $A_1, \ldots, A_m$ are *independent* if

$$(1.12) \qquad P(A_{j_1} \cap \cdots \cap A_{j_k}) = \prod_{\nu=1}^{k} P(A_{j_\nu}),$$

for all distinct $j_1, \ldots, j_k \in \{1, \ldots, m\}$.

For instance, three events A, B, C are independent if and only if all of the following conditions are met: (i) $P(A \cap B \cap C) = P(A)P(B)P(C)$; (ii) $P(A \cap B) = P(A)P(B)$; (iii) $P(A \cap C) = P(A)P(C)$; and (iv) $P(B \cap C) = P(B)P(C)$. This cannot be relaxed (Problem 1.3).

Example 1.13. A fair coin is tossed n times at random. Let H_j denote the event that the jth toss yields heads, and $T_j = H_j^c$ the event that the said toss yields tails. We assume that all of the possible outcomes of this experiment are equally likely. It follows that

$$(1.13) \qquad P(R_1 \cap \cdots \cap R_n) = \frac{1}{2^n},$$

where R_i can stand for either H_i or T_i. Therefore, the H_i's, and hence the T_i's, are independent. Moreover, $P(H_i) = P(T_i) = 1/2$ ($i = 1, \ldots, n$). Conversely, if the H_i's are independent with $P(H_i) = 1/2$ for all i, then all possible outcomes of this experiment are equally likely.

Example 1.14. An urn contains 2 black and 3 white balls. Two balls are drawn at random from the urn; all such pairs are equally likely. If B_j denotes the event that the jth draw yields a black ball, then B_1 and B_2 are *not* independent. This assertion follows from the following:

$$(1.14) \qquad P(B_1 \cap B_2) = \frac{\binom{3}{0}\binom{2}{2}}{\binom{5}{2}} = 0.1,$$

$$P(B_1) = P(B_2) = \frac{\binom{3}{0}\binom{2}{2}}{\binom{5}{2}} + \frac{\binom{2}{1}\binom{3}{1}}{\binom{5}{2}} = 0.7.$$

4. Discrete Distributions

What is a random variable? Usually, elementary probability texts define a random variable as a numerical outcome of a random experiment. Although

4. Discrete Distributions

this definition is attractive, it is flawed. For instance, suppose we performed a random experiment three times and it yielded the numbers $0.8322, -1.253$, and 0.003. Certainly, there is nothing random about any one of these real numbers! So what is the random variable that is produced?

To answer this question, one has to recognize the fact that what is random here is the procedure that led to the said numbers, and not so much the observed numbers themselves. Procedures are, of course, encoded by functions. Thus, we are led to the following.

Definition 1.15. A *random variable* X is a function from Ω into $\mathbf{R}$. For any set $E \subseteq \mathbf{R}$, the event $\{X \in E\}$ is defined as $\{\omega \in \Omega : X(\omega) \in E\}$, and we write $\mathrm{P}\{X \in E\}$ in favor of $\mathrm{P}(\{X \in E\})$.

Because Ω is a denumerable set, all random variables on Ω are *discrete* in the sense that we can find a denumerable set of points $\{x_i\}_{i=1}^{\infty}$ such that $p_j = \mathrm{P}\{X = x_j\}$ is strictly positive, and $\sum_{j=1}^{\infty} p_j = 1$.

Example 1.16. Suppose we are interested in studying a random variable that takes the values $1, \ldots, N$ with probabilities $1/N$ each. Intuitively speaking, this makes perfect sense even if we do not have a ready sample space in mind. How then does one construct such a random variable? Here is one possibility among many: Let $\Omega = \{1, \ldots, N\}$, and suppose the elements of Ω are equally likely. If we define $X(\omega) = \omega$ for all $\omega \in \Omega$, then

$$(1.15) \qquad \mathrm{P}\{X = j\} = \mathrm{P}(\{j\}) = \frac{1}{N} \qquad \forall j = 1, \ldots, N,$$

as needed.

Definition 1.17. The *distribution* of a random variable X is a collection of distinct real numbers $\{x_j\}_{j=1}^{\infty}$, and $p_1, p_2, \ldots \in [0, 1]$, such that: (i) $\sum_{j=1}^{\infty} p_j = 1$; and (ii) $\mathrm{P}\{X = x_j\} = p_j$ for all $j \geq 1$. If $p_j > 0$, then the corresponding x_j is a *possible value* of X.

Definition 1.18. The *mass function* of a discrete random variable X is the function $p : \mathbf{R} \to [0, 1]$ defined by

$$(1.16) \qquad p(x) = \mathrm{P}\{X = x\} \qquad \forall x \in \mathbf{R}.$$

The mass function is a convenient way to code the entire distribution of X. If the distinct possible values of X are $\{x_i\}_{i=1}^{\infty}$ and the corresponding probabilities are $\{p_i\}_{i=1}^{\infty}$, then $p(x_i) = p_i$, whereas $p(x) = 0$ for other values of x.

One can always construct a random variable with a pre-described distribution (Problem 1.4). However, when we discuss a random variable X, we rarely need to specify the underlying probability space. Usually, all that matters is the distribution of X. Next we present three examples to illustrate this remark. The exercises contain a few more examples.

4.1. The Binomial Distribution.
Consider a random experiment that has two possible outcomes only: *Success* versus *failure* (or heads vs. tails, female vs. male, smoker vs. non-smoker, etc.). The probability of success is denoted by $p \in (0,1)$.

We perform this experiment n times independently. The latter means that the outcomes of all the trials are independent from one another. Such experiments are called *Bernoulli trials* with *parameter* p. Let X denote the total number of successes. Then we say that X has the *binomial distribution* with parameters n and p, and X is sometimes also written as $\text{Bin}(n,p)$.

For instance, suppose the proportion of women in a certain population is p. We sample n people at random and with replacement. Then, the total number of women in the sample has the binomial distribution with parameters n and p.

Quite generally, if X has the binomial distribution with parameters n and p, then the possible values of X are $0, \ldots, n$. It remains to find the mass function; i.e., $p(k) = \text{P}\{X = k\}$ for $k = 0, \ldots, n$. This is the probability of getting exactly k successes, and also $n - k$ failures.

Consider first the problem of finding the probability that the first k trials lead to successes, and the remaining $n-k$ trials to failures. By independence, this probability is $p^k(1-p)^{n-k}$. Now if we fix any k of the n trials, then the probability that those k succeed and all remaining trials fail is the same; i.e., $p^k(1-p)^{n-k}$. The union of these events is the event $\{X = k\}$. Because the latter is a disjoint union, $p(k) = Np^k(1-p)^{n-k}$ where N is the number of ways to choose k spots for the successes among all n trials. Therefore,

$$(1.17) \qquad p(k) = \binom{n}{k} p^k (1-p)^{n-k} \qquad \forall k = 0, \ldots, n.$$

This notation tacitly implies that $\text{P}\{X = k\} = 0$ for all values of k other than $k = 0, \ldots, n$. From now on, we adopt this way of writing a probability mass function.

Binomials have the following interpretation: Let X denote the total number of successes in n independent Bernoulli trials with success probability $p \in [0,1]$ each. Let $I_j := 1$ if the jth trial leads to a success, and $I_j := 0$ otherwise ($1 \leq j \leq n$). Then, $X = I_1 + \cdots + I_n$. Among other things this means that $\text{Bin}(n,p)$ is a sum of n independent, identically distributed random variables.

4.2. The Geometric Distribution.
Consider independent Bernoulli trials, where each trial has the same success probability $p \in (0,1)$. We perform these trials until the first success appears. Let X denote the number of trials required. The distribution of X is the so-called *geometric distribution* with

parameter p, and X is sometimes written as Geom(p). Its mass function is

$$(1.18) \qquad p(x) = (1-p)^{x-1}p \qquad \forall x = 1, 2, \ldots.$$

4.3. The Poisson Distribution. Suppose $\lambda > 0$ is fixed. Then a random variable X is said to have the *Poisson distribution* with parameter λ if the mass function of X is

$$(1.19) \qquad p(x) = \frac{e^{-\lambda}\lambda^x}{x!} \qquad \forall x = 0, 1, 2, \ldots.$$

Sometimes we write Poiss(λ) for such X.

The Poisson distribution plays a natural role in a number of approximation theorems. Next is an instance in random permutations. We will see another example in Chapter 2; see the law of rare events (p. 19).

Example 1.19 (Example 1.8, Continued). Let X denote the total number of fixed points in a random permutation π of $\{1, \ldots, n\}$. In Example 1.8 we found that if n is large, then $\mathrm{P}\{X = 0\} \approx e^{-1}$. Now we go one step further and compute the mass function of X. Define $F_\varnothing = G_\varnothing = \Omega$, and for all non-empty $A \subseteq \{0, \ldots, n\}$ let

$$(1.20) \qquad F_A := \bigcap_{j \in A}\{\pi(j) = j\} \quad \text{and} \quad G_A := \bigcap_{j \notin A}\{\pi(j) \neq j\}.$$

In words, F_A denotes the event that all elements of A are fixed points, and G_A the event that there are no fixed points in A^c.

If k is an integer between 0 and n, then we can write

$$(1.21) \qquad \mathrm{P}\{X = k\} = \sum_{\substack{A \subseteq \{1, \ldots, n\}: \\ |A| = k}} \mathrm{P}(F_A)\{1 - \mathrm{P}(G_A^c \mid F_A)\}.$$

We observed in Example 1.8 that

$$(1.22) \qquad \mathrm{P}(F_A) = \frac{(n - |A|)!}{n!}.$$

To find the remaining conditional probability we first note that the conditioning reduces the sample space to all permutations for which $\pi(i) = i$ for all $i \in A$. Consequently, $\mathrm{P}(G_A^c \mid F_A)$ is the probability that there is at least one fixed point in a random permutation of $n - |A|$ distinct objects [one for each $j \notin A$]. Equation (1.8) tells us that

$$(1.23) \qquad \mathrm{P}(G_A^c \mid F_A) = \sum_{i=1}^{n-|A|} \frac{(-1)^{i+1}}{i!}.$$

In summary, for all $k = 0, \ldots, n$,

$$\text{P}\{X = k\} = \sum_{\substack{A \subseteq \{1,\ldots,n\}: \\ |A|=k}} \frac{(n-k)!}{n!} \left(1 - \sum_{i=1}^{n-k} \frac{(-1)^{i+1}}{i!}\right)$$
(1.24)
$$= \frac{1}{k!} \sum_{i=0}^{n-k} \frac{(-1)^i}{i!}.$$

Let $n \to \infty$ to find that $\text{P}\{X = k\} \to e^{-1}/k!$. It follows that, when n is large, the distribution of X is approximately that of a Poiss(1).

Next we present another example of Poisson distributions. This example highlights some of the deep connections between Poisson and Binomial random variables.

Example 1.20 (Poissonization). Choose and fix $p \in [0,1]$ and $\lambda > 0$, and let $N, I_1, I_2, \ldots$ be independent random variables such that: (i) $\text{P}\{I_j = 1\} = 1 - \text{P}\{I_j = 0\} = p$ for all $j \geq 1$; and (ii) $N = \text{Poiss}(\lambda)$. We have seen already that $S_n := \sum_{i=1}^n I_i = \text{Bin}(n, p)$ for all $n \geq 1$. Now consider the randomized sum S_N. Its distribution can be computed as follows: For all $k \geq 0$,

$$\text{P}\{S_N = k\} = \sum_{n=k}^{\infty} \text{P}\{S_n = k\} \text{P}\{N = n\}$$
(1.25)
$$= \frac{e^{-\lambda p}(\lambda p)^k}{k!}.$$

(Check!) This proves that $\text{Bin}(N, p) = I_1 + \cdots + I_N = \text{Poiss}(\lambda p)$.

5. Absolutely Continuous Distributions

It is not difficult to imagine random experiments that lead to numerical outcomes which can, in principle, take an arbitrary positive real value. One would like to call such outcomes *absolutely continuous*, or merely *continuous*, random variables. For example, the average weight of 100 randomly-selected individuals is probably best modeled by an absolutely continuous random variable.

Definition 1.21. A function f on **R** is a *density function* if it is non-negative, Riemann-integrable, and $\int_{-\infty}^{\infty} f(x)\,dx = 1$.

Define $\Omega := \{x \in \mathbf{R} : f(x) > 0\}$ and $\text{P}(A) := \int_A f(x)\,dx$ for all sets $A \subseteq \Omega$ for which the Riemann integral exists. One can then check that Proposition 1.2 continues to hold for this choice of P and Ω. Similarly, one can describe conditional probabilities for this P.

5. Absolutely Continuous Distributions

Definition 1.22. An *absolutely continuous random variable* X with density function f is a real-valued function on Ω that satisfies

$$P(\{\omega \in \Omega : X(\omega) \in A\}) = \int_A f(x)\,dx. \tag{1.26}$$

This is valid for all sets A for which the integral is defined. The displayed probability is usually written as $P\{X \in A\}$.

Frequently, one assumes further that f is *piecewise continuous*. This means that we can find points $\{a_i\}_{i=1}^\infty$ such that for all i: (a) $a_i < a_{i+1}$; and (b) f is continuous on (a_i, a_{i+1}).

Next we describe two illustrative examples. A few more are included in the exercises.

5.1. The Uniform Distribution. If $a < b$ are two fixed numbers, then the following defines a density function:

$$f(x) = \frac{1}{b-a} \qquad \forall x \in (a,b). \tag{1.27}$$

This notation implies tacitly that $f(x) = 0$ for all other values of x. The function f is called the *uniform density function* on (a,b). If the density function of X is f, then X is said to be *distributed uniformly* on (a,b). We might refer to X as Unif(a,b). Clearly, $X = $ Unif(a,b) if and only if $P\{X \in A\}$ is proportional to the length of A.

5.2. The Normal Distribution. If $\mu \in \mathbf{R}$ and $\sigma > 0$ are fixed, then X is *normally* distributed with parameters μ and σ^2 if its density function is

$$f(x) = \frac{1}{\sqrt{2\pi\sigma^2}} \exp\left(-\frac{(x-\mu)^2}{2\sigma^2}\right) \qquad \forall x \in \mathbf{R}. \tag{1.28}$$

Frequently, the symbol $N(\mu, \sigma^2)$ denotes a normal random variable with parameters μ and σ^2, and $N(0,1)$ is called a *standard normal* random variable. We define $N(\mu, 0)$ to be the non-random quantity μ. This represents the *degenerate* case.

To complete this discussion we need to verify that $\mathscr{S} := \int_{-\infty}^\infty f(x)\,dx$ is equal to one. A change of variables $[y = (x-\mu)/\sigma]$ reduces the problem to the standard normal case. We then appeal to a trick of J. Liouville, and compute $\mathscr{S}^2$ in polar coordinates, viz.,

$$\mathscr{S}^2 = \int_{-\infty}^\infty \int_{-\infty}^\infty \frac{e^{-(x^2+y^2)/2}}{2\pi}\,dx\,dy = \int_0^{2\pi} \int_0^\infty \frac{e^{-r^2/2}}{2\pi} r\,dr\,d\theta = 1. \tag{1.29}$$

6. Expectation and Variance

Suppose that X is a discrete random variable with mass function p. Its *expectation* (or *mean*), when defined, is

$$(1.30) \qquad \mathrm{E}X := \sum_{x \in \mathbf{R}} x p(x).$$

That is, $\mathrm{E}X$ is the weighted average of the possible values of X where the weights are the respective probabilities. If X denotes the amount of money that is to be gained/lost in a certain game of chance, then $\mathrm{E}X$ is a natural predictor of the as-yet-unrealized value of X.

Similarly, if X has density function f, then the expectation of X is defined as

$$(1.31) \qquad \mathrm{E}X := \int_{-\infty}^{\infty} x f(x)\, dx,$$

provided that the integral is well defined. In this regard see Problem 1.13.

It is not too difficult to check that for any reasonable function $g : \mathbf{R} \to \mathbf{R}$,

$$(1.32) \qquad \mathrm{E}g(X) = \begin{cases} \sum_{x \in \mathbf{R}} g(x) p(x) & \text{if } X \text{ is discrete,} \\ \int_{-\infty}^{\infty} g(x) f(x)\, dx & \text{if } X \text{ is absolutely continuous.} \end{cases}$$

In particular, we do not need to work out the distribution of $g(X)$ before we compute its expectation.

Define the *variance* of X by

$$(1.33) \qquad \mathrm{Var}X := \mathrm{E}\left[(X - \mu)^2\right],$$

where $\mu := \mathrm{E}X$ denotes the mean of X. Then,

$$(1.34) \qquad \mathrm{Var}X = \begin{cases} \sum_{x \in \mathbf{R}} (x - \mu)^2 p(x) & \text{if } X \text{ is discrete,} \\ \int_{-\infty}^{\infty} (x - \mu)^2 f(x)\, dx & \text{if } X \text{ is absolutely continuous.} \end{cases}$$

The following is an equivalent formulation:

$$(1.35) \qquad \mathrm{Var}X = \mathrm{E}[X^2] - (\mathrm{E}X)^2.$$

The square-root of the variance is the so-called *standard deviation* $\mathrm{SD}(X)$ of X. It gauges the best bet for the "distance" between the as-yet-unrealized value X and its predictor $\mathrm{E}X$.

Example 1.23. If $X = \mathrm{Bin}(n,p)$ then $\mathrm{E}X = np$ and $\mathrm{Var}X = np(1-p)$.

Example 1.24. If $X = \mathrm{Geom}(1/\alpha)$ then $\mathrm{E}X = \alpha$ and $\mathrm{Var}X = \alpha(1-\alpha)$.

Example 1.25. If $X = \mathrm{Unif}(a,b)$ then $\mathrm{E}X = (b+a)/2$ and $\mathrm{Var}X = (b-a)^2/12$.

Example 1.26. If $X = N(\mu,\sigma^2)$ then $\mathrm{E}X = \mu$ and $\mathrm{Var}X = \sigma^2$.

Problems

1.1. Prove Propositions 1.2, 1.10, as well as the law of total probability (p. 5).

1.2. Derive (1.32) carefully, and verify Examples 1.23 through 1.26.

1.3. Construct a sample space and three events A, B, and C in this space such that:

(1) Any two of the three events are independent, but A, B, and C are not independent.

(2) A, B, and C have strictly positive probabilites, and $P(A \cap B \cap C) = P(A)P(B)P(C)$ even though A, B, and C are not independent.

1.4. Suppose $\{p_j\}_{j=1}^\infty$ are non-negative numbers that add up to one. If $\{x_j\}_{j=1}^\infty$ are fixed and distinct, then construct a probability space and a random variable X on this probability space such that for all integers $j \geq 1$, $P\{X = x_j\} = p_j$. (HINT: Example 1.16.)

1.5 (A Bonferroni Inequality). Use the inclusion-exclusion principle to deduce that for all events $E_1, \ldots, E_n$,

$$(1.36) \qquad P\left(\bigcup_{j=1}^n E_j\right) \geq \sum_{j=1}^n P(E_j) - \sum\sum_{1 \leq i < j \leq n} P(E_i \cap E_j).$$

1.6 (Binomial Theorem). Use (1.17) to prove that for all $x, y > 0$, and all integers $n \geq 0$,

$$(1.37) \qquad (x+y)^n = \sum_{k=0}^n \binom{n}{k} x^k y^{n-k}.$$

1.7 (Distribution Functions). Let X be a random variable with a density function f. For all $a \in \mathbf{R}$ define $F(a) = P\{X \leq a\}$. F is the *distribution function* of X. Prove that $F'(a) = f(a)$ if f is continuous in a neighborhood of a.

1.8 (Hypergeometric Distribution). An urn contains $r + b$ balls; r are red, and the other b are blue. With the exception of their colors, the balls are identical. We sample, at random and without replacement, n of the $r + b$ balls. Let X denote the total number of red balls in the sample. Then, find the distribution of X, as well as EX and $\text{Var}X$. The distribution of X is called *hypergeometric*.

1.9. Compute EX and $\text{Var}X$, where $X = \text{Poiss}(\lambda)$ for some $\lambda > 0$.

1.10 (Negative Binomial Distribution). Imagine that we perform independent Bernoulli trials with success-per-trial probability $p \in (0, 1)$. We do this until the kth success appears; here, k is a fixed positive integer. Let X denote the number of trials to the kth success. Compute the distribution, expectation, and variance of X. The distribution of X is called *negative binomial*.

1.11 (Exponential Distribution). If $\lambda > 0$ is fixed, then we can define $f(x) = \lambda e^{-\lambda x}$ ($x > 0$). Check that this is a density function. A random variable X with density function f is said to have the *exponential* distribution with parameter λ. Compute the mean and variance of X.

1.12 (Gamma Distribution). Given $\alpha, \lambda > 0$, define

$$(1.38) \qquad f(x) := \frac{\lambda^\alpha}{\Gamma(\alpha)} x^{\alpha-1} e^{-\lambda x} \qquad \forall x > 0,$$

where $\Gamma(\alpha) := \int_0^\infty x^{\alpha-1} e^{-x}\, dx$ is Euler's gamma function. Verify that f is a density function. If X has density function f, then it is said to have the *gamma* distribution with parameters (α, λ). Compute the mean and variance of X in terms of Γ. Verify also that $\Gamma(n+1) = n!$ by first deriving the "duplication formula" $\Gamma(\alpha + 1) = \alpha \Gamma(\alpha)$, valid for all $\alpha > 0$.

1.13 (Cauchy Distribution). Prove that f is a density function, where

$$f(x) := \frac{1}{\pi(1+x^2)} \qquad \forall x \in \mathbf{R}. \tag{1.39}$$

If X has density function f, then it is said to have the *Cauchy distribution*. Prove that in this case $\mathrm{E}X$ is not well defined. Construct a random variable whose mean is well defined but infinite.

1.14 (Standardization). If $X = N(\mu, \sigma^2)$, then prove that $(X-\mu)/\sigma$ is standard normal. (HINT: First compute $F(x) = \mathrm{P}\{(X-\mu)/\sigma \le x\}$. Then compute F'.)

1.15 (Standard Normal Distribution). Let Z be a standard normal random variable. Compute $\mathrm{E}[Z^r]$, for all $r > 0$, using facts about the Gamma function $\Gamma(t) = \int_0^\infty x^{t-1} e^{-x}\,dx$ (Problem 1.12). What happens if r is a positive integer? Also compute $\mathrm{E}[\exp(tZ)]$ and $\mathrm{E}[\exp(tZ^2)]$ for $t \in \mathbf{R}$.

1.16 (Tails of the Normal Distribution). Prove that for all $x > 0$,

$$\left(\frac{1}{x} - \frac{1}{x^3}\right) e^{-x^2/2} \le \int_x^\infty e^{-u^2/2}\,du \le \frac{1}{x} e^{-x^2/2}. \tag{1.40}$$

(HINT: Integrate $(1 + x^{-2})\exp(-x^2/2)$ and $(1 - 3x^{-4})\exp(-x^2/2)$.)

1.17. Prove that if $X = \mathrm{Unif}(0, 1)$ and $a \in \mathbf{R}$ and $b > 0$ are fixed, then $bX + a = \mathrm{Unif}(a, a+b)$. (HINT: First compute $F(x) = \mathrm{P}\{bX + a \le x\}$. Then compute F'.)

1.18. You have n distinct keys, one of which unlocks a certain door. You select a key at random and try to unlock the door with that key. If the key works, then you are done. Else, you select another key at random and try the door again. You repeat this procedure until the door is unlocked. Let X denote the number of sampled keys needed to unlock the door.

(1) Compute the mean and variance of X if the sampling is done with replacement.

(2) Compute the mean and variance of X if the sampling is done without replacement.

1.19 (Hard). Prove that if $X = \mathrm{Bin}(n, p)$, then

$$\mathrm{E}\left[\frac{1}{\ell + X}\right] = \frac{1}{p^\ell} \int_0^p s^{\ell-1}(s+1-p)^n\,ds \qquad \forall \ell > 0. \tag{1.41}$$

For what values of ℓ can you evaluate this integral?

The following exercises explore some two-dimensional extensions of the one-dimensional theory of this chapter. The reader is encouraged to think independently about three-dimensional, or even higher-dimensional, generalizations.

If X and Y are random variables, both defined on a common sample space Ω, then (X, Y) is said to be a *random vector*. It is also known as a *random variable* in two dimensions. If X and Y are discrete, then (X, Y) is said to be *discrete*; in this case,

$$p(x, y) := \mathrm{P}\{X = x, Y = y\} \tag{1.42}$$

is its *mass function*. On the other hand, suppose there exists an integrable function f of two variables such that

$$\mathrm{P}\{(X, Y) \in A\} = \iint_A f(x, y)\,dx\,dy, \tag{1.43}$$

for all sets for which the integral can be defined. Then (X, Y) is said to be *absolutely continuous* and f is its *density function*. We say that X and Y are *independent* if

$$\mathrm{P}\{X \le x, Y \le y\} = \mathrm{P}\{X \le x\}\mathrm{P}\{Y \le y\} \qquad \forall x, y \in \mathbf{R}. \tag{1.44}$$

1.20 (Discrete Random Vectors). Let (X, Y) denote a discrete random vector with mass function p, and define the respective mass functions of X and Y as $p_X(x) = \mathrm{P}\{X = x\}$ and $p_Y(y) = \mathrm{P}\{Y = y\}$. Prove that for all $x, y \in \mathbf{R}$, $p_X(x) = \sum_z p(x, z)$ and $p_Y(y) = \sum_z p(z, y)$. Prove also that for all functions $g : \mathbf{R}^2 \to \mathbf{R}$,

$$\mathrm{E}g(X, Y) = \sum_{x,y \in \mathbf{R}} g(x, y) p(x, y), \tag{1.45}$$

provided that the sum converges.

1.21 (Absolutely Continuous Random Vectors). Let (X, Y) denote an absolutely continuous random vector with a continuous density function f. Prove that X and Y are both absolutely continuous random variables, and their respective densities f_X and f_Y are defined as follows: $f_X(x) = \int_{-\infty}^{\infty} f(x, z)\, dz$ and $f_Y(y) = \int_{-\infty}^{\infty} f(z, y)\, dz$ for all $x, y \in \mathbf{R}$. Prove also that for all bounded continuous functions $g : \mathbf{R}^2 \to \mathbf{R}$,

$$\text{E}g(X, Y) = \int_{-\infty}^{\infty} \int_{-\infty}^{\infty} g(x, y) f(x, y)\, dx\, dy. \tag{1.46}$$

1.22. Suppose (X, Y) is an absolutely continuous random vector with density function f. Suppose f is continuous. Then prove that X and Y are independent if and only if we can write f as $f(x, y) = f_1(x) f_2(y)$. Explore the case that f is piecewise continuous in each variable.

1.23 (Convolutions). Suppose (X, Y) is an absolutely continuous random vector with density function f. Prove that if f is piecewise continuous and X and Y are independent, then for all $z \in \mathbf{R}$, $\text{P}\{X + Y \le z\}$ is equal to

$$\int_{-\infty}^{\infty} \text{P}\{Y \le z - x\} f_X(x)\, dx = \int_{-\infty}^{\infty} \text{P}\{X \le z - y\} f_Y(y)\, dy. \tag{1.47}$$

Conclude that, in particular, $X + Y$ is absolutely continuous with density function

$$f_{X+Y}(z) = \int_{-\infty}^{\infty} f_Y(z - x) f_X(x)\, dx = \int_{-\infty}^{\infty} f_X(z - y) f_Y(y)\, dy. \tag{1.48}$$

The function f_{X+Y} is said to be the *convolution* of f_X and f_Y.

1.24. Let X_1 and X_2 be two independent Poisson random variables with respective parameters λ_1 and λ_2. Prove that $X_1 + X_2 = \text{Poiss}(\lambda_1 + \lambda_2)$. Compute $\text{P}\{X_1 = a \mid X_1 + X_2 = n\}$. Use your computation to argue that the conditional distribution of X_1, given that $X_1 + X_2 = n$, is binomial.

1.25. Suppose (X_1, X_2) are independent normal random variables with $\text{E}X_i = \mu_i$ and $\text{Var}X_i = \sigma_i^2$ ($i = 1, 2$). Use Problem 1.23 to prove that $X_1 + X_2$ is normal with mean $\mu_1 + \mu_2$ and variance $\sigma_1^2 + \sigma_2^2$. Generalize this to a sum of n independent normal random variables.

1.26. Let T be absolutely continuous with density $f(t) = t \exp(-t)$, $t \ge 0$. Prove that $\text{Unif}(0, T)$ has the exponential distribution with mean one.

There are instances where one may wish to mix continuous and discrete distributions as well. The following may require some loose interpretation on the part of the reader, but it should not prove to be ambiguous.

Recall that a "p-coin" is a coin that tosses heads with probability $p \in [0, 1]$.

1.27. Suppose we observe $U = \text{Unif}(0, 1)$, and then toss a U-coin n times independently. Let X denote the number of heads thus tossed. Prove that X has the [discrete] *uniform distribution* on $\{0, \ldots, n\}$. That is,

$$\text{P}\{X = k\} = \frac{1}{n+1} \qquad \forall k = 0, \ldots, n. \tag{1.49}$$

1.28. Suppose we observe $U = \text{Unif}(0, 1)$, and then independently toss a U-coin until the first head appears. Let X denote the number of tosses that are needed. Compute $\text{P}\{X = k\}$ for all $k = 1, 2, \ldots$.

Notes

(1) Independence was first formally introduced by de Moivre (1718). He called it "statistical independence."

(2) Example 1.20 is based on a very old idea which is ripened in Kac (1949).

(3) The fact that the normal density function satisfies $\int_{-\infty}^{\infty} \phi(x)\,dx = 1$ is found within the work of P.-S. Laplace in 1774. Most authors prefer Liouville's elegant proof (1.29). In this regard I add the following, which was proved in 1835 by J. Liouville: The distribution function $\Phi(y) := \int_{-\infty}^{y} \phi(x)\,dx$ cannot be expressed in terms of elementary functions (Rosenlicht, 1972). This theorem of Liouville has spurred the creation of "normal tables" for statisticians.

For a charming, though "circular" derivation of the identity "$\int_{-\infty}^{\infty} \phi(x)\,dx = 1$" see Exercise 4.20 on page 50.

(4) The material of this chapter barely skims the surface of a rich theory that is worthy of study in its own right. Feller (1957) and Poincaré (1912) are two natural starting points for learning more of the classical theory.

(5) The method suggested in Problem 1.16 is borrowed from Feller (1957, Lemma 2, pp. 166–168). But the result itself is very old. See for instance, Note 4 on page 157 below, where I mention the definitive work of Laplace (1805, pp. 490–493) on this problem. Problem 1 of Feller (1957, p. 179) contains another estimate.

Chapter 2

Bernoulli Trials

Everything would return to its pristine state after the passage of innumerable ages.

–Jacob Bernoulli

Consider the following statements:

(1) If we toss a fair coin n times for a large n, then the proportion of heads is close to one half.

(2) If we roll an unbiased die n times for a large n, then about one-sixth of the time we see one dot, approximately one-sixth two dots, etc.

It is a remarkable fact that these assertions can be made into precise mathematical theorems. We now formulate things more carefully in order to state and prove just such theorems.

Consider first the problem of the coins. Suppose, in addition, that the outcomes of the coin tosses are independent from one another. Let S_n denote the number of heads in the first n tosses. Then S_n is a binomial random variable with parameters n and $1/2$.

Next consider the die example. Suppose also that the outcomes of the rolls are independent from one another. Fix some $k \in \{1, \ldots, 6\}$ and define S_n to be the number of the rolls that led to k dots. Then S_n is binomial with parameters n and $1/6$.

These remarks suggest that, quite generally and with good probabilty, a binomial random variable is close to its expectation. We will start this chapter by proving that this is often the case.

Henceforth, in this chapter, we have in mind n Bernoulli trials all with the same parameter $p \in (0,1)$. The random variable S_n denotes the total number of successes, and has the binomial distribution with parameters n and p.

1. The Classical Theorems

The following is the celebrated law of large numbers of Bernoulli (1713):

Bernoulli's Law of Large Numbers. *If p does not depend on n, then for any $\epsilon > 0$,*

$$(2.1) \qquad \lim_{n\to\infty} P\left\{\left|\frac{S_n}{n} - p\right| > \epsilon\right\} = 0.$$

For example, consider a large independent random sample of people from a given population. Then, with high probability, the sample proportion of women is very close to the true proportion of women in the population.

Bernoulli's law of large numbers follows at once from our next result that is due to Chebyshev (1846, 1867).

Chebyshev's Inequality. *For all $n \geq 1$ and $\epsilon > 0$,*

$$(2.2) \qquad P\{|S_n - np| > n\epsilon\} \leq \frac{p(1-p)}{n\epsilon^2}.$$

Proof of Chebyshev's Inequality. We compute directly:

$$(2.3) \qquad \begin{aligned} P\{|S_n - np| > n\epsilon\} &= \sum_{\substack{0 \leq k \leq n: \\ |k-np| > n\epsilon}} \binom{n}{k} p^k (1-p)^{n-k} \\ &\leq \frac{1}{(n\epsilon)^2} \sum_{\substack{0 \leq k \leq n: \\ |k-np| > n\epsilon}} (k-np)^2 \binom{n}{k} p^k (1-p)^{n-k} \\ &\leq \frac{1}{(n\epsilon)^2} \sum_{k=0}^{n} (k-np)^2 \binom{n}{k} p^k (1-p)^{n-k}. \end{aligned}$$

The latter sum is equal to $\mathrm{Var}\, S_n = np(1-p)$; see Example 1.23 on page 12. Chebyshev's inequality follows from this. □

The second limit theorem of this chapter describes the distribution of S_n for large n where the parameter p is inversely proportional to n. Unlike the fixed-p case, S_n does not cluster around a non-random value. Instead, for large n, the distribution of S_n looks like that of a Poisson random variable; see §4.3 (p. 9).

1. The Classical Theorems

Poisson's Law of Rare Events. *If $p = p_n$ satisfies $\lim_{n \to \infty} n p_n = \lambda$ for some fixed number $\lambda > 0$, then*

$$(2.4) \qquad \lim_{n \to \infty} P\{S_n = k\} = \frac{e^{-\lambda} \lambda^k}{k!} \qquad \forall k = 0, 1, \ldots.$$

In rough terms, Bernoulli's law of large numbers states that $S_n \approx np$. Therefore, it is natural to ask about the size of $S_n - np$. The central limit theorem (CLT) of de Moivre (1733) and Laplace (1812) tells about the behavior of the distribution of $(S_n - np)/\sqrt{n}$ for large n. This behavior is described in terms of the all-mysterious "bell curve," which is the graph of the $N(0,1)$-density defined in (1.28), page 11.

The de Moivre–Laplace Central Limit Theorem. *If p does not depend on n, then for all finite numbers $a < b$,*

$$(2.5) \qquad \lim_{n \to \infty} P\left\{ a < \frac{S_n - np}{\sqrt{np(1-p)}} \le b \right\} = P\{a < N(0,1) \le b\}.$$

A by-product of the CLT is that $S_n - np$ is roughly of the order $\sqrt{n}$.

Remark 2.1. It can be shown that $(N(\mu, \sigma^2) - \mu)/\sigma$ is standard normal (Problem 1.14, p. 14). Therefore, Theorem 1 asserts that the distribution of S_n is approximately that of $N(np, np(1-p))$; i.e., a normal random variable whose mean and variance agree with those of S_n.

The following combinatorial device will be used to estimate the distribution of S_n.

de Moivre's Formula. *There exists $\beta \in (0, \infty)$ such that*

$$(2.6) \qquad n! \sim \beta n^{n+\frac{1}{2}} e^{-n} \qquad \text{as } n \to \infty.$$

Proof. Define for all integers $n \ge 1$,

$$(2.7) \qquad f(n) := \frac{n!}{n^{n+\frac{1}{2}} e^{-n}}.$$

Because $f(1) = e$, we write $f(n)$ as a telescoping product, viz.,

$$(2.8) \qquad f(n) = e \prod_{j=2}^{n} \frac{f(j)}{f(j-1)} = \exp\left\{ 1 + \sum_{j=2}^{n} [\ln f(j) - \ln f(j-1)] \right\}.$$

Evidently, as $j \to \infty$,

$$(2.9) \qquad \ln f(j) - \ln f(j-1) = 1 + \left(j - \frac{1}{2}\right) \ln\left(1 - \frac{1}{j}\right) \sim -\frac{1}{12 j^2}.$$

The result follows from this, the summability of j^{-2} and (2.8). $\square$

We can now complete our proof of the de Moivre–Laplace central limit theorem.

Proof of the de Moivre–Laplace CLT. Throughout, we write $q := 1-p$ and seek to approximate the following for large values of n:

$$
(2.10) \quad \mathrm{P}\left\{a < \frac{S_n - np}{\sqrt{npq}} \le b\right\} = \sum_{np+a\sqrt{npq}<k\le np+b\sqrt{npq}} \binom{n}{k} p^k q^{n-k}
$$

$$
= \sum_{a\sqrt{npq}<\lambda\le b\sqrt{npq}} \binom{n}{\lambda + np} p^{\lambda+np} q^{nq-\lambda}.
$$

The index k is in $\{0, \ldots, n\}$, while λ runs over all real numbers of the form $k - np$. Let us call any λ of this form an *n-admissible* number.

According to de Moivre's Formula, as $n \to \infty$,

$$
(2.11) \quad \binom{n}{\lambda + np} \sim \frac{1}{\beta\sqrt{npq}} \left(\frac{\lambda}{n} + p\right)^{-\lambda - np} \left(q - \frac{\lambda}{n}\right)^{-nq+\lambda},
$$

uniformly for all n-admissible $\lambda \in [a\sqrt{npq}, b\sqrt{npq}]$. Thus, as $n \to \infty$,

$$
(2.12) \quad \mathrm{P}\left\{a < \frac{S_n - np}{\sqrt{npq}} \le b\right\}
$$
$$
\sim \frac{1}{\beta\sqrt{npq}} \sum_{a\sqrt{npq}<\lambda\le b\sqrt{npq}} \left(1 + \frac{\lambda}{np}\right)^{-\lambda - np} \left(1 - \frac{\lambda}{nq}\right)^{-nq+\lambda}.
$$

By the Taylor expansion, $\ln(1 - x) = -x + \frac{1}{2}x^2 + o(x^2)$ if $x \approx 0$. By keeping track of the errors incurred we can prove that uniformly over all n-admissible $\lambda \in [a\sqrt{npq}, b\sqrt{npq}]$,

$$
(2.13) \quad \left(1 + \frac{\lambda}{np}\right)^{-\lambda - np} \left(1 - \frac{\lambda}{nq}\right)^{-nq+\lambda} \sim e^{-\lambda^2/(2npq)}.
$$

It follows from these remarks, and the integral test for Riemann integrals, that as $n \to \infty$,

$$
(2.14) \quad \mathrm{P}\left\{a < \frac{S_n - np}{\sqrt{npq}} \le b\right\} \sim \frac{1}{\beta\sqrt{npq}} \int_{a\sqrt{npq}}^{b\sqrt{npq}} e^{-x^2/(2npq)}\, dx.
$$

A change of variables then shows that

$$
(2.15) \quad \lim_{n\to\infty} \mathrm{P}\left\{a < \frac{S_n - np}{\sqrt{npq}} \le b\right\} = \int_a^b \frac{e^{-x^2/2}}{\beta}\, dx = \frac{\sqrt{2\pi}}{\beta} \int_a^b \phi(x)\, dx,
$$

where ϕ denotes the standard-normal density function. See (1.28) on page 11, and set $\mu = 0$ and $\sigma = 1$. Because the preceding display is valid for all finite real numbers $a < b$, it remains to prove that $\beta = \sqrt{2\pi}$.

We may note that for all $a > 0$,

$$(2.16) \qquad 1 \geq P\left\{-a < \frac{S_n - np}{\sqrt{npq}} \leq a\right\} \geq 1 - \frac{1}{a^2}.$$

Indeed, the first bound is a tautology, while the second follows from Chebyshev's inequality (p. 18). Thanks to (2.15), we can deduce that for all $a > 0$,

$$(2.17) \qquad \frac{\beta}{\sqrt{2\pi}} \geq \int_{-a}^{a} \phi(x)\, dx \geq \left(1 - \frac{1}{a^2}\right)\frac{\beta}{\sqrt{2\pi}}.$$

Let $a \uparrow \infty$ to deduce that $\beta = \sqrt{2\pi}$, as was claimed. $\square$

During the course of our proof of the CLT, we verified that $\beta = \sqrt{2\pi}$. This leads to the following celebrated result of Stirling (1730):

Stirling's Formula. $n! \sim n^{n+\frac{1}{2}} e^{-n} \sqrt{2\pi}$ as $n \to \infty$.

Problems

2.1. Derive the law of rare events (p. 19).

2.2. de Moivre's formula (p. 19) has quantitative versions as well. For instance, derive the bounds

$$(2.18) \qquad n^{n+\frac{1}{2}} e^{-n} \leq n! \leq n^{n+\frac{1}{2}} e^{-n+1} \qquad \forall n \geq 1.$$

2.3. Derive Bernoulli's law of large numbers (p. 18) from the de Moivre–Laplace central limit theorem (p. 19).

2.4. An urn contains N balls. They are all identical except for their colors: k are white and $N - k$ are black. We take a random sample of n balls, without replacement, from this urn. Let X denote the number of white balls in the sample. Suppose that $k/N \to p \in (0, 1)$ as $N \to \infty$. Then, compute $\lim_{N \to \infty} P\{X = j\}$ for all j.

2.5. Suppose X_p has the geometric distribution with parameter $p \in (0,1)$ (§4.2, page 8). Then prove that $\lim_{p \to 0} P\{pX_p \geq x\} = P\{Y \geq x\}$ for all $x > 0$, where Y has the exponential distribution with parameter one.

2.6. Suppose: (a) $\lim_{n \to \infty} P\{X_n \leq x\} = P\{X \leq x\}$ for all $x \in \mathbf{R}$; and (b) $x \mapsto P\{X \leq x\}$ is continuous. Prove that $\lim_{n \to \infty} \sup_{x \in \mathbf{R}} |P\{X_n \leq x\} - P\{X \leq x\}| = 0$. Use this to prove that (2.5) holds uniformly for all real numbers $a < b$.

2.7. Construct discrete random variables $\{X_n\}_{n=1}^{\infty}$ such that $p(x) := \lim_{n \to \infty} P\{X_n = x\}$ exists for all $x \in \mathbf{R}$, but p is not a mass function.

2.8 (Problem 2.7, continued). Suppose $\{X_n\}_{n=1}^{\infty}$ are discrete random variables with respective mass functions $\{p_n\}_{n=1}^{\infty}$. Prove that if there exists a function p such that $\lim_{n \to \infty} \sum_{z \in \mathbf{R}} |p_n(z) - p(z)| = 0$, then p is a mass function.

2.9. Derive the following "normal approximation":

$$(2.19) \qquad \int_0^t e^{-x^2}\, dx = t - \frac{1}{3} t^3 + o(t^3) \qquad \text{as } t \to 0.$$

(Laplace, 1782, vol. 10, p. 230).

2.10 (Hard). Suppose X is a discrete, non-negative random variable with mass function p. Prove that $\lim_{n \to \infty} (E[X^n])^{1/n} = \sup\{x \in \mathbf{R} : p(x) > 0\}$.

2.11 (Hard). Suppose X has the Poisson distribution with parameter $\lambda > 0$. Prove that for all $a \le b$,

$$\lim_{\lambda \to \infty} \mathrm{P}\left\{ a \le \frac{X-\lambda}{\sqrt{\lambda}} \le b \right\} = \mathrm{P}\{a \le N(0,1) \le b\}. \tag{2.20}$$

2.12 (Problem 2.8, continued; Hard). Consider two discrete random variables X and Y, and denote their respective mass functions by p_X and p_Y.

(1) Prove that $\sup_{A \subseteq \mathbf{R}} |\mathrm{P}\{X \in A\} - \mathrm{P}\{Y \in A\}| = \frac{1}{2} \sum_{z \in \mathbf{R}} |p_X(z) - p_Y(z)|$.

(2) Let X_n denote the number of fixed points of a random permutation of $\{1, \ldots, n\}$ and $Y = \mathrm{Poiss}(1)$. Use the preceding, together with Example 1.19 (p. 9), to deduce that $\lim_{n \to \infty} \sup_{A \subseteq \mathbf{R}} |\mathrm{P}\{X_n \in A\} - \mathrm{P}\{Y \in A\}| = 0$.

2.13 (Hard). Let f be defined by (2.7). The proof of de Moivre's formula (pp. 19–21) asserts that as n tends to infinity, $f(n) \sim \sqrt{2\pi}$. Refine this by proving that $(2\pi)^{-1/2} f(n) = 1 + (12n)^{-1} + o(1/n)$ as $n \to \infty$.

2.14 (Harder). Prove that if $\zeta(s) = \sum_{j=1}^{\infty} j^{-s}$ for all $s > 1$, then

$$\pi = \frac{e^2}{2} \exp\left\{ -\sum_{k=2}^{\infty} \frac{(k-1)(\zeta(k) - 1)}{k(k+1)} \right\}. \tag{2.21}$$

Notes

(1) Items (1) and (2) (p. 17) both rely implicitly on physical assumptions whose validity can, and should, be questioned. In this regard see Keller (1986); see also Diaconis and Keller (1989).

(2) Bernoulli's original proof of his law of large numbers (1713) was much more involved than the present, modern argument. But it also yielded much more than the stated theorem of this chapter.

(3) Our presentation of De Moivre's Formula (p. 19) is motivated by the exposition of Poincaré (1912, pp. 84–87).

(4) Some authors incorrectly ascribe the de Moivre–Laplace CLT, in the $p \ne 1/2$ case, to P.-S. Laplace. For a detailed clarification see Dudley (2002, pp. 330–331). de Moivre's innovation was, in fact, a normal approximation to the binomial expansion of $(a+b)^n$, although he placed special emphasis on the case that $a = b$. This is equivalent to the de Moivre–Laplace CLT in its entirety. However, Laplace (1812, Chapter III) seems to be the progenitor of the modern probabilistic interpretation of this theorem, and so deserves a part of the credit (Adams, 1974, Chapter 2). In the words of Adams (1974, p. 27), "I can find no evidence to support the view that de Moivre thought of what we now call the normal law as itself being a probability law of errors. A concept of probability law of errors, due to Thomas Simpson, first appeared in 1755, the year after de Moivre's death."

(5) de Moivre (1738) proved the following refinement to our "de Moivre's Formula" (p. 19):

$$\ln(n!) = \ln \beta + \left(n + \frac{1}{2}\right) \ln n - n + \frac{1}{12n} - \frac{1}{360n^3} + \frac{1}{1260n^4} + \cdots,$$

and evaluated β numerically. Problem 2.13 is a start in this direction.

(6) There are a vast number of [mostly overlapping] proofs of the Stirling formula. Here is a synopsis of a novel derviation due to Wong (1977): Let $S_n = \mathrm{Poiss}(n)$ and $Z = N(0,1)$, so that $(S_n - n)/\sqrt{n}$ converges in distribution to Z by Problem 2.11 above. One can prove then that $\lim_{n \to \infty} n^{-1/2} \mathrm{E}[(S_n - n)^-] = \mathrm{E}[Z^-] = (2\pi)^{-1/2}$. See Problem 7.34 on page 115. Stirling's formula ensues because of the following computation:

$$\mathrm{E}\left[(S_n - n)^-\right] = \sum_{j=0}^{n} \frac{e^{-n}(n-j)n^j}{j!} = \frac{e^{-n} n^{n+1}}{n!}.$$

Chapter 3

Measure Theory

If you are afraid of something, measure it, and you will realize it is a mere trifle.

–Renato Caccioppoli

Modern probability is spoken in the language of measure theory, and the latter is the main subject of this chapter. To understand why measure theory is needed consider a random variable X which is distributed uniformly on $(0,1)$; thus, $\mathrm{P}\{X \in A\}$ is the length of A, as long as this length is defined. A primary goal of measure theory is to describe when A has a well-defined length. This matter arises because if we adopt just about any reasonable model of logic, then there exist subsets of $\mathbf{R}$ which do not have a definable length (Solovay, 1970; Freiling, 1986).

We have other uses for measure theory as well. For example, the measure-theoretic approach allows us to study distributions that are not of the types discussed in Chapter 1. We will see later on (Theorem 6.20, p. 71) examples of distributions that are neither absolutely continuous, nor discrete; nor are they simple combinations of the two types mentioned.

The present chapter addresses our forthcoming measure-theoretic needs.

1. Measure Spaces

Throughout, Ω denotes an abstract set that we call the *sample space*.

Definition 3.1. A collection $\mathscr{F}$ of subsets of Ω is a σ-*algebra* if: (i) $\Omega \in \mathscr{F}$; (ii) $\mathscr{F}$ is closed under complementation; i.e., if $A \in \mathscr{F}$ then $A^c \in \mathscr{F}$; and (iii) $\mathscr{F}$ is closed under countable unions; i.e., if $A_1, A_2, \ldots \in \mathscr{F}$, then $\cup_{n=1}^{\infty} A_n \in \mathscr{F}$. It is an *algebra* if, instead of (iii), $\mathscr{F}$ is merely closed under finite unions.

Definition 3.2. Elements of a σ-algebra $\mathscr{F}$ are said to be $\mathscr{F}$-*measurable*, or *measurable with respect to* $\mathscr{F}$. When it is clear from the context that $\mathscr{F}$ is the σ-algebra under study, its elements are referred to as *measurable*.

Of course, σ-algebras (respectively, algebras) are also closed under countable (respectively, finite) intersections, and they also contain the empty set. Furthermore, any σ-algebra is an algebra but the converse is false: The collection of all finite unions of subintervals of $[0,1]$ is an algebra but not a σ-algebra.

Example 3.3. $\{\Omega, \varnothing\}$ is called the *trivial* σ-algebra; it is the smallest σ-algebra of subsets of Ω. The largest such σ-algebra is the *power set* of Ω; it is the collection of all subsets of Ω, and is written as $\mathscr{P}(\Omega)$.

Lemma 3.4. *If A is a set, and if $\mathscr{F}_\alpha$ is a σ-algebra of subsets of Ω for each $\alpha \in A$, then $\cap_{\alpha \in A} \mathscr{F}_\alpha$ is a σ-algebra. Consequently, given any algebra $\mathscr{A}$, there exists a smallest σ-algebra that contains $\mathscr{A}$.*

Definition 3.5. If $\mathscr{A}$ is a collection of subsets of Ω, then we write $\sigma(\mathscr{A})$ for the smallest σ-algebra that contains $\mathscr{A}$; this is the σ-*algebra generated by* $\mathscr{A}$.

It might help to note that $\sigma(\mathscr{A}) = \cap \mathscr{F}$, where the intersection is taken over all σ-algebras $\mathscr{F} \supseteq \mathscr{A}$, and is non-empty (Example 3.3).

Definition 3.6. If Ω is a topological space, then the σ-algebra $\mathscr{B}(\Omega)$ generated by the open subsets of Ω is called the *Borel σ-algebra* of Ω. The elements of $\mathscr{B}(\Omega)$ are called *Borel sets*.

Definition 3.7. Let $\mathscr{F}$ be a σ-algebra of subsets of Ω. A set function $\mu: \mathscr{F} \to [0, \infty]$ is said to be a *measure* on $(\Omega, \mathscr{F})$ if: (i) $\mu(\varnothing) = 0$; and (ii) for any denumerable collection $\{A_n\}_{n=1}^\infty$ of disjoint measurable sets,

$$(3.1) \qquad \mu\left(\bigcup_{n=1}^\infty A_n\right) = \sum_{n=1}^\infty \mu(A_n).$$

By virtue of their definition, measures are non-negative though possibly infinite. Of course, real-valued (often called signed) or complex measures can be defined just as easily.

Definition 3.8. Let $\mathscr{G}$ denote a collection of subsets of Ω, and let μ be a set function on Ω. We say that μ is *countably additive* on $\mathscr{G}$ if (3.1) holds for all disjoint sets $A_1, A_2, \ldots \in \mathscr{G}$ that satisfy $\cup_{n=1}^\infty A_n \in \mathscr{G}$. We say that μ is *countably subadditive* on $\mathscr{G}$ if for all $A_1, A_2, \ldots \in \mathscr{G}$ that satisfy $\cup_{n=1}^\infty A_n \in \mathscr{G}$,

$$(3.2) \qquad \mu\left(\bigcup_{n=1}^\infty A_n\right) \leq \sum_{n=1}^\infty \mu(A_n).$$

The following is simple but not entirely obvious if you read the definitions carefully.

Lemma 3.9. *Countably additive set functions are countably subadditive.*

Definition 3.10. If $\mathscr{F}$ is a σ-algebra of subsets of Ω, then $(\Omega, \mathscr{F})$ is called a *measurable space*. If, in addition, μ is a measure on $(\Omega, \mathscr{F})$, then $(\Omega, \mathscr{F}, \mu)$ is called a *measure space*.

Listed below are some of the elementary properties of measures:

Lemma 3.11. *Let $(\Omega, \mathscr{F}, \mu)$ denote a measure space.*
  (i) *(Continuity from below) If $A_1 \subseteq A_2 \subseteq \cdots$ are all measurable, then $\mu(A_n) \uparrow \mu(\cup_{m=1}^\infty A_m)$ as $n \to \infty$.*
  (ii) *(Continuity from above) If $A_1 \supseteq A_2 \supseteq \cdots$ are all measurable and $\mu(A_n) < \infty$ for some n, then $\mu(A_n) \downarrow \mu(\cap_{m=1}^\infty A_m)$ as $n \to \infty$.*

Definition 3.12. A measure space $(\Omega, \mathscr{F}, \mu)$ is a *probability space* if $\mu(\Omega) = 1$. In this case μ is called a *probability measure*. If $\mu(\Omega) < \infty$, then μ is called a *finite measure* and $(\Omega, \mathscr{F}, \mu)$ a *finite measure space*. Finally, μ and/or $(\Omega, \mathscr{F}, \mu)$ are called *σ-finite* if there exist measurable sets $\Omega_1 \subseteq \Omega_2 \subseteq \cdots$ such that $\cup_{n=1}^\infty \Omega_n = \Omega$ and $\mu(\Omega_n) < \infty$.

Often we denote probability measures by $P, Q, \ldots$ rather than $\mu, \nu, \ldots$.

The following shows that the preceding abstract set-up includes the discrete theory of Chapter 1.

Lemma 3.13. *If $\Omega = \{\omega_i\}_{i=1}^\infty$ is denumerable, then we can find $p_1, p_2, \ldots \in [0, 1]$ such that $P(A) = \sum_{i:\ \omega_i \in A} p_i$ for all $A \subseteq \Omega$. Conversely, any non-negative sequence $\{p_i\}_{i=1}^\infty$ that adds to one defines a probability measure P on the power set of Ω via the assignment $P(\{\omega_i\}) = p_i$ $(i \geq 1)$.*

Definition 3.14. For all measurable spaces $(\Omega, \mathscr{F})$ and $x \in \Omega$ define the *point-mass* δ_x at x as $\delta_x(A) := \mathbf{1}_A(x)$. This is a probability measure.

In the notation of point-masses, the probability measure of Lemma 3.13 can be written as $P(A) = \sum_{j=1}^\infty p_j \delta_{\omega_j}(A)$.

2. Lebesgue Measure

Lebesgue measure on $(0, 1]$ describes a notion of length for various subsets of $(0, 1]$. To construct the Lebesgue measure we first define a set function m on half-closed finite intervals of the form $(a, b] \subseteq (0, 1]$ that evaluates the lengths of the said intervals: $m((a, b]) = b - a$.

Let $\mathscr{A}$ denote the collection of all finite unions of half-closed subintervals of $(0,1]$. It is easy to see that $\mathscr{A}$ is an algebra. We extend the definition of m to $\mathscr{A}$ as follows: For all disjoint half-closed intervals $I_1, \ldots, I_n \subseteq (0,1]$,

$$(3.3) \qquad m\left(\bigcup_{i=1}^n I_i\right) = \sum_{i=1}^n m(I_i).$$

So far, we have defined $m(E)$ for all $E \in \mathscr{A}$. Of course, we need to insure that this definition is consistent. Consistency follows from induction and the following obvious fact: For all $0 < a < b < c$,

$$(3.4) \qquad m\left((a,c]\right) = m\left((a,b]\right) + m\left((b,c]\right).$$

More importantly, we have

Lemma 3.15. *The set function m is countably additive on the algebra $\mathscr{A}$.*

In other words, suppose $A_1, A_2, \ldots$ are disjoint elements of $\mathscr{A}$ such that $\cup_{n=1}^\infty A_n \in \mathscr{A}$. Then, $m(\cup_{n=1}^\infty A_n) = \sum_{n=1}^\infty m(A_n)$. (If there are only finitely many non-empty A_n's, then this is an obvious consequence of (3.3).)

Proof. Because $\cup_{n=1}^\infty A_n$ and $\cup_{n=1}^{N-1} A_n$ are both in $\mathscr{A}$, so is $\cup_{n=N}^\infty A_n$. Also,

$$(3.5) \qquad m\left(\bigcup_{n=1}^\infty A_n\right) = \sum_{n=1}^{N-1} m(A_n) + m\left(\bigcup_{n=N}^\infty A_n\right).$$

Therefore, it suffices to prove that $\lim_{N \to \infty} m(\cup_{n=N}^\infty A_n) = 0$. Hence, our goal is to establish the following: Given a sequence of sets $B_n \downarrow \varnothing$, all in $\mathscr{A}$, $m(B_n)$ decreases to zero.

Suppose to the contrary that there exists $\epsilon > 0$ such that $m(B_n) \geq \epsilon$ for all $n \geq 1$. We will derive a contradiction from this.

Write B_n as a finite union of half-open intervals, $B_n = \cup_{j=1}^{k_n}(a_j^n, b_j^n]$, where $0 \leq a_j^n < b_j^n \leq 1$. Also recall that the B_n's are decreasing. Choose some $\alpha_j^n \in (a_j^n, b_j^n)$ such that

$$(3.6) \qquad \alpha_j^n \leq a_j^n + \frac{\epsilon}{2^{n+2}k_n},$$

and define $C_n = \cup_{j=1}^{k_n}(\alpha_j^n, b_j^n]$. Then,

$$(3.7) \qquad m\left(\bigcup_{i=1}^n (B_i \setminus C_i)\right) \leq \sum_{j=1}^n \sum_{i=1}^{k_j} \left(\alpha_i^j - a_i^j\right) \leq \frac{\epsilon}{2}.$$

In particular, $m(C_n) \geq m(B_n) - (\epsilon/2) \geq (\epsilon/2)$. But the closure $\overline{C_n}$ of every C_n is closed, bounded, and non-empty. If we knew that the $\overline{C_n}$'s were decreasing, then this and the Heine–Borel property of $[0,1]$ together would imply that $\cap_n \overline{C_n} \neq \varnothing$. Because $\overline{C_n} \subseteq B_n$ and $B_n \downarrow \varnothing$, this would yield the desired contradiction.

Unfortunately, the $\overline{C_n}$'s need not be decreasing. So instead consider $D_n = \cap_{j=1}^n \overline{C_j}$. Then, $\overline{D_1}, \overline{D_2}, \ldots$ are closed, bounded, and decreasing. It suffices to show that they are non-empty. Because $B_n = D_n \cup (B_n \cap D_n^c)$,

$$(3.8) \qquad B_n \subseteq D_n \cup \left(\bigcup_{j=1}^n B_j \cap D_n^c \right) \subseteq D_n \cup \bigcup_{j=1}^n (B_j \setminus C_j),$$

since the B_n's are decreasing. Therefore, $\epsilon \leq m(B_n) \leq m(D_n) + (\epsilon/2)$, thanks to (3.7). This shows that $m(D_n) \geq \epsilon/2$, so that $D_n \neq \varnothing$; this completes the proof of countable additivity. The remainder of the proof is smooth sailing. □

Lemma 3.15 and the following result together extend the domain of the definition of m to $\sigma(\mathscr{A})$; the latter is obviously equal to $\mathscr{B}((0,1])$.

The Carathéodory Extension Theorem (Carathéodory, 1948). *Suppose Ω is a set, and $\mathscr{A}$ denotes an algebra of subsets of Ω. Given a countably additive set function μ on $\mathscr{A}$, there exists a measure $\bar{\mu}$ on $(\Omega, \sigma(\mathscr{A}))$ such that $\mu(E) = \bar{\mu}(E)$ for all $E \in \mathscr{A}$. Suppose, in addition, that there exist $\Omega_1 \subseteq \Omega_2 \subseteq \cdots$ in $\mathscr{A}$ such that $\cup_{n=1}^\infty \Omega_n = \Omega$ and $\mu(\Omega_i) < \infty$ for all $i \geq 1$. Then, the extension $\bar{\mu}$ of μ is unique, and $\bar{\mu}$ is a σ-finite measure with $\bar{\mu}(\Omega) = \mu(\Omega)$.*

This result is proved in §4 at the end of this chapter (p. 30). Note that the method of this section also yields Lebesgue's measure on $\mathbf{R}$, $[0,1]$, etc.

In order to construct Lebesgue's measure on $(0,1]^d$ for $d \geq 2$, we proceed as in the case $d = 1$, except start by defining the measure of a *hypercube* $\{x \in (0,1]^d : a_j < x_j \leq b_j\}$ as $\prod_{j=1}^d (b_j - a_j)$. Since the collection of all finite unions of hypercubes is an algebra that generates $\mathscr{B}((0,1]^d)$, we can appeal to the Carathéodory extension theorem to construct Lebesgue's measure on $\mathscr{B}((0,1]^d)$. Further extensions to $[0,1]^d, \mathbf{R}^d$, etc. are made similarly. It is also possible to construct Lebesgue measure on $\mathscr{B}(\mathbf{R}^d)$ as a *product measure*; see Chapter 5.

Thanks to the Carathéodory extension theorem, we can easily construct many measures on the Borel-measurable subsets of $\mathbf{R}^d$ as the following shows. This example will be generalized greatly in the next chapter where we introduce the abstract integral.

Theorem 3.16. *Suppose $f : \mathbf{R}^d \to \mathbf{R}_+$ is a continuous function such that the Riemann integral $\int_{\mathbf{R}^d} f(x)\, dx$ is equal to one. Given $a, b \in \mathbf{R}^d$ with $a_j \leq b_j$ for all $j \leq d$, consider the hypercube $C := (a_1, b_1] \times \cdots \times (a_d, b_d]$,* and define $\mu(C) = \int_C f(x)\,dx$. Then μ extends uniquely to a probability measure on $(\mathbf{R}^d, \mathscr{B}(\mathbf{R}^d))$, and f is called the density function of μ.

Proof. (Sketch) Let $\mathscr{A}$ denote the collection of all finite unions of disjoint hypercubes of the type mentioned. Then $\mathscr{A}$ is an algebra of subsets of $\mathbf{R}^d$. If $A \in \mathscr{A}$, then we can write $A = \cup_{i=1}^m C_i$, where the C_i's are hypercubes of the type considered here, and define

$$\tag{3.9} \mu\left(\bigcup_{i=1}^m C_i\right) = \sum_{i=1}^m \mu(C_i).$$

This is a consistent definition. Now we proceed as we did when we constructed the Lebesgue measure. In this way, we can show that μ is countably additive on $\mathscr{A}$. Finally, we appeal to the Carathéodory extension theorem (p. 27) to finish. □

As an instructive exercise, one can verify that the preceding includes the distributions of §5.1 (p. 11) and §5.2 (p. 11). Below are two more examples.

Example 3.17 (Uniform). Suppose $X \in \mathscr{B}(\mathbf{R}^n)$ has finite and strictly positive n-dimensional Lebesgue measure $m(X)$. Then, the *uniform distribution* ν is the measure defined by

$$\tag{3.10} \nu(A) = \frac{m(A \cap X)}{m(X)} \qquad \forall A \in \mathscr{B}(\mathbf{R}^n).$$

This generalizes §5.1 on page 11.

Example 3.18 (Normal/Gaussian in $\mathbf{R}^n$). Let $\mu \in \mathbf{R}^n$ be a column vector, and Σ a symmetric, invertible, and positive-definite matrix with n rows and n columns. Then, the *normal (or Gaussian) distribution* with parameters μ and Σ corresponds to the density function

$$\tag{3.11} f(x) = \frac{1}{(2\pi)^{n/2}\sqrt{\det \Sigma}} \exp\left(-\frac{1}{2}(x-\mu)'\Sigma^{-1}(x-\mu)\right) \qquad \forall x \in \mathbf{R}^n.$$

See §5.2 (p. 11) for the case $n = 1$.

3. Completion

In the previous section we constructed the Lebesgue measure on $\mathscr{B}((0,1]^d)$, and this is good enough for most of our needs. However, one can extend the definition of m further by defining $m(E)$ for a slightly larger class of sets E.

Definition 3.19. Given a measure space $(\Omega, \mathscr{F}, \mu)$, a measurable set E is *null* if $\mu(E) = 0$. The σ-algebra $\mathscr{F}$ is said to be *complete* if all subsets of null sets are themselves measurable and null. When $\mathscr{F}$ is complete, we say also that $(\Omega, \mathscr{F}, \mu)$ is complete.

3. Completion

We can always ensure completeness, as the following demonstrates.

Theorem 3.20. *Given a measure space $(\Omega, \mathscr{F}, \mu)$, there exists a complete σ-algebra $\mathscr{F}' \supseteq \mathscr{F}$ and a measure μ' on $(\Omega, \mathscr{F}')$ such that μ and μ' agree on $\mathscr{F}$.*

Definition 3.21. The measure space $(\Omega, \mathscr{F}', \mu')$ is called the *completion* of $(\Omega, \mathscr{F}, \mu)$.

On one or two occasions, in particular when discussing Brownian motion, we will complete a measure space or two in order to handle some technical points. However, for the most part, we need not worry about such issues overly much.

Proof of Theorem 3.20. (Sketch) For any two sets A and B define

$$(3.12) \quad \mathscr{F}' = \Big\{ A \subseteq \Omega : \exists B, N \in \mathscr{F} \text{ such that } \mu(N) = 0 \text{ and } A \triangle B \subseteq N \Big\}.$$

Step 1. $\mathscr{F}'$ is a σ-algebra. Since $A^c \triangle B^c = A \triangle B$, $\mathscr{F}'$ is closed under complementation. If $A_1, A_2, \ldots \in \mathscr{F}'$, then we can find $B_1, B_2, \ldots \in \mathscr{F}$ and null sets $N_1, N_2, \ldots$ such that $A_i \triangle B_i \subseteq N_i$ for all $i \geq 1$. But

$$(3.13) \quad \left(\bigcup_{i=1}^{\infty} A_i \right) \triangle \left(\bigcup_{i=1}^{\infty} B_i \right) \subseteq \bigcup_{i=1}^{\infty} (A_i \triangle B_i) \subseteq \bigcup_{i=1}^{\infty} N_i,$$

and the latter is null, thanks to countable subadditivity. Thus, $\mathscr{F}'$ is a σ-algebra.

Step 2. The measure μ'. For any $A \in \mathscr{F}'$ define $\mu'(A) := \mu(B)$, where $B \in \mathscr{F}$ is a set such that for a null set $N \in \mathscr{F}$, $A \triangle B \subseteq N$. It is not hard to see that this is well defined, as it does not depend on the representation (B, N) of A. Clearly, $\mu' = \mu$ on $\mathscr{F}$; we need to show that μ' is a measure on $(\Omega, \mathscr{F}')$. The only interesting portion is countable additivity.

Step 3. Countable Additivity. Suppose $A_1, A_2, \ldots \in \mathscr{F}'$ are disjoint. Find $B_i, N_i \in \mathscr{F}$ as before and note that whenever $j \leq i$,

$$(3.14) \quad B_{i+1} \cap B_j \subseteq (A_{i+1} \cup N_*) \cap (A_j \cup N_*) = N_*,$$

where $N_* := \cup_{i=1}^{\infty} N_i$ is a null set. Define $C_1 = B_1$ and iteratively define

$$(3.15) \quad C_{i+1} := B_{i+1} \setminus (C_1 \cup \cdots \cup C_i).$$

The C_i's are disjoint, and $B_{i+1} \cap C_i \subseteq N_*$ thanks to the previous display; in particular,

$$(3.16) \quad \mu'(B_{i+1}) = \mu'(B_{i+1} \setminus C_i) = \mu'(C_{i+1}).$$

Since $B_1 = C_1$, this shows that $\mu(B_i) = \mu(C_i)$ for all i; we have used the fact that $\mu' = \mu$ on $\mathscr{F}$. Because the C_j's are disjoint, and since

$\cup_{j=1}^\infty C_j = \cup_{j=1}^\infty B_j$, we obtain $\sum_{j=1}^\infty \mu(B_j) = \mu(\cup_{j=1}^\infty B_j)$. It follows readily that $\mu'(\cup_{j=1}^\infty A_j) = \sum_{j=1}^\infty \mu'(A_j)$, and our task is done. □

Let m denote the Lebesgue measure on $(0,1]^d$. We can complete the probability space $((0,1]^d, \mathcal{B}((0,1]^d), m)$ in order to obtain the probability space $((0,1]^d, \mathcal{L}((0,1]^d), \lambda)$. Here, $\mathcal{L}((0,1]^d)$ denotes the completion of $\mathcal{B}((0,1]^d)$, and is called the collection of all *Lebesgue-measurable sets* in $(0,1]^d$. Likewise, we could define $\mathcal{L}([0,1]^d)$, $\mathcal{L}(\mathbf{R}^d)$, etc. We have now defined the Lebesgue measure $\lambda(E)$ of $E \subset \mathbf{R}^d$ for a large class of sets E. Problem 3.17 shows that one cannot define $\lambda(E)$ for all $E \subset \mathbf{R}^d$ and preserve the all-important translation-invariance of the Lebesgue measure.

4. Proof of Carathéodory's Theorem

The proof of Carathéodory's theorem (p. 27) is somewhat lengthy, and relies on a set of ingenious ideas that are also useful elsewhere. Throughout, Ω is a set, and $\mathscr{A}$ is an algebra of subsets of Ω.

Definition 3.22. A collection of subsets of Ω is a *monotone class* if it is closed under increasing countable unions and decreasing countable intersections.

Lemma 3.23. *An arbitrary intersection of monotone classes is a monotone class. In particular, there exists a smallest monotone class containing $\mathscr{A}$.*

Definition 3.24. The smallest monotone class that contains $\mathscr{A}$ is written as $\mathrm{mc}(\mathscr{A})$, and is called the monotone class *generated* by $\mathscr{A}$.

The following result is of paramount use in abstract measure theory:

The Monotone Class Theorem. *Any monotone class that contains $\mathscr{A}$ also contains $\sigma(\mathscr{A})$. In other words, $\mathrm{mc}(\mathscr{A}) = \sigma(\mathscr{A})$.*

Before we prove this let us use it to establish the uniqueness assertion of Carathéodory's extension theorem.

Proof of Carathéodory's Theorem (Uniqueness). Suppose there were two extensions $\bar{\mu}$ and ν, and define

(3.17) $$\mathscr{C} := \{E \in \sigma(\mathscr{A}) : \nu(E) = \bar{\mu}(E)\}.$$

One can check directly that $\mathscr{C}$ is a monotone class that contains $\mathscr{A}$, and hence the theorem follows [decode this!]. □

Proof of the Monotone Class Theorem. Because $\sigma(\mathscr{A})$ is a monotone class, this implies that $\sigma(\mathscr{A}) \supseteq \mathrm{mc}(\mathscr{A})$. Therefore, it suffices to prove that $\mathrm{mc}(\mathscr{A}) \supseteq \sigma(\mathscr{A})$. The proof is non-constructive. First, note that the following are monotone classes:

$$\begin{aligned}
\text{(3.18)} \quad \mathscr{C}_1 &:= \{E \in \sigma(\mathscr{A}) : \ E^c \in \mathrm{mc}(\mathscr{A})\}, \\
\mathscr{C}_2 &:= \left\{E \in \sigma(\mathscr{A}) : \ {}^\forall F \in \sigma(\mathscr{A}), \ E \cup F \in \mathrm{mc}(\mathscr{A})\right\}.
\end{aligned}$$

Since $\mathscr{C}_1$ is a monotone class that contains $\mathscr{A}$, we have $\mathscr{C}_1 \supseteq \mathrm{mc}(\mathscr{A})$. This means that $\mathrm{mc}(\mathscr{A})$ is closed under complementation. If we knew also that $\mathscr{A} \subseteq \mathscr{C}_2$, then we could deduce similarly that $\mathscr{C}_2 \supseteq \mathrm{mc}(\mathscr{A})$. This would imply that $\mathrm{mc}(\mathscr{A})$ is closed under finite unions and is therefore a σ-algebra. Consequently, we would have $\mathrm{mc}(\mathscr{A}) \supseteq \sigma(\mathscr{A})$ and complete our proof. However, the proof that $\mathscr{A} \subseteq \mathscr{C}_2$ requires one more idea. Consider

$$\text{(3.19)} \quad \mathscr{C}_3 := \left\{E \in \sigma(\mathscr{A}) : \ {}^\forall F \in \mathscr{A}, \ E \cup F \in \mathrm{mc}(\mathscr{A})\right\}.$$

Because it is a monotone class that contains $\mathscr{A}$, $\mathscr{C}_3 \supseteq \mathrm{mc}(\mathscr{A})$. By reversing the roles of E and F in the definition of $\mathscr{C}_3$ we can see that $\mathscr{C}_2 \supseteq \mathscr{A}$ as well, and this is what we needed to prove. $\square$

Proof of Carathéodory's Theorem (Existence). It takes too much effort to produce a completely rigorous proof. Therefore, we outline a proof only. The main idea is to try and prove more. Namely, we will define, in a natural way, $\bar{\mu}(E)$ for all $E \subseteq \Omega$. This defines a set function $\bar{\mu}$ on the power set $\mathscr{P}(\Omega)$ of Ω which may be too big a σ-algebra in the sense that $\bar{\mu}$ may fail to be countably additive on $\mathscr{P}(\Omega)$. However, it will be countably additive on $\sigma(\mathscr{A})$. Now, we fill in more details.

For all $E \subseteq \Omega$, define

$$\text{(3.20)} \quad \bar{\mu}(E) := \inf\left\{\sum_{n=1}^\infty \mu(E_n) : \ {}^\forall j \geq 1, \ E_j \in \mathscr{A} \text{ and } E \subseteq \bigcup_{n=1}^\infty E_n\right\}.$$

In the jargon of measure theory, $\bar{\mu}$ is a *Carathéodory outer measure* on $(\Omega, \mathscr{P}(\Omega))$. This defines a natural extension of μ. The proof proceeds in three steps.

Step 1. Countable Subadditivity of $\bar{\mu}$. First, we want to prove that $\bar{\mu}$ is countably subadditive on $\mathscr{P}(\Omega)$. Indeed, we wish to show that $\bar{\mu}(\cup_{n=1}^\infty A_n) \leq \sum_{n=1}^\infty \bar{\mu}(A_n)$ for all $A_1, A_2, \ldots \subseteq \Omega$. To this end, consider any collection $\{A_{j,n}\}$ of elements of $\mathscr{A}$ such that $A_n \subseteq \cup_{j=1}^\infty A_{j,n}$ for all n. By the definition of $\bar{\mu}$,

$$\text{(3.21)} \quad \bar{\mu}\left(\bigcup_{n=1}^\infty A_n\right) \leq \sum_{n=1}^\infty \sum_{j=1}^\infty \mu(A_{j,n}).$$

A second appeal to the definition of $\bar{\mu}$ implies that for any $\epsilon > 0$ we could choose the $A_{j,n}$'s such that

$$(3.22) \qquad \sum_{j=1}^{\infty} \mu(A_{j,n}) \leq \frac{\epsilon}{2^n} + \bar{\mu}(A_n),$$

whence

$$(3.23) \qquad \bar{\mu}\left(\bigcup_{n=1}^{\infty} A_n\right) \leq \epsilon + \sum_{n=1}^{\infty} \bar{\mu}(A_n).$$

Because $\epsilon > 0$ is arbitrary, this yields the countable subadditivity of $\bar{\mu}$.

Step 2. $\bar{\mu}$ extends μ. Next, we plan to prove that $\bar{\mu}$ and μ agree on $\mathscr{A}$ so that $\bar{\mu}$ is indeed an extension of μ. Because $\bar{\mu}(E) \leq \mu(E)$ for all $E \in \mathscr{A}$, we seek to prove the converse inequality. Consider a collection $E_1, E_2, \ldots$ of elements of $\mathscr{A}$ that cover E. For any $\epsilon > 0$ we can arrange things so that $\sum_{n=1}^{\infty} \mu(E_n) \leq \bar{\mu}(E) + \epsilon$. Since μ is countably additive on $\mathscr{A}$,

$$(3.24) \qquad \mu(E) \leq \mu\left(\bigcup_{n=1}^{\infty} E_n\right) \leq \sum_{n=1}^{\infty} \mu(E_n) \leq \bar{\mu}(E) + \epsilon.$$

Because $\epsilon > 0$ is arbitrary, Step 2 is completed.

Step 3. Countable Additivity. We now complete our proof by showing that the restriction of $\bar{\mu}$ to $\sigma(\mathscr{A})$ is countably additive. Thanks to Step 1, it suffices to show that $\sum_{n=1}^{\infty} \bar{\mu}(A_n) \leq \bar{\mu}(\cup_{n=1}^{\infty} A_n)$ for all disjoint $A_1, A_2, \ldots \in \sigma(\mathscr{A})$. With this in mind consider

$$(3.25) \qquad \mathscr{M} = \left\{ E \subseteq \Omega : \,^\forall F \in \mathscr{A}, \, \bar{\mu}(E) = \bar{\mu}(E \cap F) + \bar{\mu}(E \cap F^c) \right\}.$$

According to Step 2, $\mathscr{M}$ contains $\mathscr{A}$. Thus, thanks to the monotone class theorem, if $\mathscr{M}$ were a monotone class then $\sigma(\mathscr{A}) \subseteq \mathscr{M}$. This proves that $\bar{\mu}$ is finitely additive on $\sigma(\mathscr{A})$. Since the A_n's were disjoint, it follows that

$$(3.26) \qquad \bar{\mu}\left(\bigcup_{n=1}^{\infty} A_n\right) \geq \bar{\mu}\left(\bigcup_{n=1}^{N} A_n\right) = \sum_{n=1}^{N} \bar{\mu}(A_n),$$

for every $N \geq 1$. Step 3, whence the Carathéodory extension theorem, follows from this upon letting $N \uparrow \infty$. Define

$$(3.27) \qquad \mathscr{N} := \left\{ E \subseteq \Omega : \, \bar{\mu}(E) \geq \bar{\mu}(E \cap F) + \bar{\mu}(E \cap F^c) \quad \,^\forall F \in \mathscr{A} \right\}.$$

Owing to Step 1, it suffices to show that $\mathscr{N}$ is a monotone class. This is proved by appealing to similar covering arguments that we used in Steps 1 and 2. □

Problems

3.1. Prove Lemma 3.4.

3.2. Construct an example of a countable family $\{\mathscr{F}_i\}_{i=1}^\infty$ of σ-algebras such that $\sigma(\cup_{i=1}^\infty \mathscr{F}_i) \neq \cup_{i=1}^\infty \mathscr{F}_i$. Can you do this so that $\mathscr{F}_i \subseteq \mathscr{F}_{i+1}$ for all i? Typically, one writes $\vee_{i=1}^\infty \mathscr{F}_i$ for $\sigma(\cup_{i=1}^\infty \mathscr{F}_i)$.

3.3. Construct a σ-algebra $\mathscr{F}$ of subsets of $\mathbf{R}$ such that no open interval is measurable with respect to $\mathscr{F}$, although any singleton $\{x\}$ is $(x \in \mathbf{R})$.

3.4. Prove that $\mathscr{B}(\mathbf{R}^k)$ is generated by the collection of all balls whose center and radius are both rational. This implies that $\mathscr{B}(\mathbf{R}^k)$ is "countably generated," i.e., generated by a countable family of sets. Prove also that any singleton $\{x\}$ is $\mathscr{B}(\mathbf{R}^k)$-measurable.

3.5. Prove Lemma 3.11.

3.6. Prove Lemma 3.13.

3.7. Prove Lemma 3.23.

3.8 (Counting Measure). Suppose Ω is a set. For any $A \subseteq \Omega$ define $\mu(A)$ to be the cardinality of A. Prove that μ is a measure on $(\Omega, \mathscr{P}(\Omega))$, where $\mathscr{P}(\Omega)$ denotes the power set of Ω.

3.9 (Distribution Functions). A function $F: \mathbf{R} \to [0,1]$ is a (cumulative) *distribution function* on $\mathbf{R}$ if: (i) It is non-decreasing and right-continuous [this means that $F(a+) := \lim_{b \downarrow a} F(b) = F(a)$]; (ii) $\lim_{a \to -\infty} F(a) = 0$; and (iii) $\lim_{a \to \infty} F(a) = 1$.

Prove that if μ is a probability measure on $(\mathbf{R}, \mathscr{B}(\mathbf{R}))$, then $F(a) := \mu((-\infty, a])$ defines a distribution function on $\mathbf{R}$. Conversely, prove that if F is a distribution function on $\mathbf{R}$, then there exists a unique probability measure μ on $(\mathbf{R}, \mathscr{B}(\mathbf{R}))$ such that $\mu((-\infty, a]) := F(a)$ for all $a \in \mathbf{R}$.

3.10. If $x, r \in \mathbf{R}$ and $A \subseteq \mathbf{R}$, then consider the sets

$$(3.28) \qquad x + A := \{x + a : a \in A\} \quad \text{and} \quad rA := \{ra : a \in A\}.$$

Prove that Lebesgue measure on $\mathbf{R}$ is *translation invariant*. That is, the measure of $x + A$ is the same as the measure of A, provided that $x + A$ and A are Borel measurable. Furthermore, prove that if m_α denotes the Lebesgue measure on $([0, \alpha], \mathscr{B}([0, \alpha]))$ for a given $\alpha > 0$, then $\alpha^{-1} A \in \mathscr{B}([0,1])$ for all measurable $A \subseteq [0, \alpha]$, and $m_\alpha(A) = \alpha m_1(\alpha^{-1} A)$. In other words, prove that Lebesgue measure is also scale invariant.

3.11 (Problem 3.10, Continued). Let μ be a translation invariant σ-finite measure on $(\mathbf{R}, \mathscr{B}(\mathbf{R}))$. Prove that there exists a $c \in (0, \infty)$ such that $c^{-1} \mu$ is Lebesgue measure.

3.12 (Lebesgue Measure on the Circle; Problem 3.10, Continued). Let $S^1 = \{z \in \mathbf{C} : |z| = 1\}$ denote the unit circle in the plane. We say that $A \subseteq S^1$ is an *open* subset of S^1 if A is an open subset of $\mathbf{C}$; this defines $\mathscr{B}(S^1)$ unambiguously.

Prove that $f(\theta) = \exp(i 2\pi\theta)$ defines a homeomorphism from $(0, 1]$ onto S^1; that is, f^{-1} exists, and f and f^{-1} are both continuous. Let m denote the Lebesgue measure on $(0, 1]$ and define $\mu(A) = m(f^{-1}(A))$ for all $A \in \mathscr{B}(S^1)$. Prove that μ is a probability measure on $(S^1, \mathscr{B}(S^1))$. Prove also that μ is "rotation invariant." That is, $\mu(rA) = \mu(A)$ for all $A \in \mathscr{B}(S^1)$ and $r \in \mathbf{C}$ with $|r| = 1$. Frequently, μ is called the *Lebesgue measure* on S^1.

3.13. Suppose $(\Omega, \mathscr{B}(\Omega), \mu)$ is a topological (Borel-) measure space. Define $\text{supp}(\mu)$ to be the smallest closed set whose complement has μ-measure zero; this is called the *support* of μ. Prove that $\text{supp}(\mu)$ is well defined. Prove also that a point $x \in \Omega$ is not in $\text{supp}(\mu)$ if and only if there exists an open neighborhood U of x such that $\mu(U) = 0$.

3.14. Consider a finite measure space $(\Omega, \mathscr{F}, \mu)$, and suppose $\mathscr{A} \subseteq \mathscr{F}$ is a monotone class. Prove that $\mathscr{M}$ is a monotone class, where

$$(3.29) \qquad \mathscr{M} := \bigcap_{n=1}^\infty \left\{ A \in \mathscr{F} : \exists B \in \mathscr{A} : \mu(A \triangle B) \leq \frac{1}{n} \right\}.$$

3.15 (Relative Measures). If μ is a σ-finite measure on $(\mathbf{R}, \mathscr{B}(\mathbf{R}))$, then define $\mathscr{A}$ to be the collection of all $A \in \mathscr{B}(\mathbf{R})$ such that the following limit exists and is finite:

$$(3.30) \qquad (D\mu)(A) = \lim_{n \to \infty} \frac{\mu(A \cap [-n, n])}{n}.$$

Is $\mathscr{A}$ an algebra? What if $\mathbf{R} \in \mathscr{A}$? Is $D\mu$ countably additive on $(\mathbf{R}, \mathscr{A})$?

3.16. Let μ be a measure on $(\mathbf{R}, \mathscr{B}(\mathbf{R}))$ such that the following limit exists for all $x \in \mathbf{R}$:

$$(3.31) \qquad (L\mu)(x) = \lim_{T \to \infty} \frac{\mu([x-T, x+T])}{T}.$$

Prove that $L\mu$ is a constant (Plancherel and Pólya, 1931).

3.17 (Hard). In this exercise we construct a set in the circle S^1 that is not Borel measurable. As usual, we can think of S^1 as the unit circle in $\mathbf{C}$. That is, $S^1 = \{e^{i\theta} : \theta \in (0, 2\pi]\}$.

(1) Given any $z = e^{i\alpha}, w = e^{i\beta} \in S^1$, we write $z \sim w$ if $\alpha - \beta$ is a rational number. Show that this defines an equivalence relation on S^1.

(2) Use the axiom of choice to construct a set Λ whose elements are one from each $\sim$-equivalence class of S^1. Λ is often written as $S^1/\sim$.

(3) For any rational $\alpha \in (0, 2\pi]$ let $\Lambda_\alpha = e^{i\alpha}\Lambda$ denote the rotation of Λ by angle α, and check that if $\alpha, \beta \in (0, 2\pi] \cap \mathbf{Q}$ are distinct, then $\Lambda_\alpha \cap \Lambda_\beta = \varnothing$.

(4) Let μ denote the Lebesgue measure on $(S^1, \mathscr{B}(S^1))$ (Problem 3.12), and show that $\mu(\Lambda)$ is not defined. (HINT: $S^1 = \cup_{\alpha \in (0, 2\pi] \cap \mathbf{Q}} \Lambda_\alpha$.)

(5) Conclude that Λ is not Borel measurable.

3.18 (Harder). For any compact $E \subset [0, 1]$ and $\epsilon, \beta > 0$ define

$$(3.32) \qquad \mathrm{H}_\beta^\epsilon(E) := \inf \sum_{i=1}^\infty |E_i|^\beta,$$

where $|A|$ denotes the Lebesgue measure of A, and the infimum is computed over all sequences $\{E_i\}_{i=1}^\infty$ of closed intervals such that $\sup_i |E_i| \le \epsilon$ and $\cup_{i=1}^\infty E_i \supseteq E$. Prove that

$$(3.33) \qquad \mathrm{H}_\beta(E) := \lim_{\epsilon \to 0} \mathrm{H}_\beta^\epsilon(E)$$

exists and defines a measure on $\mathscr{B}([0, 1])$. The set function H_β is called the β-dimensional *Hausdorff measure*. Can you identify H_1 and H_β for $\beta > 1$? (HINT: You may wish to consult the book of Falconer (1986, §1.1 and §1.2).)

Notes

(1) The theorem of Solovay (1970), referred to in the preamble of this chapter, states that there are non-measurable subsets of the real line if and only if Cantor's axiom of denumerable choice [ADC] holds. Note that ADC lies at the very heart of nearly all of real analysis.

(2) Textbook expositions of Lemma 3.4 have a long tradition; see, for example, Hausdorff (1927, p. 85). Similarly, we can refer to Hausdorff (1927, pp. 177–181) for Definition 3.6.

Chapter 4

Integration

Nature laughs at the difficulties of integration.

–Pierre-Simon de Laplace

We are ready to define nearly household terms such as "random variables," "expectation," "standard deviation," and "correlation." Next follows a brief preview:

- A random variable X is a measurable function.
- The expectation $\mathrm{E}X$ is the integral $\int X\,d\mathrm{P}$ of the function X with respect to the underlying probability measure P.
- The standard deviation is the distance, in $L^2(\mathrm{P})$, between X and its expectation. Correlation is related to an expectation of a certain function of two random variables.

Thus, in this chapter we describe measurable functions, as well as the abstract integral $\int X\,d\mathrm{P}$. Throughout, $(\Omega, \mathscr{F}, \mu)$ denotes a measure space.

1. Measurable Functions

Definition 4.1. A function $f : \Omega \to \mathbf{R}^n$ is (Borel) *measurable* if $f^{-1}(E) \in \mathscr{F}$ for all $E \in \mathscr{B}(\mathbf{R}^n)$. Measurable functions on probability spaces are often referred to as *random variables,* and written as $X, Y, \ldots$ instead of $f, g, \ldots$. Measurable subsets of probability spaces are called *events*.

Because $f^{-1}(E) = \{\omega \in \Omega : f(\omega) \in E\}$, f is measurable (equivalently, f is a random variable) if and only if the pre-images of measurable sets under f are themselves measurable.

Example 4.2. The *indicator function* of $A \subseteq \Omega$ is

(4.1) $$\mathbf{1}_A(\omega) = \begin{cases} 1 & \text{if } \omega \in A, \\ 0 & \text{if } \omega \in A^c. \end{cases}$$

If $A \in \mathscr{F}$, then $\mathbf{1}_A : \Omega \to \{0,1\}$ is a measurable function.

Checking the measurability of a function can be a painful chore. The following alleviates some of the pain most of the time.

Lemma 4.3. *If $\mathscr{A}$ is an algebra that generates $\mathscr{B}(\mathbf{R}^n)$ and $f^{-1}(A) \in \mathscr{F}$ for all $A \in \mathscr{A}$, then $f : \Omega \to \mathbf{R}^n$ is measurable.*

Proof. The lemma follows from the monotone class theorem (p. 30), because $\{A \in \mathscr{B}(\mathbf{R}^n) : f^{-1}(A) \in \mathscr{F}\}$ is a monotone class that contains $\mathscr{A}$. □

The following shows how to use this to produce measurable functions.

Lemma 4.4. *Consider functions $f, f_1, f_2, \ldots : \Omega \to \mathbf{R}^n$ and $g : \mathbf{R}^n \to \mathbf{R}^m$.*

  (i) *If g is continuous, then it is measurable.*
  (ii) *If f, f_1, f_2 are measurable, then so are αf and $f_1 + f_2$ for all $\alpha \in \mathbf{R}$. If $n = 1$, then $f_1 \times f_2$ is measurable too.*
  (iii) *If $n = 1$ and $f_1, f_2, \ldots$ are measurable, then so are $\sup_k f_k$, $\inf_k f_k$, $\limsup_k f_k$, and $\liminf_k f_k$.*
  (iv) *If g and f are measurable, then so is their composition $(g \circ f)(x) = g(f(x))$.*

Proof. By definition, if g is continuous then for all open sets $G \subseteq \mathbf{R}^m$, $g^{-1}(G)$ is open and hence Borel measurable. Because $g^{-1}(G^c) = (g^{-1}(G))^c$ and $g^{-1}(G_1 \cup G_2) = g^{-1}(G_1) \cup g^{-1}(G_2)$, (i) follows from Lemma 4.3. The functions $g(x) = \alpha x$ and $g(x,y) = x+y$ and $g(x,y) = xy$ are all continuous on the appropriate Euclidean spaces. So if we proved (iv), then (ii) would follow from (i) and (iv). But (iv) is an elementary consequence of the identity $(g \circ f)^{-1}(A) = f^{-1}(g^{-1}(A))$. It remains to prove (iii). From now on, we assume that the values of the f_k's are one-dimensional.

Let $S(\omega) = \sup_k f_k(\omega)$ and note that

(4.2) $$S^{-1}((-\infty, x]) = \bigcap_{k=1}^{\infty} f_k^{-1}((-\infty, x]) \in \mathscr{F}$$

for all $x \in \mathbf{R}$. Because $S^{-1}((x,y]) = S^{-1}((-\infty,y]) \setminus S^{-1}((\infty,x])$ for all reals $x < y$, it follows that $S^{-1}((x,y]) \in \mathscr{F}$. The collection of finite disjoint unions of sets of the form $(x,y]$ is an algebra that generates $\mathscr{B}(\mathbf{R})$. Therefore, $\sup_k f_k$ is measurable by Lemma 4.3. Apply (iv) to $g(x) = -x$ to deduce that $\inf_k f_k = -\sup_k(-f_k)$ is also measurable. But we have $\limsup_k f_k = \inf_m \sup_{m \geq n} f_k = \inf_k h_k$ where $h_k = \sup_{m \geq k} f_m$. Since denumerable suprema and infima preserve measurability, $\limsup_k f_k$ is measurable. Finally, the lim inf is measurable because $\liminf_k f_k = -\limsup_k(-f_k)$. □

2. The Abstract Integral

Throughout this section $(\Omega, \mathscr{F}, \mu)$ is a finite measure space unless we explicitly specify that μ is σ-finite.

We now wish to define the integral $\int f\, d\mu$ for measurable functions $f : \Omega \to \mathbf{R}$. Much of what we do here works for σ-finite measure spaces using the following localization method: Find disjoint measurable $K_1, K_2, \ldots$ such that $\cup_n K_n = \Omega$ and $\mu(K_n) < \infty$. Define μ_n to be the restriction of μ to K_n; i.e., $\mu_n(A) = \mu(K_n \cap A)$ for all $A \in \mathscr{F}$. It is easy to see that μ_n is a finite measure on $(\Omega, \mathscr{F})$. Apply the integration theory of this module to μ_n, and define $\int f\, d\mu = \sum_n \int f\, d\mu_n$. For us the details are not worth the effort. After all probability measures are finite!

The abstract integral is derived in three steps.

2.1. Elementary and Simple Functions.

When f is a nice function, $\int f\, d\mu$ is easy to define. Indeed, suppose $f = c\mathbf{1}_A$ where $A \in \mathscr{F}$ and $c \in \mathbf{R}$. Such functions are called *elementary functions*. Then, we define $\int f\, d\mu = c\mu(A)$.

More generally, suppose $A_1, \ldots, A_n \in \mathscr{F}$ are disjoint, $\alpha_1, \ldots, \alpha_n \in \mathbf{R}$, and $f = \sum_{j=1}^n \alpha_j \mathbf{1}_{A_j}$. Then f is measurable by Lemma 4.4, and such functions are called *simple functions*. For them we define $\int f\, d\mu = \sum_{j=1}^n \alpha_j \mu(A_j)$. This notion is well defined; in other words, writing a simple function f in two different ways does not yield two different integrals. One proves this first in the case where f is an elementary function. Indeed, suppose $f = a\mathbf{1}_A = b\mathbf{1}_B + c\mathbf{1}_C$, where B, C are disjoint. It follows easily from this that $a = b = c$ and $A = B \cup C$. Therefore, by the finite additivity of μ, $a\mu(A) = b\mu(B) + c\mu(C)$. This is another way of saying that our integral is well defined in this case. The general case follows from this, the next lemma, and induction.

Lemma 4.5. *If f is a simple function, then so is $|f|$. If $f \geq 0$ pointwise, then $\int f\, d\mu \geq 0$. Furthermore, if f, g are simple functions, then for $a, b \in \mathbf{R}$,*

$$(4.3) \qquad \int (af + bg)\, d\mu = a \int f\, d\mu + b \int g\, d\mu.$$

In other words, $\Lambda(f) := \int f\, d\mu$ defines a non-negative linear functional on simple functions. A consequence of this is that $\int f\, d\mu \leq \int g\, d\mu$ whenever

$f \leq g$ are simple functions. In particular, we have also the following important consequence: $|\int f\,d\mu| \leq \int |f|\,d\mu$. This is called the *triangle inequality*.

2.2. Bounded Measurable Functions. Suppose $f : \Omega \to \mathbf{R}$ is bounded and measurable. To define $\int f\,d\mu$ we use the following to approximate f by simple functions.

Lemma 4.6. *If $f : \Omega \to \mathbf{R}$ is bounded and measurable, then we can find simple functions $\underline{f}_n, \overline{f}_n$ ($n = 1, 2, \ldots$) such that as $n \to \infty$: $\underline{f}_n \uparrow f$; $\overline{f}_n \downarrow f$; and $\overline{f}_n \leq \underline{f}_n + 2^{-n}$ pointwise.*

We can deduce the following by simply combining Lemmas 4.5 and 4.6:

- $\int \underline{f}_n\,d\mu \leq \int \overline{f}_n\,d\mu \leq \int \underline{f}_n\,d\mu + 2^{-n}\mu(\Omega)$ for all $n \geq 1$; and
- $\int f\,d\mu := \lim_{n\to\infty} \int \underline{f}_n\,d\mu = \lim_{n\to\infty} \int \overline{f}_n\,d\mu$ exists and is finite.

This produces an integral $\int f\,d\mu$ that inherits the properties of $\int \overline{f}_n\,d\mu$ and $\int \underline{f}_n\,d\mu$ that were described by Lemma 4.5. That is,

Lemma 4.7. *If f is a bounded measurable function, then so is $|f|$. If f is a pointwise-nonnegative measurable function, then $\int f\,d\mu \geq 0$. Furthermore, if f, g are bounded and measurable functions, then for $a, b \in \mathbf{R}$,*

$$(4.4) \qquad \int (af + bg)\,d\mu = a\int f\,d\mu + b\int f\,d\mu.$$

2.3. The General Case. Let $\mathbf{R}_+ := [0, \infty)$, and consider a non-negative measurable $f : \Omega \to \mathbf{R}_+$. For all $n \geq 1$, the function $f_n(\omega) := \min(f(\omega), n)$ is measurable [Lemma 4.4] and $0 \leq f_n \leq f$. Because $f_n \uparrow f$ as $n \to \infty$, Lemma 4.7 insures that $\int f_n\,d\mu$ increases with n, and hence has a limit, which is denoted by $\int f\,d\mu$. This "integral" inherits the properties of the integrals for bounded measurable integrands, but may be infinite.

In order to define the most general integral of this type let us consider an arbitrary measurable function $f : \Omega \to \mathbf{R}$ and write $f = f^+ - f^-$, where

$$(4.5) \qquad f^+(\omega) := \max(f(\omega), 0) \quad \text{and} \quad f^-(\omega) := -\min(f(\omega), 0).$$

The functions f^+ and f^- are respectively called the *positive* and the *negative* parts of f. Both $f^\pm$ are measurable (Lemma 4.5), and if $\int |f|\,d\mu < \infty$, then we can define $\int f\,d\mu = \int f^+\,d\mu - \int f^-\,d\mu$. This integral has the following properties.

Proposition 4.8. *Let f be a measurable function such that $\int |f|\,d\mu < \infty$. If $f \geq 0$ pointwise, then $\int f\,d\mu \geq 0$. If g is another measurable function such that $\int |g|\,d\mu < \infty$, then for $a, b \in \mathbf{R}$,*

$$(4.6) \qquad \int (af + bg)\,d\mu = a\int f\,d\mu + b\int g\,d\mu.$$

Our arduous construction is over and gives us an "indefinite integral." We can get "definite integrals" as follows: For all $A \in \Omega$ define

$$\text{(4.7)} \qquad \int_A f \, d\mu = \int f \mathbf{1}_A \, d\mu.$$

This is well defined as long as $\int_A |f| \, d\mu < \infty$. In particular, note that $\int_\Omega f \, d\mu = \int f \, d\mu$.

Definition 4.9. We say that f is *integrable* (with respect to μ) if $\int |f| \, d\mu < \infty$. On occasion, we will write $\int f(\omega) \, \mu(d\omega)$ for the integral $\int f \, d\mu$. This will be useful later when f will have other variables in its definition, and the $\mu(d\omega)$ reminds us to only "integrate out the variable ω."

Definition 4.10. When $(\Omega, \mathscr{F}, \mathrm{P})$ is a probability space and $X : \Omega \to \mathbf{R}$ is a random variable, we write $\mathrm{E}X = \int X \, d\mathrm{P}$ and call this integral the *expectation* or *mean* of X. When $A \in \mathscr{F}$ [i.e., when A is an event], we may write $\mathrm{E}[X; A]$ in place of the more cumbersome $\mathrm{E}[X\mathbf{1}_A]$ or $\int_A X \, d\mathrm{P}$.

3. L^p-Spaces

Throughout this section $(\Omega, \mathscr{F}, \mathrm{P})$ is a probability space.

We can define for all $p \in (0, \infty)$ and all random variables $X : \Omega \to \mathbf{R}$,

$$\text{(4.8)} \qquad \|X\|_p := (\mathrm{E}\{|X|^p\})^{1/p},$$

provided that the integral exists; i.e., that $|X|^p$ is P-integrable.

Definition 4.11. The space $L^p(\mathrm{P})$ is the collection of all random variables $X : \Omega \to \mathbf{R}$ that are p times P-integrable. More precisely, these are the random variables X such that $\|X\|_p < \infty$.

Remark 4.12. More generally, if $(\Omega, \mathscr{F}, \mu)$ is a σ-finite measure space, then $L^p(\mu)$ will denote the collection of all measurable functions $f : \Omega \to \mathbf{R}$ such that $\|f\|_p < \infty$. Occasionally we write respectively $\|f\|_{L^p(\mu)}$ and $L^p(\Omega, \mathscr{F}, \mu)$ in place of $\|f\|_p$ and $L^p(\mu)$ in order to emphasize that the underlying measure space is $(\Omega, \mathscr{F}, \mu)$.

Next we list some of the elementary properties of L^p-spaces. Note that the following properties do not rely on the finiteness of μ.

Theorem 4.13. *The following hold for a σ-finite measure μ:*

  (i) $L^p(\mu)$ *is a linear space. That is,* $\|af\|_p = |a| \cdot \|f\|_p$ *for all* $a \in \mathbf{R}$ *and* $f \in L^p(\mu)$, *and* $f + g \in L^p(\mu)$ *if* $f, g \in L^p(\mu)$.
  (ii) *(Hölder's Inequality) If* $p > 1$ *and* $p^{-1} + q^{-1} = 1$, *then*

$$\|fg\|_1 \le \|f\|_p \|g\|_q \qquad \forall f \in L^p(\mu), \, g \in L^q(\mu).$$

(iii) (Minkowski's Inequality) If $p \geq 1$, then
$$\|f+g\|_p \leq \|f\|_p + \|g\|_p \qquad \forall f,g \in L^p(\mu).$$

Proof. It is clear that $\|af\|_p = |a| \cdot \|f\|_p$, and $|x+y|^p \leq 2^p\{|x|^p + |y|^p\}$ for $x, y \in \mathbf{R}$. Hence, $\|f+g\|_p^p \leq 2^p\{\|f\|_p^p + \|g\|_p^p\}$, which proves (i).

Hölder's inequality holds trivially if $\|f\|_p$ or $\|g\|_p$ is equal to 0. Thus, we can assume without loss of generality that $\|f\|_p, \|g\|_p > 0$. For all $x, y \geq 0$,

(4.9) $$xy \leq \frac{x^p}{p} + \frac{y^q}{q};$$

this follows by minimizing $\phi(x) = p^{-1}x^p + q^{-1}y^q - xy$. Replace x and y respectively by $F(\omega) = |f(\omega)|/\|f\|_p$ and $G(\omega) = |g(\omega)|/\|g\|_q$, and integrate $[d\mu]$ to obtain Hölder's inequality, viz.,

(4.10) $$\frac{\|fg\|_1}{\|f\|_p \cdot \|g\|_q} = \int |FG|\, d\mu \leq \frac{\|F\|_p^p}{p} + \frac{\|G\|_q^q}{q} = \frac{1}{p} + \frac{1}{q} = 1.$$

Minkowski's inequality follows from Hölder's inequality as follows: Since $|x+y|^p \leq |x| \cdot |x+y|^{p-1} + |y| \cdot |x+y|^{p-1}$,

(4.11) $$\int |f+g|^p\, d\mu \leq \int |f| \cdot |f+g|^{p-1}\, d\mu + \int |g| \cdot |f+g|^{p-1}\, d\mu$$
$$\leq \{\|f\|_\alpha + \|g\|_\alpha\} \cdot \left(\int |f+g|^{\beta(p-1)}\, d\mu\right)^{1/\beta},$$

where $\alpha \geq 1$ and $\alpha^{-1} + \beta^{-1} = 1$. Choose $\alpha = p$ and note that $\beta = q$ solves $\beta(p-1) = p$. This yields $\|f+g\|_p^p \leq \{\|f\|_p + \|g\|_p\}\|f+g\|_p^{p-1}$. If $\|f+g\|_p = 0$, then Minkowski's inequality holds trivially. Else, we can solve the preceding display to finish. $\square$

The following is a consequence of Hölder's inequality with $p = 2$, but it is sufficiently important that it deserves special mention.

The Cauchy–Schwarz–Bunyakovsky Inequality. *For all $f, g \in L^2(\mu)$,*

(4.12) $$\left|\int fg\, d\mu\right| \leq \|f\|_2 \|g\|_2.$$

Definition 4.14. A function $\psi : \mathbf{R} \to \mathbf{R}$ is *convex* if

(4.13) $$\psi(\lambda x + (1-\lambda) y) \leq \lambda \psi(x) + (1-\lambda)\psi(y) \qquad \forall \lambda \in [0,1], x, y \in \mathbf{R}.$$

See Problem 4.9 for a useful criterion of convexity.

Jensen's Inequality. *Suppose μ is a probability measure. If $\psi : \mathbf{R} \to \mathbf{R}$ is convex and $\psi(f)$ and f are integrable, then*

(4.14) $$\int \psi(f)\, d\mu \geq \psi\left(\int f\, d\mu\right).$$

3. L^p-Spaces

Example 4.15. Since $\psi(x) = |x|$ is convex, Jensen's inequality extends the *triangle inequality*: $\int |f|\,d\mu \ge |\int f\,d\mu|$. A second noteworthy example is the inequality $\int e^f\,d\mu \ge \exp(\int f\,d\mu)$, valid because $\psi(x) = e^x$ is convex. These examples do not presuppose any integrability (why?).

Proof of Jensen's inequality. We will soon see that because ψ is convex, there are linear functions $\{L_z\}_{z \in \mathbf{R}}$ such that

$$(4.15) \qquad \psi(x) = \sup_{z \in \mathbf{R}} L_z(x) \qquad \forall x \in \mathbf{R}_+.$$

Therefore, by Proposition 4.8,

$$(4.16) \qquad \int \psi(f)\,d\mu \ge \sup_{z \in \mathbf{R}} \int L_z(f)\,d\mu = \sup_{z \in \mathbf{R}} L_z\left(\int f\,d\mu\right) = \psi\left(\int f\,d\mu\right).$$

[Here is where we need μ to be a probability measure.] It is easy to describe L_z pictorially: L_z describes the line "tangent" to the graph of ψ at the point $(z, \psi(z))$. Nonetheless (4.15) merits an honest proof.

Consider three points $x \le z \le y$. We can write $z = \lambda x + (1-\lambda)y$ where $\lambda = (y-z)/(y-x)$. Because $\lambda \in [0,1]$ and ψ is convex,

$$(4.17) \qquad \psi(z) \le \left(\frac{y-z}{y-x}\right)\psi(x) + \left(\frac{z-x}{y-x}\right)\psi(y).$$

Subtract $\psi(z)(z-x)/(y-x)$ from both sides and then solve to obtain

$$(4.18) \qquad \frac{\psi(z) - \psi(x)}{z-x} \le \frac{\psi(y) - \psi(z)}{y-z} \qquad \forall x \le z \le y.$$

Now choose and fix some $z \in \mathbf{R}$. Appeal to (4.18) to deduce that $\psi(z) - \psi(x) \le A_1(z-x)$ for all $x \le z$, where $A_1 := \inf_{w \ge z}(\psi(w) - \psi(z))/(w-z)$ depends only on z. That is,

$$(4.19) \qquad \psi(x) \ge \psi(z) + A_1(x-z) \qquad \forall x \le z.$$

Similiarly,

$$(4.20) \qquad \psi(y) \ge \psi(z) + A_2(y-z) \qquad \forall y \ge z,$$

where $A_2 := \sup_{w \le z}(\psi(z) - \psi(w))/(z-w)$ depends only on z. Because $A_1 \le A_2$, we can define $\alpha_z := (A_1 + A_2)/2$ and observe that

$$(4.21) \qquad \psi(x) \ge \psi(z) + \alpha_z(x-z) := L_z(x) \qquad \forall x \in \mathbf{R}.$$

For each $z \in \mathbf{R}$ fixed: L_z is a linear function; $L_z(z) = \psi(z)$; and $\psi(x) \ge L_z(x)$ for all x. This implies (4.15) and thence Jensen's inequality. $\square$

An important consequence of this is that $L^p(\mu)$-spaces are nested when μ is finite.

Proposition 4.16. If $\mu(\Omega) < \infty$ and $r > p \geq 1$, then $L^r(\mu) \subseteq L^p(\mu)$. In fact,

$$\|f\|_p \leq [\mu(\Omega)]^{\frac{1}{p} - \frac{1}{r}} \|f\|_r \qquad \forall f \in L^r(\mu). \tag{4.22}$$

Proof. The proposition follows from the displayed inequality. Since this is a result that involves only the function $|f|$, we can assume without loss of generality that $f \geq 0$. Consider simple functions S_n that converge upward to f. Suppose we could prove the proposition for each S_n. Then we can let $n \uparrow \infty$ and then appeal to our construction of integrals to derive the theorem for f. In particular, we can assume without loss of generality that f is bounded and hence in $L^v(\mu)$ for all $v > 0$.

The function $\phi(x) = |x|^s$ is convex for all $s \geq 1$. Let $s = (r/p)$ and apply Jensen's inequality to deduce that when μ is a probability measure,

$$\|f\|_p^r = \phi\left(\int |f|^p \, d\mu\right) \leq \int \phi(|f|^p) \, d\mu = \|f\|_r^r. \tag{4.23}$$

This is the desired result. If $\mu(\Omega) > 0$ is finite but not equal to 1, then define $\bar{\mu}(A) := \mu(A)/\mu(\Omega)$. This is a probability measure, and according to what we have shown thus far, $\|f\|_{L^p(\bar{\mu})} \leq \|f\|_{L^r(\bar{\mu})}$. Solve for $\|f\|_{L^p(\mu)}$ and $\|f\|_{L^p(\mu)}$ to finish. Finally, if $\mu(\Omega) = 0$ then the result holds vacuously. □

Fix some $p \geq 1$ and define $d(f, g) := \|f - g\|_p$ for all $f, g \in L^p(\mu)$. According to Minkowski's inequality (Theorem 4.13), d has the following properties:

(1) $d(f, f) = 0$;
(2) $d(f, g) \leq d(f, h) + d(h, g)$; and
(3) $d(f, g) = d(g, f)$.

In other words, if it were the case that "$d(f, g) = 0 \Longrightarrow f = g$," then $d(\cdot, \cdot)$ would metrize $L^p(\mu)$. Unfortunately, the latter property does not hold in general. For an example consider $g = f\mathbf{1}_A$ where $A \neq \emptyset$ and $\mu(A^c) = 0$. Evidently then $g \neq f$ but $d(f, g) = 0$.

Nonetheless, if we can identify the elements of $L^p(\mu)$ that are equal to each other outside a null set, then the resulting collection of equivalence classes—endowed with the usual quotient topology and Borel σ-algebra—is indeed a metric space. It is also complete; i.e., every Cauchy sequence converges.

Theorem 4.17. Let $(\Omega, \mathscr{F}, \mu)$ denote a σ-finite measure space. For any $f, g \in L^p(\mu)$, write $f \sim g$ iff $f = g$ μ-a.e. That is, $\mu(\{\omega : f(\omega) \neq g(\omega)\}) = 0$. Then $\sim$ is an equivalence relation on $L^p(\mu)$. Let $[f]$ denote the $\sim$-orbit of f; i.e., $f \in [f]$ iff $f \sim g$. Let $\mathscr{L}^p(\mu) = \{[f] : f \in L^p(\mu)\}$ and define

$\|[f]\|_p = \|f\|_p$. Then, $\mathscr{L}^p(\mu)$ is a complete normed linear space. Moreover, $\mathscr{L}^2(\mu)$ is a Hilbert space.

We will prove this in Section 5 below; see page 46.

4. Modes of Convergence

There are many ways in which a sequence of functions can converge. We will be primarily concerned with the following. Throughout, $(\Omega, \mathscr{F}, \mu)$ is a measure space, and $f, f_1, f_2, \ldots : \Omega \to \mathbf{R}$ are measurable.

Definition 4.18. We say that f_n converges to f μ-*almost everywhere* (written μ-a.e., a.e. $[\mu]$, or even a.e.) if

$$(4.24) \quad \mu\left(\left\{\omega \in \Omega : \limsup_{n\to\infty} |f_n(\omega) - f(\omega)| > 0\right\}\right) = 0.$$

Frequently, we write $\{f \in A\}$ for $\{\omega \in \Omega : f(\omega) \in A\}$ and $\mu\{f \in A\}$ for $\mu(\{f \in A\})$. In this way, f_n converges to f a.e. iff $\mu\{f_n \not\to f\} = 0$. When $(\Omega, \mathscr{F}, P)$ is a probability space and $X, X_1, X_2, \ldots$ are random variables on this space, we say instead that X_n converges to X *almost surely* (written a.s.).

Definition 4.19. We say that $f_n \to f$ in $L^p(\mu)$ if $\lim_{n\to\infty} \|f_n - f\|_p = 0$. Also, $f_n \to f$ *in measure* if $\lim_{n\to\infty} \mu\{|f_n - f| \geq \epsilon\} = 0$ for all $\epsilon > 0$. If $X, X_1, X_2, \ldots$ are random variables on the probability space $(\Omega, \mathscr{F}, P)$, then we say that X_n converges to X in *probability* when $X_n \to X$ in P-measure; that is, if $\lim_n P\{|X_n - X| \geq \epsilon\} = 0$ for all $\epsilon > 0$. We write this as $X_n \xrightarrow{P} X$.

Theorem 4.20. *Either a.e.-convergence or L^p-convergence implies convergence in measure. Conversely, if $\sup_{j\geq n} |f_j| \to 0$ in measure, then $f_n \to 0$ almost everywhere.*

The interesting portion of this relies on the following result:

Markov's Inequality. *If $f \in L^1(\mu)$, then for all $\lambda > 0$,*

$$(4.25) \quad \mu\{|f| \geq \lambda\} \leq \frac{1}{\lambda} \int_{\{f \geq \lambda\}} |f|\, d\mu \leq \frac{\|f\|_1}{\lambda}.$$

Proof. Set $\Lambda := \{|f| \geq \lambda\}$ and note that $\int_\Lambda |f|\, d\mu \geq \int_\Lambda \lambda\, d\mu = \lambda\mu(\Lambda)$. This yields the first inequality. The second one is even more transparent. $\square$

We can apply the preceding to the function $|f|^p$ to deduce the following:

Chebyshev's Inequality (1846; 1867). *For all $p, \lambda > 0$ and $f \in L^p(\mu)$,*

$$(4.26) \quad \mu\{|f| \geq \lambda\} \leq \frac{1}{\lambda^p} \int_{\{|f|\geq\lambda\}} |f|^p\, d\mu \leq \frac{\|f\|_p^p}{\lambda^p}.$$

Proof of Theorem 4.20. By the Chebyshev inequality, $L^p(\mu)$-convergence implies convergence in measure. In order to prove that a.e.-convergence implies convergence in measure we first need to understand a.e.-convergence a little better.

Note that $f_n \to f$ a.e. if and only if $\mu(\cap_{N=1}^\infty \cup_{n=N}^\infty \{|f_n - f| \geq \epsilon\}) = 0$ for all $\epsilon > 0$. Since μ is continuous from above,

$$(4.27) \qquad f_n \to f \text{ a.e. iff } \lim_{N \to \infty} \mu\left(\bigcup_{n=N}^\infty \{|f_n - f| \geq \epsilon\}\right) = 0 \quad \forall \epsilon > 0.$$

Because $\mu\{|f_N - f| \geq \epsilon\} \leq \mu(\cup_{n \geq N}\{|f_n - f| \geq \epsilon\})$, if $f_N \to f$ a.e. then $f_N \to f$ in measure.

Finally, if $\sup_{j \geq n} |f_j| \to 0$ in measure then

$$(4.28) \qquad \mu\left(\bigcap_{N=1}^\infty \bigcup_{n=N}^\infty \left\{\sup_{j \geq n} |f_j| \geq \epsilon\right\}\right) = \lim_{N \to \infty} \mu\left(\left\{\sup_{j \geq N} |f_j| \geq \epsilon\right\}\right) = 0.$$

Thus, $\limsup_m |f_m| \leq \limsup_n \sup_{j \geq n} |f_j| < \epsilon$ a.e. $[\mu]$. If $N(\epsilon)$ denotes the set of ω's for which this inequality fails, then $\cup_{\epsilon \in \mathbf{Q}_+} N(\epsilon)$ is a null set off which $\lim_m |f_m| < \epsilon$ for every rational $\epsilon > 0$; i.e., off $\cup_{\epsilon \in \mathbf{Q}_+} N(\epsilon)$ we have $\lim_m |f_m| = 0$. □

Here are two examples to test the strength of the relations between the various modes of convergence. The first involves the Steinhaus probability space which was a starting-point of modern probability theory.

Example 4.21 (The Steinhaus Probability Space). The *Steinhaus probability space* is the probability space $(\Omega, \mathscr{B}(\Omega), P)$ where Ω is either $(0,1)$, $[0,1)$, $(0,1]$, or $[0,1]$; P denotes the Lebesgue measure on Ω. On this space consider

$$(4.29) \qquad X_n(\omega) := n^a \mathbf{1}_{[0,1/n)}(\omega),$$

where $a > 0$ is fixed and $\omega \in \Omega$. Then $X_n \to 0$ almost surely (in fact for all $\omega \in \Omega$). And yet if $p \geq a^{-1}$, then $\|X_n\|_p^p = n^{ap-1}$ is bounded away from 0. Therefore, a.s.-convergence does not imply L^p-convergence. The trouble comes from the fact that $\sup_n |X_n|$ is not in $L^p(P)$; compare with the dominated convergence theorem (p. 46).

Example 4.22. Let $((0,1], \mathscr{B}((0,1]), P)$ be the Steinhaus probability space of the previous example. Now we construct random variables $\{X_n\}_{n=1}^\infty$ such that $\lim_n X_n(\omega)$ does not exist for any $\omega \in (0,1]$, and yet $\lim_n \|X_n\|_p = 0$ for all $p > 0$.

Define a "triangular array" of functions $f_{i,j}$ ($\forall i \geq 1$, $j \leq 2^{i-1}$) as follows: First let $f_{1,1}(\omega) = 1$ for all $\omega \in (0,1]$. Then define

(4.30) $\quad f_{2,1}(\omega) = \begin{cases} 2, & \text{if } \omega \in \left(0, \frac{1}{2}\right] \\ 0, & \text{otherwise} \end{cases}$, $f_{2,2}(\omega) = \begin{cases} 0, & \text{if } \omega \in \left(\frac{1}{2}, 1\right] \\ 2, & \text{otherwise} \end{cases}$, ...

In general, for all $i \geq 1$ and $j = 1, \ldots, 2^{i-1}$, we can define $f_{i,j}$ to be i on $((j-1)2^{-i-1}, j2^{-i-1}]$, and zero elsewhere. Let us enumerate the $f_{i,j}$'s according to the dictionary ordering, and call the resulting relabeling (X_k); i.e., $X_1 = f_{1,1}$, $X_2 = f_{2,1}$, $X_3 = f_{2,2}$, $X_4 = f_{3,1}$, $X_5 = f_{3,2}, \ldots$. Evidently, $\limsup_{k \to \infty} X_k(\omega) = \infty$ whereas $\liminf_{k \to \infty} X_k(\omega) = 0$ for all $\omega \in (0,1]$. In particular, the $X_n(\omega)$'s do not converge for any ω. On the other hand, X_n converges to zero in $L^p(P)$ for all $p > 0$ because $\|f_{i,j}\|_p = i^p 2^{-(i-1)}$.

5. Limit Theorems

Proposition 4.8 expresses two of the essential properties of the abstract integral: (i) Integration is a positive operation (i.e., if $f \geq 0$ then $\int f \, d\mu \geq 0$); and (ii) it is a linear operation (i.e., equation (4.6)). We now turn to some of the important properties that involve limiting operations.

Throughout this section, we let $(\Omega, \mathscr{F}, \mu)$ denote a finite measure space, and address the following question: If f_n converges to f, then does $\int f_n \, d\mu$ converge to $\int f \, d\mu$?

The Bounded Convergence Theorem. *Suppose $f_1, f_2, \ldots$ are measurable functions on $(\Omega, \mathscr{F})$ such that $\sup_n |f_n|$ is bounded by a constant K. If $f_n \to f$ in measure $[\mu]$, then $\lim_{n \to \infty} \int f_n \, d\mu = \int f \, d\mu$.*

Proof. For all $n \geq 1$, f_n is integrable because $|f_n(\omega)| \leq K$ and μ is a finite measure. Now fix an $\epsilon > 0$ and let $E_n := \{\omega \in \Omega : |f(\omega) - f_n(\omega)| > \epsilon\}$. According to Proposition 4.8,

(4.31) $\quad \left| \int f \, d\mu - \int f_n \, d\mu \right| \leq \int_{E_n^c} |f - f_n| \, d\mu + \int_{E_n} |f - f_n| \, d\mu$
$\leq \epsilon \mu(\Omega) + 2K \mu(E_n).$

Since $\lim_{n \to \infty} \mu(E_n) = 0$, we can then let $\epsilon \downarrow 0$ to finish. $\square$

Fatou's Lemma. *If $\{f_i\}_{i=1}^\infty$ is a collection of non-negative integrable functions on $(\Omega, \mathscr{F}, \mu)$, then*

(4.32) $\quad \int \liminf_{n \to \infty} f_n \, d\mu \leq \liminf_{n \to \infty} \int f_n \, d\mu.$

Proof. Let $g_n = \inf_{j \geq n} f_j$ and observe that $g_n \uparrow f := \liminf_k f_k$ as $n \to \infty$. In particular, for any constant $K > 0$, $(f \wedge K - g_n \wedge K)$ is a bounded measurable function that converges to 0 as $n \to \infty$. Because $g_n \le f_n$, the bounded convergence theorem implies that

$$(4.33) \qquad \liminf_{n\to\infty} \int f_n \, d\mu \ge \lim_{n\to\infty} \int (g_n \wedge K) \, d\mu = \int (f \wedge K) \, d\mu.$$

Therefore, it suffices to prove that

$$(4.34) \qquad \lim_{K\uparrow\infty} \int (f \wedge K) \, d\mu = \int f \, d\mu.$$

For all $\epsilon > 0$ we can find a simple function S such that: (i) $0 \le S \le f$; (ii) there exists $C > 0$ such that $S(\omega) \le C$; and (iii) $\int S \, d\mu \ge \int f \, d\mu - \epsilon$. Now $\int (f \wedge K) \, d\mu \ge \int (S \wedge K) \, d\mu = \int S \, d\mu \ge \int f \, d\mu - \epsilon$ if $K > C$. This proves (4.34), whence follows the result. □

The Monotone Convergence Theorem. *Suppose $\{f_n\}_{n=1}^\infty$ is a sequence of non-negative integrable functions on $(\Omega, \mathscr{F}, \mu)$ such that $f_n(x) \le f_{n+1}(x)$ for all $n \ge 1$ and $x \in \Omega$, and $f(x) := \lim_{n\to\infty} f_n(x)$ exists for all $x \in \Omega$. Then, $\lim_{n\to\infty} \int f_n \, d\mu = \int f \, d\mu$.*

Proof. By monotonicity $L := \lim_{n\to\infty} \int f_n \, d\mu$ exists and is $\le \int f \, d\mu$. Apply Fatou's lemma to deduce the complementary inequality. □

The Dominated Convergence Theorem. *Suppose $\{f_i\}_{i=1}^\infty$ is a sequence of measurable functions on $(\Omega, \mathscr{F})$ such that $\sup_m |f_m|$ is integrable $[d\mu]$. Then, $\lim_{n\to\infty} \int f_n \, d\mu = \int \lim_{n\to\infty} f_n \, d\mu$ provided that $f(x) := \lim_{n\to\infty} f_n(x)$ exists for all $x \in \Omega$.*

Proof. Thanks to Fatou's lemma $f \in L^1(\mu)$. Also, $F := \sup_{i \ge 1} |f_i| \in L^1(\mu)$ by assumption. We can apply Fatou's lemma to the non-negative function $g_n := 2F - |f_n - f|$ to deduce that $\lim_{n\to\infty} \int |f_n - f| \, d\mu = 0$. The dominated convergence theorem follows from this and the bound

$$(4.35) \qquad \left| \int f_n \, d\mu - \int f \, d\mu \right| \le \int |f_n - f| \, d\mu,$$

which is merely the triangle inequality for integrals. □

We can now prove our Theorem 4.17 (p. 42) on completions of L^p spaces.

Proof of Theorem 4.17. The fact that $L^p(\mu)$, and hence $\mathscr{L}^p(\mu)$, is a linear space has already been established in Theorem 4.13. As we argued a few paragraphs earlier, $d(f,g) := \|f - g\|_p$ is a norm (now on $\mathscr{L}^p(\mu)$) as soon as we prove that $d(f,g) = 0 \Rightarrow [f] = [g]$; but this is obvious.

In order to establish completeness suppose $\{f_n\}_{n=1}^\infty$ is a Cauchy sequence in $L^p(\mu)$. It suffices to show that f_n converges in $L^p(\mu)$. (Translate this to a statement about $[f_n]$'s.) Recall that "$\{f_n\}_{n=1}^\infty$ is Cauchy" means that $\lim_{m,n\to\infty} \|f_n - f_m\|_p \to 0$. Thus, we can find a subsequence $\{n_k\}_{k=1}^\infty$

such that $\|f_{n_{k+1}} - f_{n_k}\|_p \leq 2^{-k}$. Consequently, $\sum_k \|f_{n_{k+1}} - f_{n_k}\|_p < \infty$. Thanks to Minkowski's inequality and the monotone convergence theorem, $\|\sum_{k=1}^\infty |f_{n_{k+1}} - f_{n_k}|\|_p < \infty$. In particular, $\sum_k (f_{n_{k+1}} - f_{n_k})$ converges μ-almost everywhere (why?).

If $f := \sum_k (f_{n_{k+1}} - f_{n_k})$ then $f \in L^p(\mu)$ by Fatou's lemma. By the triangle inequality for L^p-norms,

$$(4.36) \qquad \|f - f_{n_k}\|_p \leq \sum_{j=k+1}^\infty \|f_{n_{j+1}} - f_{n_j}\|_p \to 0 \qquad \text{as } k \to \infty.$$

Minkowski's inequality implies that

$$(4.37) \qquad \|f - f_N\|_p \leq \|f - f_{n_k}\|_p + \|f_{n_k} - f_N\|_p \qquad \forall N, k \geq 1.$$

Therefore, we can let N and k tend to infinity to see that $f_n \to f$ in $L^p(\mu)$.

Finally, we can recognize that by Hölder's inequality $\langle f, g \rangle := \int fg \, d\mu$ is an inner product. Therefore, $L^2(\mu)$ is a Hilbert space. This completes the proof. □

6. The Radon–Nikodým Theorem

Given two measures μ and ν one can ask, "*When can we find a function $\pi_\star$ such that for all measurable sets A, $\nu(A) = \int_A \pi_\star \, d\mu$?*" If μ denotes the Lebesgue measure, then the function $\pi_\star$ is a probability density function, and the prescription $\nu(A) := \int_A \pi_\star \, d\mu$ defines a probability measure ν. For instance, the standard-normal distribution is precisely the measure ν when $\pi_\star(x) = (2\pi)^{-1/2} \exp(-x^2/2)$ and μ is the Lebesgue measure on the line.

Definition 4.23. Given two measures μ and ν on $(\Omega, \mathscr{F})$, we say that ν is *absolutely continuous* with respect to μ (written $\nu \ll \mu$) if $\nu(A) = 0$ for all $A \in \mathscr{F}$ such that $\mu(A) = 0$.

For instance, suppose that $(\Omega, \mathscr{F}, \mu)$ is a finite measure space and $f \in L^1(\mu)$. Then $\nu(A) = \int_A f \, d\mu$ defines a finite measure, and $\nu \ll \mu$. The following states that what we have just seen is the only example of its kind.

The Radon–Nikodým Theorem. *If $\nu \ll \mu$ are two finite measures on $(\Omega, \mathscr{F})$, then there exists a non-negative $\pi_\star \in L^1(\mu)$ such that $\int f \, d\nu = \int f \pi_\star \, d\mu$ for all bounded measurable functions $f : \Omega \to \mathbf{R}$. Furthermore, $\pi_\star$ is unique up to a μ-null set.*

Remark 4.24. Frequently we write $\pi_* := d\nu/d\mu$. The function π_* is called the *Radon–Nikodým derivative* of ν with respect to μ. The notation is suggestive because for all bounded measurable $f : \Omega \to \mathbf{R}$,

$$(4.38) \qquad \int f \, d\nu = \int f \left(\frac{d\nu}{d\mu}\right) d\mu.$$

Remark 4.25. Suppose $(\Omega, \mathscr{F}, \mu)$ is a measure space, $f \in L^1(\mu)$ is nonnegative, and $\nu(E) = \int_E f \, d\mu$ for all $E \in \mathscr{F}$. Then ν is a measure also and $(d\nu/d\mu) = f$ a.e. $[\mu]$. Furthermore, if $\|f\|_1 = 1$ then ν is a probability measure with density function f. We can now see that Examples 3.17–3.18 (p. 28) contain a series of Radon–Nikodým derivatives that arise in elementary probability theory.

The Radon–Nikodým theorem is a simple but deep result. Next is a "geometric proof," due to von Neumann (1940, Lemma 3.2.3, p. 127).

Proof. First, we prove the theorem under the stronger domination condition that $\nu(A) \le \mu(A)$ for all $A \in \mathscr{F}$.

Step 1. The Case $\nu \le \mu$. Consider the linear functional $\mathscr{L}(f) := \int f \, d\nu$ that acts on all $f \in L^1(\nu)$. By the Cauchy–Bunyakovsky–Schwarz inequality,

$$(4.39) \qquad |\mathscr{L}(f)|^2 \le \nu(\Omega) \int |f|^2 \, d\mu.$$

Hence, $\mathscr{L}$ is a bounded linear functional on $L^2(\mu)$. By completeness (Theorem 4.17), and the general theory of Hilbert spaces (Theorem A.4, p. 204), $\mathscr{L}$ is obtained by an inner product. That is, there exists a μ-almost everywhere unique $\pi \in L^2(\mu)$ such that $\int f \, d\nu = \int f\pi \, d\mu$ for all $f \in L^2(\mu)$. Choose an $\alpha > 0$ and replace f by the indicator of the measurable set $\{\pi \le -\alpha\}$ to deduce that $\mu\{\pi \le -\alpha\} = 0$. It follows from the right-continuity of μ that $\pi \ge 0$ a.e. $[\mu]$. That is, we have established the theorem for all $f \in L^2(\mu)$. By the monotone convergence theorem this fact holds for all measurable $f \ge 0$, and the entire theorem follows, with $\pi_\star = \pi$, in the case of dominated measures.

Step 2. General ν, μ. Because $\nu \le (\mu + \nu)$, Step 1 extracts a μ-a.e. unique (in fact, $(\mu+\nu)$-a.e. unique) and non-negative $\pi \in L^2(\mu+\nu)$ such that $\int f(1-\pi) \, d\nu = \int f\pi \, d\mu$ for all $f \in L^2(\mu+\nu)$. Replace f by the indicator of $\{\pi \ge 1\}$ to deduce that $\mu\{\pi \ge 1\} = 0$. Consequently,

$$(4.40) \qquad \int_{\{\pi<1\}} f(1-\pi) \, d\nu = \int_{\{\pi<1\}} f\pi \, d\mu \qquad \forall f \in L^2(\mu+\nu).$$

According to the monotone convergence theorem, the preceding holds for all non-negative measurable functions f. Replace f by $f(1-\pi)^{-1}\mathbf{1}_{\{\pi<1\}}$ and consider $\Pi := \pi(1-\pi)^{-1}\mathbf{1}_{\{\pi<1\}}$ to deduce that there exists a non-negative measurable function Π such that $\int_{\{\pi<1\}} f \, d\nu = \int f\Pi \, d\mu$ for all measurable $f \ge 0$. In general, one cannot go further and remove the $\{\pi < 1\}$ from the integral. However, if $\nu \ll \mu$, then the already-proven fact that $\mu\{\pi \ge 1\} = 0$ implies that $\nu\{\pi \ge 1\} = 0$. Hence, for all non-negative measurable f,

$$(4.41) \qquad \int f \, d\nu = \int f\Pi \, d\mu.$$

Plug in $f(x) \equiv 1$ to deduce that $\int \Pi \, d\mu = \nu(\Omega) < \infty$. It follows that $\Pi \in L^1(\mu)$, and the theorem concludes with $\pi_* = \Pi$.

Step 3. Uniqueness. We conclude the proof by proving that if there exist $\pi_*, \pi^* \in L^1(\mu)$ such that $\int f \pi_* \, d\mu = \int f \pi^* \, d\mu$ for all bounded measurable $f : \Omega \to \mathbf{R}$, then $\pi_* = \pi^*$ a.e. $[\mu]$.

Let us posit the existence of such π_*, π^*. Fix $\epsilon > 0$ and define $f = \mathbf{1}_{A(\epsilon)}$, where $A(\epsilon) := \{\omega \in \Omega : \pi^*(\omega) \geq \pi_*(\omega) + \epsilon\}$. Then,

$$(4.42) \quad \int f \pi_* \, d\mu = \int f \pi^* \, d\mu \geq \int f(\pi_* + \epsilon) \, d\mu = \int f \pi_* \, d\mu + \epsilon \mu(A(\epsilon)).$$

Because $\int f \pi_* \, d\mu \leq \|\pi_*\|_{L^1(\mu)} < \infty$, this proves that $\mu(A(\epsilon)) = 0$ for all $\epsilon > 0$. By the continuity properties of measures (Lemma 3.11, page 25),

$$(4.43) \quad 0 = \lim_{\substack{\epsilon \downarrow 0 \\ \epsilon \in \mathbf{Q}}} \mu(A(\epsilon)) = \mu\left(\bigcup_{\substack{\epsilon > 0: \\ \epsilon \in \mathbf{Q}}} A(\epsilon) \right).$$

But the right-hand side is the μ-measure of the set where $\pi^* > \pi_*$. This proves that $\pi^* \leq \pi_*$ a.e. $[\mu]$. Reverse the roles of π^* and π_* to find that they are equal almost everywhere $[\mu]$. □

Problems

4.1. Let Ω be a set and $(A, \mathscr{A})$ a measurable space. For any function $X : \Omega \to A$ define $\sigma(X)$ to be the collection of all inverse images of X; i.e., $\{X^{-1}(B); B \in \mathscr{A}\}$. Prove that $\sigma(X)$ is a σ-algebra, and is the smallest σ-algebra with respect to which X is measurable.

4.2. Prove Lemma 4.6. (HINT: If $f(\omega) \in [j 2^{-n}, (j+1)2^{-n})$, then set $\overline{f}_n(\omega) := (j+1)2^{-n}$.)

4.3. Let μ denote the counting measure on the measure space $(\Omega, \mathscr{F})$; see Problem 3.8 on page 33. Prove, carefully, that $\int f \, d\mu = \sum_{x \in \Omega} f(x)$ whenever f is absolutely summable.

4.4 (Distributions). Suppose $(\Omega, \mathscr{F}, \mu)$ is a measure space. Let Ω' be a set, and let $\mathscr{F}'$ denote a σ-algebra of subsets of Ω'. Prove that if $f : \Omega \to \Omega'$ is measurable, then $\mu \circ f^{-1}$ is a measure on $(\Omega', \mathscr{F}')$, where $(\mu \circ f^{-1})(A) = \mu(\{\omega \in \Omega : f(\omega) \in A\})$ is the so-called *distribution* of f.

4.5 (Riesz Representation Theorem). Let $C(0,1]$ denote the collection of all continuous functions that map $(0,1]$ to $\mathbf{R}$. A map $T : C(0,1] \to \mathbf{R}$ is a *positive, bounded linear functional* if: (i) For all $a, b \in \mathbf{R}$ and $f, g \in C(0,1]$, $T(af + bg) = aT(f) + bT(g)$; (ii) there exists a finite constant A such that $|T(f)| \leq A \sup_{y \in (0,1]} |f(y)|$ for all $f \in C(0,1]$; and (iii) $T(f) \geq 0$ whenever $f \geq 0$. The smallest such A is the *norm* $\|T\|$ of T.

(1) Given any finite measure μ on $\mathscr{B}((0,1])$, check that $T(f) = \int f \, d\mu$ defines a positive, bounded linear functional. Compute $\|T\|$.

(2) Conversely, prove that for any positive, bounded, linear functional T on $(0,1]$ there exists a finite measure μ such that for all $f \in C(0,1]$, $T(f) = \int f \, d\mu$. (HINT: For any closed set C define $\mu(C) := \inf\{T(f) : f \geq \mathbf{1}_C\}$. For a general set G define $\mu(G) := \sup\{\mu(C) : C \subset G, C \text{ closed}\}$.)

This is due to F. Riesz. (HINT: Examine Carathéodory's theorem on page 27. Also, see the proof of Lemma 3.15, p. 26.)

4.6. Consider two σ-finite measures μ and ν, both defined on $(\mathbf{R}^k, \mathscr{B}(\mathbf{R}^k))$. Prove that if $\int f \, d\mu = \int f \, d\nu$ for all continuous functions $f : \mathbf{R}^k \to \mathbf{R}$, then $\mu = \nu$.

4.7. Choose and fix $p_1, \ldots, p_n > 0$ such that $\sum_{\nu=1}^n p_\nu = 1$. Prove that if $x_1, \ldots, x_n$ are positive, then $\prod_{\nu=1}^n x_\nu^{p_\nu} \leq \sum_{\nu=1}^n p_\nu x_\nu$. Prove also that the inequality is strict unless $x_1 = \cdots = x_n$.

4.8. Define $\mathscr{A} = \sum_{i=1}^n a_i/n$ to be the arithmetic mean, $\mathscr{G} = (\prod_{i=1}^n a_i)^{1/n}$ the geometric mean, and $\mathscr{H} = n/\sum_{i=1}^n a_i^{-1}$ the harmonic mean of $\{a_i\}_{i=1}^n$, where $a_i \geq 0$. Prove that $\mathscr{H} \leq \mathscr{G} \leq \mathscr{A}$.

4.9. Choose and fix an interval $I \subseteq \mathbf{R}$. A function $f : I \to \mathbf{R}$ is said to be *convex on I* if $f(\lambda x + (1-\lambda)y) \leq \lambda f(x) + (1-\lambda)f(y)$ for all $x,y \in I$ and $\lambda \in [0,1]$. Prove that if $f'' \geq 0$ on I, then f is convex on I. Use this to prove that:

(1) The Euler gamma function $\Gamma(t) = \int_0^\infty x^{t-1} e^{-x}\,dx$ is convex on $(0,\infty)$.

(2) The function $x^{-1}\exp(-x^2/2)$ is convex on $\mathbf{R}_+$.

4.10 (Problem 4.9, Continued). Suppose $f : \mathbf{R} \to \mathbf{R}$ is convex. Prove that f has right and left derivatives everywhere; i.e., prove that $\partial_+ f(x) = \lim_{\epsilon \downarrow 0}(f(x+\epsilon)-f(x))/\epsilon$ and $\partial_- f(x) = \lim_{\epsilon \downarrow 0}(f(x)-f(x-\epsilon))/\epsilon$ exist. Prove, in addition, that $\partial_+ f$ and $\partial_- f$ are both non-decreasing.

4.11. Prove that if f is convex, then it is continuous. Conversely, prove that if f is continuous and $2f((a+b)/2) \leq f(a)+f(b)$ for all $a,b \in \mathbf{R}$, then f is convex.

4.12. Prove that if $\sup_n |X_n| \leq Y$, where $Y \in L^1(\mathrm{P})$, then, $\limsup_{n\to\infty} \mathrm{E}X_n \leq \mathrm{E}[\limsup_{n\to\infty} X_n]$. Construct an example to show that the domination condition on the X_n's cannot be altogether dropped.

4.13. Let $\phi(a) := \min(|a|, 1)$, and given any two random variables X and Y, define $d_\mathrm{P}(X, Y) = \mathrm{E}\phi(X-Y)$. Prove that d_P is a metric, and X_n converges to X in probability if and only if $d_\mathrm{P}(X_n, X) \to 0$. That is, d_P metrizes convergence in probability.

4.14. Prove that if $X \in L^p(\mathrm{P})$ for some $p > 0$, then $\lim_{t\to\infty} t^p \mathrm{P}\{|X| > t\} = 0$, and the latter condition implies that $X \in L^r(\mathrm{P})$ for all $r \in (0, p)$. (HINT: Apply Fubini–Tonelli to $\int_0^\infty t^{r-1} \mathrm{P}\{|X| > t\}\,dt$.)

4.15 (Slutsky's Theorem). Suppose $\{X_n\}_{n=1}^\infty$ and $\{Y_n\}_{n=1}^\infty$ are two sequences of random variables such that $X_n \xrightarrow{\mathrm{P}} X$ and $Y_n \xrightarrow{\mathrm{P}} Y$. Prove that if f is a continuous function of two variables, then $f(X_n, Y_n) \xrightarrow{\mathrm{P}} f(X, Y)$.

4.16. Prove that X_n converges to X in probability iff for any subsequence $\{n_k\}_{k=1}^\infty$ there exists a further sub-subsequence $\{n_{k(i)}\}_{i=1}^\infty$ such $X_{n_{k(i)}} \to X$ a.s.

4.17. Consider the measure space $([0,1], \mathscr{B}([0,1]), \mu)$ where μ is a finite measure. Prove that if $f : [0,1] \to \mathbf{R}$ is continuous and $\{\lambda_j\}_{j=1}^\infty$ is a sequence of numbers in $[0,1]$, then

$$\text{(4.44)} \qquad \lim_{n\to\infty} \sum_{j=1}^n f\left(\frac{j-\lambda_j}{n}\right) \mu\left(\left(\frac{j-1}{n}, \frac{j}{n}\right]\right) = \int f\,d\mu - f(0)\mu(\{0\}).$$

Use this to prove that if μ denotes Lebesgue measure, then the Riemann integral of f agrees with its Lebesgue integral. Can you extend this to σ-finite measure spaces $(\mathbf{R}^d, \mathscr{B}(\mathbf{R}^d), \mu)$ and integrable continuous functions $f : \mathbf{R}^d \to \mathbf{R}$?

4.18. Let μ denote the Lebesgue measure on $([0,1]^d, \mathscr{B}([0,1]^d))$. Prove that continuous functions are dense in $L^p(\mu)$ for every $p \geq 1$. That is, prove that given $\epsilon > 0$ and $f \in L^p(\mu)$ we can find a continuous function $g : [0,1]^d \to \mathbf{R}$ such that $\|f-g\|_p \leq \epsilon$.

4.19. If $f : \mathbf{R}^k \to \mathbf{R}$ satisfies $\int |f(x)|^p\,dx < \infty$ for some $p > 1$, then prove that f is *continuous in $L^p(\mathbf{R}^k)$*; i.e., $\lim_{\epsilon \to 0} \int_{\mathbf{R}^k} |f(x+\epsilon) - f(x)|^p\,dx = 0$. (HINT: Problem 4.18.)

4.20. Prove that the following exists, and compute its value:

$$\text{(4.45)} \qquad \lim_{n\to\infty} \int_{-\sqrt{n}}^{\sqrt{n}} \left(1 - \frac{x^2}{2n}\right)^n dx.$$

4.21. Construct a σ-finite measure space $(\Omega, \mathscr{F}, \mu)$ such that $L^2(\mu) \not\subset L^1(\mu)$.

4.22 (Mixtures). Suppose $(\Omega, \mathscr{F})$ and $(\Theta, \mathscr{G})$ are two measure spaces. Assume that ν is a probability measure on $\mathscr{G}$ and P_θ a probability measure on $\mathscr{F}$ for each $\theta \in \Theta$. Then prove that $\mu(A) = \int_\Theta P_\theta(A)\,\nu(d\theta)$ $(A \in \mathscr{F})$ defines a probability measure on $\mathscr{F}$, provided that $\theta \mapsto P_\theta(A)$ is $\mathscr{G}$-measurable for each $A \in \mathscr{F}$. The probability measure μ is said to be a *mixture* of the P_θ's; the *mixing measure* is the probability measure ν.

4.23 (Generalized Hölder Inequality). Let $\{X_i\}_{i=1}^n$ be non-negative random variables. Prove that for all $\rho_1, \ldots, \rho_n > 1$ that satisfy $\sum_{i=1}^n \rho_i^{-1} = 1$, $\mathrm{E}(X_1 \cdots X_n) \leq \prod_{i=1}^n \|X_i\|_{\rho_i}$. (HINT: Problem 4.7.)

4.24 (Chernoff's Inequality). Prove that for any random variable X and all $t > 0$, $\mathrm{P}\{X \geq t\} \leq \inf_{\xi \geq 0} \exp\{-t\xi + \ln \mathrm{E}e^{\xi X}\}$, and $\mathrm{P}\{X \leq t\} \leq \inf_{\xi \geq 0} \exp\{t\xi + \ln \mathrm{E}e^{-\xi X}\}$.

4.25 (Hadamard's Inequality). Suppose f is a convex and integrable function on (a, b). Then prove that $(b - a)f((a+b)/2) \leq \int_a^b f(x)\,dx$.

4.26 (Young's Inequality). Suppose f is a continuous, strictly increasing function on $[0, a]$, and $f(0) = 0$. Prove that $ab \leq \int_0^a f(x)\,dx + \int_0^b f^{-1}(x)\,dx$ for all $b \geq f(a)$, with equality iff $b = f(a)$. Here, f^{-1} is the inverse function to f. Use this to find another proof of (4.9). (HINT: Plot f, and consider the areas under and over f, respectively.)

4.27 (An Uncertainty Principle). Prove that all continuously differentiable functions $f : \mathbf{R} \to \mathbf{R}$ that have compact support satisfy the inequality

$$(4.46) \qquad \int_{-\infty}^\infty |f(x)|^2\,dx \leq 2\left(\int_{-\infty}^\infty x^2 |f(x)|^2\,dx\right)^{1/2} \left(\int_{-\infty}^\infty |f'(x)|^2\,dx\right)^{1/2}.$$

Use the preceding to relax the compact-support assumption. (HINT: Integrate by parts to find that $\int_{-\infty}^\infty xf(x)f'(x)\,dx = -\frac{1}{2}\int_{-\infty}^\infty f^2(x)\,dx$.)

4.28 (Uniform Integrability). Let $\{X_n\}_{n=1}^\infty$ denote a collection of random variables on a common probability space $(\Omega, \mathscr{F}, \mathrm{P})$. The X's are *uniformly integrable* (UI) if they are integrable and

$$(4.47) \qquad \lim_{t \to \infty} \limsup_{n \to \infty} \mathrm{E}\{|X_n|; |X_n| \geq t\} = 0.$$

Prove that $\{X_n\}_{n=1}^\infty$ is UI iff $\lim_{t \to \infty} \sup_{n \geq 1} \mathrm{E}\{|X_n|; |X_n| \geq t\} = 0$. Also prove:

(1) If $|X_n| \leq |Y_n|$ $(n \geq 1)$ and $\{Y_n\}_{n=1}^\infty$ is UI, then so is $\{X_n\}_{n=1}^\infty$.

(2) If $\{X_n\}_{n=1}^\infty$ and $\{Y_n\}_{n=1}^\infty$ are UI, then so is $\{X_n + Y_n\}_{n=1}^\infty$.

(3) $\{X_n\}_{n=1}^\infty$ is UI as long as $\sup_n \|X_n\|_p < \infty$ for some $p > 1$.

(4) $X_n \to X$ in $L^1(\mathrm{P})$ if and only if: (a) $X_n \xrightarrow{\mathrm{P}} X$; and (b) $\{X_n\}_{n=1}^\infty$ is UI.

4.29 (Problem 4.28, Continued). Let $p \geq 1$, and consider $X, X_1, X_2, \ldots \in L^p(\mathrm{P})$ such that $X_n \to X$, in probability. Prove that either one of the following is equivalent to the uniform integrability of $\{|X_i|^p\}_{i=1}^\infty$: (i) $X_n \to X$ in $L^p(\mathrm{P})$; or (ii) $\mathrm{E}\{|X_n|^p\} \to \mathrm{E}\{|X|^p\}$ as $n \to \infty$.

4.30 (Hoeffding's Inequality). Suppose $\mathrm{E}X = 0$ and $\mathrm{P}\{|X| \leq c\} = 1$ for some non-random constant $c > 0$. Prove that for all $\xi \in \mathbf{R}$, $\mathrm{E}e^{\xi X} \leq \exp(\xi^2 c^2/2)$ and $\mathrm{E}e^{\xi |X|} \leq 2\exp(\xi^2 c^2/2)$. (HINT: $e^{\xi x} \leq e^{\xi c}(c+x)/(2c) + e^{-\xi c}(c-x)/(2c)$ for all $x \in [-c, c]$.)

4.31 (Hard). For all $\alpha \geq 0$ compute

$$(4.48) \qquad \lim_{n \to \infty} n \int_0^{n^\alpha} \frac{\exp(x/n) - 1}{x + x^3}\,dx.$$

4.32 (The Good-Lambda Inequality; Hard). Let X and Y be two non-negative random variables, and $p \geq 1$ a fixed constant. Suppose there exist $\beta > 1$, $\gamma \in (0, 1)$, and $\delta < \beta^{-p}$ such that $\mathrm{P}\{X > \beta\lambda, Y < \gamma\lambda\} \leq \delta \mathrm{P}\{X > \lambda\}$ for all $\lambda > 0$. Then prove that $\mathrm{E}[X^p] \leq \alpha \gamma^{-p} \mathrm{E}[Y^p]$, where $\alpha = (\beta^{-p} - \delta)^{-1}$.

4.33 (Harder). Let X and Y be two non-negative random variables such that $X, Y, \log Y \in L^1(\mathrm{P})$. Suppose for all measurable sets A, $\mathrm{E}[X; A] \geq |\mathrm{E}[Y; A]|^2$. Then prove that $\mathrm{E}[\log X] > -\infty$. (HINT: Set $A_{-1} = \{X > Y\}$, and for all $n \geq 0$ define $A_n = \{e^{-n-1}Y < X \leq e^{-n}Y\}$ and $B_n = A_n \cap \{Y \leq e^{-n/4}\}$. Prove that $\sum_n n\mathrm{P}(A_n \setminus B_n) < \infty$. Use this to prove that $\sum_n n\mathrm{P}(A_n) < \infty$. Alternatively, see Dudley (1967).)

Notes

(1) The modern notions of abstract random variables—as measurable functions—and expectations—as integrals—seem to be due to Fréchet (1930). In concrete settings, these notions have been around for quite a long time. See, for example, the classic by Borel (1909).

(2) Kolmogorov (1933) created the modern, axiomatic theory of probability in his landmark book. Among other things, Kolmogorov's work is said to have solved a main part of Hilbert's sixth problem (Gnedenko, 1969).

(3) Much of the material of Section 5 is due to Lebesgue (1910). Notable exceptions to this remark are Fatou's lemma (1906) and the monotone convergence theorem of Levi (1906).

(4) Problem 4.15 is due, in its essence, to Slutsky (1925).

(5) Problem 4.24 is due to Chernoff (1952).

(6) Problem 4.27 is a disguised form of the Heisenberg uncertainty principle. In this form, it is due to H. Weyl. Another form will be discussed in Problem 7.36 on page 115.

(7) Problem 4.30 is due to Hoeffding (1963, Lemma 1).

(8) The good-λ inequality (Problem 4.32) is a fundamental tool in probability and harmonic analysis. It was invented by Burkholder and Gundy (1970) and explored further by Coifman (1972) and Burkholder, Davis, and Gundy (1972). See also the expository account by Jones (1998).

Chapter 5

Product Spaces

Nature is an infinite sphere, whose center is everywhere and whose circumference is nowhere.

–Blaise Pascal

If A_1 and A_2 are sets, then their product $A_1 \times A_2$ is defined to be the collection of all ordered pairs (a_1, a_2) where $a_1 \in A_1$ and $a_2 \in A_2$. In a like manner, we define $A_1 \times A_2 \times A_3$, etc. We can even define infinite-product spaces of the type $A_1 \times A_2 \times \ldots$.

We have two main reasons for studying the measure theory of product spaces. The first one is that an understanding of product spaces allows for the construction and analysis of several random variables simultaneously; a theme that is essential to nearly all of probability theory.

Our second reason for learning more about product spaces is less obvious at this point: We will need the so-called Fubini–Tonelli theorem that allows us to interchange the order of various multiple integrals. This is a central fact, and it leads to a number of essential computations.

1. Finite Products

Suppose $(\Omega_1, \mathscr{F}_1, \mu_1)$ and $(\Omega_2, \mathscr{F}_2, \mu_2)$ are two finite measure spaces. There is a natural σ-algebra $\mathscr{F}_1 \times \mathscr{F}_2$ and a measure $\mu_1 \times \mu_2$ that correspond to the product set $\Omega_1 \times \Omega_2$.

First consider the collection

(5.1) $$\mathscr{A}_0 := \{A_1 \times A_2 : A_1 \in \mathscr{F}_1, \ A_2 \in \mathscr{F}_2\}.$$

This is closed under finite (in fact arbitrary) intersections, but not under finite unions. For example, let $A_1 = A_2 = [0, 1]$ and $B_1 = B_2 = [1, 2]$ to see

53

that $(A_1 \times A_2) \cup (B_1 \times B_2)$ is not of the form $C_1 \times C_2$ for any C_1 and C_2. So $\mathscr{A}_0$ is not an algebra. We correct this by adding to $\mathscr{A}_0$ all finite disjoint unions of elements of $\mathscr{A}_0$, and call the resulting collection $\mathscr{A}$.

Lemma 5.1. *The collection $\mathscr{A}$ is an algebra, and $\sigma(\mathscr{A}) = \sigma(\mathscr{A}_0)$.*

Definition 5.2. We write $\mathscr{F}_1 \times \mathscr{F}_2$ in place of $\sigma(\mathscr{A})$.

Define μ on $\mathscr{A}_0$ as follows:

$$(5.2) \qquad \mu(A_1 \times A_2) := \mu_1(A_1)\mu_2(A_2) \qquad \forall A_1 \in \mathscr{F}_1, A_2 \in \mathscr{F}_2.$$

If $A^1, \ldots, A^n \in \mathscr{A}_0$ are disjoint, then we define $\mu(\cup_{i=1}^n A^i) := \sum_{i=1}^n \mu(A^i)$. This constructs μ on the algebra $\mathscr{A}$ in a well-defined manner. Indeed, suppose $\cup_{i=1}^n A^i = \cup_{j=1}^m B^j$ where the A^i's are disjoint and the B^j's are also disjoint. Then, $\cup_{i=1}^n A^i = \cup_{i=1}^n \cup_{j=1}^m (A^i \cap B^j)$ is a disjoint union of nm sets. Therefore,

$$(5.3) \qquad \mu\left(\bigcup_{i=1}^n A^i\right) = \sum_{i=1}^n \sum_{j=1}^m \mu(A^i \cap B^j) = \mu\left(\bigcup_{j=1}^m B^j\right),$$

by symmetry.

Theorem 5.3. *There exists a unique measure $\mu_1 \times \mu_2$ on $(\Omega_1 \times \Omega_2, \mathscr{F}_1 \times \mathscr{F}_2)$ such that $\mu_1 \times \mu_2 = \mu$ on $\mathscr{A}$.*

Definition 5.4. The measure $\mu_1 \times \mu_2$ is called the *product measure* of μ_1 and μ_2; the space $\Omega_1 \times \Omega_2$ is the corresponding *product space*, and $\mathscr{F}_1 \times \mathscr{F}_2$ is the *product σ-algebra*. The measure space $(\Omega_1 \times \Omega_2, \mathscr{F}_1 \times \mathscr{F}_1, \mu_1 \times \mu_2)$ is the *product measure space*.

Remark 5.5. By induction, we can construct a product measure space $(\Omega, \mathscr{F}, \mu)$ based on any finite number of measure spaces $(\Omega_i, \mathscr{F}_i, \mu_i)$, $i = 1, \ldots, n$: Define $\Omega := \Omega_1 \times \cdots \times \Omega_n$, $\mathscr{F} := \mathscr{F}_1 \times \cdots \times \mathscr{F}_n$, and $\mu := \mu_1 \times \cdots \times \mu_n$.

Proof of Theorem 5.3. By Carathéodory's extension theorem (p. 27), it suffices to prove that $(\mu_1 \times \mu_2)$ is countably additive on the algebra $\mathscr{A}$. We accomplish this in three successive steps.

Step 1. Sections of Measurable Sets are Measurable. For all $E \subseteq \Omega_1 \times \Omega_2$ and $\omega_2 \in \Omega_2$ define

$$(5.4) \qquad E_{\omega_2} := \{\omega_1 \in \Omega_1 : (\omega_1, \omega_2) \in E\}.$$

This is the *section* of E along ω_2. In the first step of the proof we demonstrate that if E is measurable, then for every $\omega_2 \in \Omega_2$, E_{ω_2} is measurable too: Fix $\omega_2 \in \Omega_2$ and consider the collection

$$(5.5) \qquad \mathscr{M} := \{E \in \mathscr{F}_1 \times \mathscr{F}_2 : E_{\omega_2} \in \mathscr{F}_1\}.$$

Because $\mathscr{M}$ is a monotone class that contains $\mathscr{A}$, the monotone class theorem (p. 30) implies that $\mathscr{M} = \mathscr{F}_1 \times \mathscr{F}_2$. This concludes Step 1.

Step 2. Disintegration. Because E_{ω_2} is measurable, $\mu_1(E_{\omega_2})$ is well defined. We now show that $\Omega_2 \ni \omega_2 \mapsto \mu_1(E_{\omega_2})$ is measurable. First suppose $E \in \mathscr{A}_0$, so that $E = A_1 \times A_2$ where $A_i \in \mathscr{F}_i$. Then $E_{\omega_2} = A_1$ if $\omega_2 \in A_2$, and $E_{\omega_2} = \varnothing$ if $\omega_2 \in A_2^c$. Consequently,

$$(5.6) \qquad \mu_1(E_{\omega_2}) = \mu_1(A_1)\mathbf{1}_{A_2}(\omega_2).$$

It follows that $\mu_1(E_{\omega_2})$ is a measurable function of $\omega_2 \in \Omega_2$. Furthermore, $(\mu_1 \times \mu_2)(E) = \mu_1(A_1)\mu_2(A_2)$, and hence

$$(5.7) \qquad (\mu_1 \times \mu_2)(E) = \int \mu_1(E_{\omega_2})\, \mu_2(d\omega_2).$$

Equation (5.7) is called a *disintegration formula*.

Step 3. Countable Additivity. By finite additivity, (5.7) extends the definition of $\mu_1 \times \mu_2$ to finite disjoint unions of elements of $\mathscr{A}_0$; i.e., (5.7) holds for all $E \in \mathscr{A}$. Furthermore, the dominated convergence theorem shows that $\mu_1 \times \mu_2$ is countably additive on the algebra $\mathscr{A}$. [It suffices to prove that if $E^N \in \mathscr{A}$ satisfy $E^N \downarrow \varnothing$, then $(\mu_1 \times \mu_2)(E^N) \downarrow 0$. But this follows from (5.7) and the monotone convergence theorem.] Therefore, owing to the Carathéodory extension theorem (p. 27), $\mu_1 \times \mu_2$ can be extended uniquely to a measure on all of $\mathscr{F}_1 \times \mathscr{F}_2$. This proves the theorem. In addition, it shows that (5.7) holds for all $E \in \mathscr{F}_1 \times \mathscr{F}_2$. (The fact that $\omega_2 \mapsto \mu_1(E_{\omega_2})$ is measurable is proved implicitly here; why?) □

The following shows that the two possible ways of constructing Lebesgue's measure coincide.

Corollary 5.6. *If m^d denotes the Lebesgue measure on $((0,1]^d, \mathscr{B}((0,1]^d))$, then $m^d = m^1 \times \cdots \times m^1$ (d times.)*

Proof. If $E = (a_1, b_1] \times \cdots \times (a_d, b_d]$ is a d-dimensional hypercube, then

$$(5.8) \qquad m^d(E) = \prod_{j=1}^{d}(b_j - a_j) = (m^1 \times \cdots \times m^1)(E).$$

By finite additivity, m^d and $(m^1 \times \cdots \times m^1)$ agree on the smallest algebra that contains hypercubes. By Carathéodory's extension theorem, m^d and $(m^1 \times \cdots \times m^1)$ agree on the σ-algebra generated by hypercubes. □

The following is an important consequence of this development.

The Fubini–Tonelli Theorem. *If $f : \Omega_1 \times \Omega_2 \to \mathbf{R}$ is product measurable, then for each $\omega_1 \in \Omega_1$, $\omega_2 \mapsto f(\omega_1, \omega_2)$ is $\mathscr{F}_2$-measurable, and by symmetry,* for each $\omega_2 \in \Omega_2$, $\omega_1 \mapsto f(\omega_1, \omega_2)$ is $\mathscr{F}_1$-measurable. If in addition $f \in L^1(\mu_1 \times \mu_2)$, then the following are a.e.-finite measurable functions:

$$\text{(5.9)} \qquad \omega_1 \mapsto \int f(\omega_1, \omega_2)\, \mu_2(d\omega_2), \quad \omega_2 \mapsto \int f(\omega_1, \omega_2)\, \mu_1(d\omega_1).$$

Finally, the following change-of-variables formula is valid:

$$\text{(5.10)} \qquad \begin{aligned} \int f\, d(\mu_1 \times \mu_2) &= \int \left(\int f(\omega_1, \omega_2)\, \mu_1(d\omega_1) \right) \mu_2(d\omega_2) \\ &= \int \left(\int f(\omega_1, \omega_2)\, \mu_2(d\omega_2) \right) \mu_1(d\omega_1). \end{aligned}$$

Proof. (Sketch) If $f = \mathbf{1}_E$ for some $E \in \mathscr{F}_1 \times \mathscr{F}_2$, then (5.7) contains (5.9) and (5.10). By linearity, these equations continue to hold for all simple functions f. Finally, we take limits to prove the result for every function $f \in L^1(\mu_1 \times \mu_2)$. $\square$

The following is an important corollary of the proof of Fubini–Tonelli's theorem. (Proof: Approximate f from below by simple functions; then appeal to the monotone convergence theorem.)

Corollary 5.7. *If $f : \Omega_1 \times \Omega_2 \to \mathbf{R}$ is measurable and non-negative, then (5.10) holds in the sense that all three double integrals converge and diverge together, and are equal in the convergent case.*

Remark 5.8. In fact, the proof shows that $f \in L^1(\mu_1 \times \mu_2)$ as long as one of the three integrals in (5.10) is finite when f is replaced by $|f|$.

The Fubini–Tonelli theorem is deceptively delicate: We cannot always interchange the order of double integrals. Our next examples highlight this fact.

Example 5.9. For all $n \geq 0$ and $x \in (0, 1]$ define

$$\text{(5.11)} \qquad \psi_n(x) := \begin{cases} 2^{n+1} & \text{if } x \in (2^{-n-1}, 2^{-n}], \\ 0 & \text{otherwise.} \end{cases}$$

Note, in particular, that $\int_0^1 \psi_n(x)\, dx = 1$. Define the measurable function $f : (0, 1]^2 \to \mathbf{R}$ as follows:

$$\text{(5.12)} \qquad f(x, y) = \sum_{n=0}^{\infty} \big[\psi_n(x) - \psi_{n+1}(x) \big] \psi_n(y) \qquad \forall x, y \in (0, 1].$$

All but possibly one of these terms are zero, so the function f is well defined. Nonetheless, we argue next that the Fubini–Tonelli theorem is not applicable to the function f.

1. Finite Products

If $y \in (2^{-n-1}, 2^{-n}]$ then $f(x,y) = 2^{n+1}[\psi_n(x) - \psi_{n+1}(x)]$. It follows that $\int_0^1 f(x,y)\,dx = 0$, whence we have $\int_0^1 \int_0^1 f(x,y)\,dx\,dy = 0$. On the other hand,

$$(5.13) \qquad \int_{2^{-n-1}}^{2^{-n}} f(x,y)\,dy = \psi_n(x) - \psi_{n+1}(x).$$

Sum this from $n = 0$ to $n = m-1$ to find that

$$(5.14) \qquad \int_{2^{-m}}^1 f(x,y)\,dy = \psi_0(x) - \psi_m(x).$$

But if $x > 0$, then $\lim_{m \to \infty} \psi_m(x) = 0$ for all $x \in (0,1]$. Thus,

$$(5.15) \qquad \int_0^1 f(x,y)\,dy = \psi_0(x).$$

We integrate this over all values of $x \in (0,1]$ to find that

$$(5.16) \qquad \int_0^1 \int_0^1 f(x,y)\,dx\,dy = 0 \neq 1 = \int_0^1 \int_0^1 f(x,y)\,dy\,dx.$$

Thus, Fubini–Tonelli's theorem does not apply, and the reason is that f is not absolutely integrable. (Prove it!)

The preceding example is slightly complicated because we had to work with finite measures. But, in fact, the Fubini–Tonelli theorem is valid for sigma-finite measures as well (Problem 5.5). If we admit this, then we can greatly simplify the preceding example.

Example 5.10. Define $f : \mathbf{R}_+^2 \to \{-1, 0, 1\}$ as follows:

$$(5.17) \quad f(x,y) = \sum_{n=0}^\infty \left[\mathbf{1}_{[n,n+1) \times [n,n+1)}(x,y) - \mathbf{1}_{[n,n+1) \times [n+1,n+2)}(x,y) \right].$$

One can check directly that $\int_0^\infty \int_0^\infty f\,dx\,dy = 1$ whereas $\int_0^\infty \int_0^\infty f\,dy\,dx = 0$. As was the case in the preceding example, the Fubini–Tonelli theorem fails to apply here because $\int_0^\infty \int_0^\infty |f|\,dx\,dy = \infty$.

Our next example, due to W. Sierpiński, illustrates that Fubini–Tonelli's theorem need not hold when f is not product-measurable. Throughout, we will rely on the axiom of choice as well as the continuum hypothesis, and let $(\Omega, \mathscr{F}, P)$ designate the Steinhaus probability space.

Example 5.11. Define c to be the first uncountable ordinal; by the axiom of choice c exists. Next, define S to be the collection of all ordinal numbers strictly less than c; S is called Hartog's c-*section* of ordinal numbers. By the continuuum hypothesis, S has the power of the continuum. That is, we can find a one-to-one map $\phi : [0,1] \to S$.

Now consider the set

(5.18) $$E := \{(x,y) \in [0,1]^2 : \phi(x) < \phi(y)\}.$$

For all $x \in [0,1]$ consider the x-section $_xE$ of E,

(5.19) $\quad _xE := \{y \in [0,1] : (x,y) \in E\} = \{y \in [0,1] : \phi(x) < \phi(y)\}.$

Both E and $_xE$ are non-empty, because ϕ is one-to-one. Moreover, $_xE^c$ is denumerable because $\phi(_xE^c)$ is. [This follows from the definition of S.] Consequently, $_xE$ is Borel measurable and $P(_xE) = 1$ for all $x \in [0,1]$.

We can also define the y-section $E_y := \{x \in [0,1] : (x,y) \in E\}$ for any $y \in [0,1]$. Since E_y is denumerable, $E_y \in \mathscr{F}$ and $P(E_y) = 0$. Hence,

(5.20) $$\int_0^1 P(_xE)\,P(dx) = 1 \neq 0 = \int_0^1 P(E_y)\,P(dy).$$

So there is no disintegration formula (5.7). This, in turn, implies that (5.10) does not hold for the bounded function $f(x,y) = \mathbf{1}_E(x,y)$. Since P is a probability measure on the Borel subsets of $[0,1]$, all bounded measurable functions are P-integrable. Thus we see that the source of the difficulty is that f is not product-measurable although $x \mapsto \int f(x,y)\,P(dy)$ and $y \mapsto \int f(x,y)\,P(dx)$ are measurable (in fact constants).

2. Infinite Products

So far, our only nontrivial example of a measure is Lebesgue measure on $(\mathbf{R}^n, \mathscr{B}(\mathbf{R}^n))$. We have also seen that we can create other interesting product measures once we know some nice measures. We now wish to add to our repertoire of nontrivial measures by defining measures on infinite-product spaces that we take to be $(0,1]^\infty$ or $\mathbf{R}^\infty$, where for any Ω the set Ω^∞ is defined as the collection of all infinite sequences of the form $(\omega_1, \omega_2, \ldots)$ where $\omega_i \in \Omega$.

In order to construct measures on $(0,1]^\infty$, or more generally $\mathbf{R}^\infty$, we first need a topology in order to have a Borel σ-algebra $\mathscr{B}((0,1]^\infty)$.

Definition 5.12. Given a topological set Ω, a set $A \subseteq \Omega^\infty$ is called a *cylinder set* if either $A = \varnothing$, or it has the form $A = A_1 \times A_2 \times \cdots$ where $A_i = \Omega$ for all but a finite number of i's. A cylinder set $A = A_1 \times A_2 \times \cdots$ is *open* if every A_i is open in Ω. The *product topology* on Ω^∞ is the smallest topology that contains all open cylinder sets. This, in turn, gives us the Borel σ-algebra $\mathscr{B}(\Omega^\infty)$.

Suppose we wanted to construct the Lebesgue measure on $(0,1]^\infty$. Note that any cylinder set has a perfectly well-defined Lebesgue measure. For example, let $I_\ell = (0, \ell^{-1}]$ for $\ell = 1, 2, 3$, and $I_\ell = (0,1]$ for $\ell \geq 4$. Then,

$I = I_1 \times I_2 \times I_3 \times I_4 \times \cdots$ is a cylinder set, and it would make perfectly good sense that the "Lebesgue measure" of I should be $1 \times \frac{1}{2} \times \frac{1}{3} = \frac{1}{6}$.

It stands to reason that if m denotes one-dimensional Lebesgue measure then one should be able to define the Lebesgue measure $m^\infty = m \times m \times \cdots$ on $((0,1]^\infty, \mathscr{B}((0,1]^\infty))$ as the (or perhaps a) "projective limit" of the n-dimensional Lebesgue measure $m^n = m \times \cdots \times m$ on $((0,1]^n, \mathscr{B}((0,1]^n))$. This argument can be made rigorous not only for m^∞, but for a large class of other measures as well. But first we need some notation for projections.

For all $1 \leq n \leq \infty$ let $I^n := (0,1]^n$ and $\mathscr{B}^n := \mathscr{B}(I^n)$. If $A = A_1 \times A_2 \times \cdots$ is a subset of I^∞, then for any integer $n \geq 1$ we can project A onto its first n coordinates as follows:

$$(5.21) \qquad \pi_n(A) = A_1 \times \cdots \times A_n.$$

More precisely, the function $\pi_n : I^\infty \to I^n$ is defined by

$$(5.22) \qquad \pi_n(x) := (x_1, \ldots, x_n) \qquad \forall x = (x_1, x_2, \ldots) \in I^\infty.$$

We will use this notation throughout.

Definition 5.13. If $A = A_1 \times A_2 \times \cdots \subseteq I^\infty$ is a nonempty cylinder set, then we define its *dimension* as

$$(5.23) \qquad \dim A := \sup\{n \geq 1 : A_n \neq (0,1]\},$$

where $\sup \varnothing := 0$. In addition, $\dim \varnothing := \infty$.

Note that: (i) $A \in I^\infty$ is a non-empty cylinder set iff all but a finite number of its coordinates are equal to $(0,1]$; and (ii) all cylinder sets are finite dimensional. We also remark that this "dimension" has the property that if $A \subseteq B$, then $\dim A \geq \dim B$.

If $\dim A = n$ then $A_n \neq (0,1]$. But $A_{i+n} = (0,1]$ for all $i \geq 1$, and you should think of $A \in I^\infty$ as the natural embedding, or lifting, of the n-dimensional Euclidean set $\pi_n(A) = A_1 \times \cdots \times A_n \in I^n$ onto I^∞.

Definition 5.14. A family $\{(I^n, \mathscr{B}^n, \mathrm{P}^n)\}_{n=1}^\infty$ of probability spaces is *consistent* if

$$(5.24) \qquad \mathrm{P}^n(A_1 \times \cdots \times A_n) = \mathrm{P}^{n+1}(A_1 \times \cdots \times A_n \times (0,1]),$$

for all $n \geq 1$ and all $A_1, \ldots, A_n \in \mathscr{B}^1$. We will also say that $\{\mathrm{P}^n\}_{n=1}^\infty$ is consistent.

Remark 5.15. There is another way to think of a consistent family $\{\mathrm{P}^n\}_{n=1}^\infty$: If $1 \leq n := \dim A < \infty$ then

$$(5.25) \qquad \mathrm{P}^m(\pi_m(A)) = \mathrm{P}^n(\pi_n(A)) \qquad \forall m \geq n.$$

The notation is admittedly heavy-handed, but once you understand it you are ready for the beautiful theorem of A. N. Kolmogorov, the proof of which is spelled out later in §3.

The Kolmogorov Extension Theorem. *Suppose $\{P^n\}_{n=1}^\infty$ is a consistent family of probability measures on each of the spaces $(I^n, \mathscr{B}^n)$. Then, there exists a unique probability measure P^∞ on $(I^\infty, \mathscr{B}^\infty)$ such that $P^\infty(B) = P^n(\pi_n(B))$ for all finite n and all n-dimensional sets $B \in \mathscr{B}^\infty$.*

Remark 5.16. One can use Kolmogorov's extension theorem to construct the Lebesgue measure on $(I^\infty, \mathscr{B}^\infty)$.

Remark 5.17. One can just as easily prove Kolmogorov's extension theorem on the measurable space $(\mathbf{R}^\infty, \mathscr{B}(\mathbf{R}^\infty))$, where $\mathbf{R}^\infty$ is endowed with the product topology.

3. Complement: Proof of Kolmogorov's Extension Theorem

First we establish the asserted uniqueness of P^∞. Indeed suppose there were two such measures P^∞ and Q^∞, both on $(I^\infty, \mathscr{B}\infty)$. It follows immediately that $P^\infty(A) = Q^\infty(A)$ for all n-dimensional cylinder sets $A = A_1 \times \cdots \times A_n \times (0,1] \times (0,1] \times \cdots$. Therefore, P^∞ and Q^∞ agree on the algebra $\mathscr{A}$ generated by all cylinder sets; this is the smallest algebra that contains all cylinder sets. The monotone class theorem (p. 30) implies that $P^\infty = Q^\infty$ on all of $\mathscr{B}^\infty$.

Here is the strategy of the remainder of the proof, in a nutshell: Let $\mathscr{A}$ denote the collection of all finite unions of cylinder sets of the form

$$(5.26) \qquad (a_1, b_1] \times (a_2, b_2] \times \cdots \times (a_k, b_k] \times (0,1] \times (0,1] \times \cdots,$$

where $0 \leq a_i < b_i \leq 1$ for all i, and $k \geq 1$. We also add $\varnothing$ to $\mathscr{A}$, so that $\mathscr{A}$ becomes an algebra that generates $\mathscr{B}^\infty$. Our goal is to construct a countably additive measure on $\mathscr{A}$ and then appeal to Carathéodory's theorem (p. 27) to finish.

Our definition of P^∞ is both simple and intuitively appealing: First, define $P^\infty(\varnothing) = 0$ and $P^\infty(I^\infty) = 1$. This takes care of the trivial elements of $\mathscr{A}$. If $A \in \mathscr{A}$ is such that $1 \leq n := \dim A < \infty$, then we let $P^\infty(A) = P^n(\pi_n(A))$. [Check that this is well defined.]

Step 1. Finite Additivity. Let us first prove that P^∞ is finitely additive on $\mathscr{A}$. We want to show that if $A, B \in \mathscr{A}$ are disjoint, then $P^\infty(A \cup B) = P^\infty(A) + P^\infty(B)$. If $A = \varnothing$ or I^∞ then $B = A^c$, and finite additivity holds trivially from the fact that $P^\infty(\varnothing) = 1 - P^\infty(I^\infty) = 0$.

If neither A nor B is I^∞, then $n = \dim A$ and $m = \dim B$ are nontrivial natural numbers. We may assume without loss of generality that $n \geq m$. It follows that

(5.27) $\quad P^\infty(A \cup B) = P^n(\pi_n(A) \cup \pi_n(B)) = P^n(\pi_n(A)) + P^n(\pi_n(B)).$

This follows, since $\pi_n(A) \cap \pi_n(B) = \varnothing$ and P^n is a measure. On the other hand, $P^n(\pi_n(A)) = P^\infty(A)$, and $P^n(\pi_n(B)) = P^m(\pi_m(B)) = P^\infty(B)$ since $\{P^k\}_{k=1}^\infty$ is a consistent family (Remark 5.15). This verifies finite additivity.

Step 2. Countable Additivity. Suppose P^∞ is countably additive on $\mathscr{A}$. Then, the Carathéodory's extension theorem implies that P^∞ can be extended uniquely to a countably additive measure on $\sigma(\mathscr{A}) = \mathscr{B}^\infty$. This extension, still written as P^∞, is the probability measure on $(I^\infty, \mathscr{B}^\infty)$ that is stated in the present theorem. Thus, it suffices to establish the countable additivity of P^∞ on $\mathscr{A}$. In order to do this, we appeal to an argument that is similar to the proof that the Lebesgue measure on $(0,1]$ is countably additive on finite unions of intervals of the form $(a,b]$. Make certain that you understand the proof of Lemma 3.15 (p. 26) before proceeding with the present proof.

Let $A^1, A^2, \ldots$ denote disjoint sets in $\mathscr{A}$ such that $\cup_{j=1}^\infty A^j$ is also in $\mathscr{A}$; we need to verify that $P^\infty(\cup_{j=1}^\infty A^j) = \sum_{j=1}^\infty P^\infty(A^j)$. We write $\cup_{j=1}^\infty A^j = (\cup_{j=1}^N A^j) \cup (\cup_{j=N+1}^\infty A^j)$, and note that $\cup_{j=1}^N A^j$ and $\cup_{j=N+1}^\infty A^j$ are disjoint elements of $\mathscr{A}$. By Step 1,

(5.28) $\quad P^\infty\left(\bigcup_{j=N+1}^\infty A^j\right) = \sum_{j=1}^N P^\infty(A^j) + P^\infty\left(\bigcup_{j=N+1}^\infty A^j\right).$

Thus, it suffices to show that if $B^N \downarrow \varnothing$—all in $\mathscr{A}$—then $P^\infty(B^N) \downarrow 0$. We assume the contrary and derive a contradiction. That is, we suppose that there exists $\epsilon > 0$ such that $P^\infty(B^n) \geq \epsilon$ for all $n \geq 1$. These remarks make it clear that $\dim B^n$ is strictly positive (i.e., $B^n \neq \varnothing$) and finite (i.e., $B^n \neq I^\infty$) for all n large. Henceforth, let $\gamma(n) := \dim B^n$, and note that the condition $B^n \downarrow \varnothing$ forces $\gamma(n)$ to be non-decreasing. Thus,

(5.29) $\quad B^n = B_1^n \times \cdots \times B_{\gamma(n)}^n \times (0,1] \times (0,1] \times \cdots,$

where $B_m^n := \cup_{j=1}^{k(n,m)} (a_j^{n,m}, b_j^{n,m}]$ $(m \leq \gamma(n))$. Now define C^n to be an approximation from inside to B via closed intervals, viz.,

(5.30) $\quad C^n = C_1^n \times \cdots \times C_{\gamma(n)}^n \times (0,1] \times (0,1] \times \cdots,$

where $C_m^n = \cup_{i=1}^{k(n,m)} [\alpha_i^{n,m}, b_i^{n,m}]$ $(m \leq \gamma(n))$, and the $\alpha_i^{n,m} \in (a_i^{n,m}, b_i^{n,m})$ are so close to the a's that

(5.31) $\quad P^\infty(B^j \setminus C^j) \leq \dfrac{\epsilon}{2^j} \quad \forall j \geq 1.$

Proof. This can always be done because $P^\infty(B^j \setminus C^j)$ is

$$P^{\gamma(j)}\left(\bigcup_{i=1}^{k(j,1)}\left[a_i^{j,1},\alpha_i^{j,1}\right]\times\cdots\times\bigcup_{i=1}^{k(j,\gamma(j))}\left[a_i^{j,\gamma(j)},\alpha_i^{j,\gamma(j)}\right]\right),$$

and $P^{\gamma(j)}$ is a measure on $(I^{\gamma(j)},\mathscr{B}^{\gamma(j)})$. □

Therefore, thanks to (5.31), $P^\infty(D^n) \geq (\epsilon/2)$, where $D^n = \cap_{j=1}^n C^j$ is a sequence of decreasing sets with $D^n \subseteq [0,1]^{\gamma(n)} \times (0,1] \times (0,1] \times \cdots$. Now we argue that $\cap_{n=1}^\infty D^n \neq \emptyset$; since $D^n \subseteq B^n$, this contradicts $B^n \downarrow \emptyset$ and our task is done.

We know that $D^n \neq \emptyset$ for any finite n since $P^\infty(D^n) \geq (\epsilon/2)$. Moreover, we can write $D^n = D_1^n \times D_2^n \times \cdots$ where: (a) $D_j^n = (0,1]$ for all $j > \gamma(n)$; and (b) D_j^n is closed in $[0,1]$ for all $j \leq \gamma(n)$. Therefore, we can choose $x^n \in D^n$ of the following form:

$$(5.32) \qquad x^n := \left(x_1^n, x_2^n, \ldots, x_{\gamma(n)}^n, \tfrac{1}{2}, \tfrac{1}{2}, \ldots\right) \qquad \forall n \geq 1.$$

Because $\gamma(n)$ is non-decreasing and $D_1^1 \supseteq D_1^2 \supseteq D_1^3 \supseteq \cdots$ is a decreasing sequence of closed subsets of $[0,1]$, $z_1 := \lim_{\ell \to \infty} x_1^\ell$ is an element of $\cap_{n=1}^\infty D_1^n$. Similarly, $z_j = \lim_{\ell \to \infty} x_j^\ell \in \cap_{n=1}^\infty D_j^n$ for all $j \geq 1$. Thus, we have found a point $z = (z_1, z_2, z_3, \ldots)$ in $\cap_{n=1}^\infty D^n$. This proves that $\cap_{n=1}^\infty D^n \neq \emptyset$, whence the theorem.

Problems

5.1. Let Ω be an uncountable set and $\mathscr{F}$ the collection of all subsets $A \subseteq \Omega$ such that either A or A^c is denumerable.

  (1) Prove that $\mathscr{F}$ is a σ-algebra.
  (2) Define the set function $P : \mathscr{F} \to \{0,1\}$ by: $P(A) = 1$ if A is uncountable, and $P(A) = 0$ if A is denumerable. Prove that P is a probability measure on $(\Omega, \mathscr{F})$.
  (3) Use only the axiom of choice to construct a set Ω, and an $E \subseteq \Omega \times \Omega$ such that for all $x, y \in \Omega$, $_xE$ and E_y^c are denumerable, where $_xE := \{y \in \Omega : (x,y) \in E\}$ and $E_y := \{x \in \Omega : (x,y) \in E\}$.

5.2. Consider a finite measure μ on $(\mathbf{R}^n, \mathscr{B}(\mathbf{R}^n))$ such that $\mu(\{x\}) = 0$ for all $x \in \mathbf{R}^n$. Define the *diagonal* D of $\mathbf{R}^{2n} = \mathbf{R}^n \times \mathbf{R}^n$ to be $\{(x,x) : x \in \mathbf{R}^n\}$. Then prove that $(\mu \times \mu)(D) = 0$.

5.3 (Problem 5.2, Continued). Let (X,Y) be a random variable that takes values in $\mathbf{R}^2$. We say that X and Y are *independent* if

$$(5.33) \qquad \mathrm{E}[f(X)g(Y)] = \mathrm{E}f(X)\,\mathrm{E}g(Y),$$

for all bounded measurable functions $f, g : \mathbf{R} \to \mathbf{R}$. Prove that if $P\{X = a\} = 0$ for all $a \in \mathbf{R}$, then $P\{X = Y\} = 0$. Why does this generalize Problem 5.2? (HINT: $\mu(A) := P\{X \in A\}$ and $\nu(B) := P\{Y \in B\}$ are probability measures.)

5.4. Prove that if $\{a_{i,j}\}_{i,j=1}^\infty$ is a sequence indexed by $\mathbf{N}^2$, then

$$(5.34) \qquad \sum_{i=1}^\infty \sum_{j=1}^\infty a_{i,j} = \sum_{j=1}^\infty \sum_{i=1}^\infty a_{i,j},$$

provided that $\sum_{i,j} |a_{i,j}| < \infty$. Construct an example of $\{a_{i,j}\}_{i,j=1}^\infty$ for which (5.34) fails to hold.

5.5. Prove that the Fubini–Tonelli theorem (p. 55) remains valid when μ is σ-finite.

5.6. If $\{\mu_i\}_{i=1}^\infty$ are probability measures on $([0,1],\mathscr{B}([0,1]))$, then carefully make sense of the probability measure $\prod_{i=1}^\infty \mu_i$. Use this to construct the Lebesgue measure m on $[0,1]^\infty$ endowed with its product σ-algebra. Finally, if $1 > a_1 > a_2 > \cdots > a_n \downarrow 0$, then prove that

$$m\left(\prod_{i=1}^\infty [a_i,1]\right) > 0 \quad \text{iff} \quad \sum_{i=1}^\infty a_i < \infty. \tag{5.35}$$

We will do much more on this. Consult the Borel–Cantelli lemma on page 73.

5.7. Define $f(x,y) := (x^2 - y^2)(x^2 + y^2)^{-2}$ for all $x,y \in (0,1]$, and verify that

$$\int_0^1 \int_0^1 f(x,y)\,dx\,dy \neq \int_0^1 \int_0^1 f(x,y)\,dy\,dx. \tag{5.36}$$

Why does the Fubini–Tonelli theorem not apply?

5.8. Define $f(x,y) := xy(x^2+y^2)^{-2}$ for all $x,y \in [-1,1]$, and prove that $f \notin L^1([-1,1]^2)$ and yet

$$\int_{-1}^1 \int_{-1}^1 f(x,y)\,dx\,dy = \int_{-1}^1 \int_{-1}^1 f(x,y)\,dy\,dx. \tag{5.37}$$

5.9. Define, for all $x,y \in \mathbf{R}^2$,

$$f(x,y) = \begin{cases} 1 & \text{if } x^2 + y^2 = 1, \\ 0 & \text{otherwise.} \end{cases} \tag{5.38}$$

Respectively define μ and ν to be the Lebesgue measure and the counting measure on $\mathbf{R}$. Prove that $\int_{-1}^1 \int_{-1}^1 f\,d\mu\,d\nu \neq \int_{-1}^1 \int_{-1}^1 f\,d\nu\,d\mu$. Why does the Fubini–Tonelli theorem not apply?

5.10 (Monotone Rearrangements). Let $f : \mathbf{R} \to \mathbf{R}_+$ be integrable, and define

$$A_z := \{x \in \mathbf{R} : f(x) \geq z\} \qquad \forall z \geq 0. \tag{5.39}$$

Prove that A_z is measurable for all $z \geq 0$. Let $f^\downarrow(z)$ denote the Lebesgue measure of A_z, and prove that: (i) $f^\downarrow$ is non-increasing and measurable; and (ii) $\int_0^\infty f^\downarrow(z)\,dz = \int_{-\infty}^\infty f(x)\,dx$.

5.11. Compute explicitly the numerical value of

$$\int_0^\infty \int_0^\infty \exp\left(-\frac{x^2}{y} - y\right)\,dy\,dx. \tag{5.40}$$

5.12. For all functions $f : [a,b] \to \mathbf{R}_+$ define the set

$$A(f) := \{(x,y) \in \mathbf{R}^2 : 0 \leq y \leq f(x)\}. \tag{5.41}$$

Prove that if f is measurable, then $A(f) \in \mathscr{B}(\mathbf{R}^2)$. Compute the two-dimensional Lebesgue measure of $A(f)$.

5.13 (The FKG Inequality). Suppose μ denotes a probability measure on $\mathscr{B}(\mathbf{R})$ and $f,g : \mathbf{R} \to \mathbf{R}$ are non-decreasing and measurable. Prove that

$$\int fg\,d\mu \geq \left(\int f\,d\mu\right)\left(\int g\,d\mu\right), \tag{5.42}$$

provided that the integrals converge. Use this to derive the *Chebyshev inequality for sums*: If $a_1 \leq a_2 \leq \cdots \leq a_n$ and $b_1 \leq b_1 \leq \cdots \leq b_n$, then

$$\left(\frac{1}{n}\sum_{i=1}^n a_i\right)\left(\frac{1}{n}\sum_{i=1}^n b_i\right) \leq \frac{1}{n}\sum_{i=1}^n a_i b_i. \tag{5.43}$$

(HINT: $[f(x) - f(y)] \cdot [g(x) - g(y)] \geq 0$.)

5.14 (Problem 5.13, Continued). Suppose $\mu_1, \ldots, \mu_n$ are probability measures on $\mathscr{B}(\mathbf{R})$, and define $\mu = \mu_1 \times \cdots \times \mu_n$. If $f,g : \mathbf{R}^n \to \mathbf{R}$ are non-decreasing in each coordinate and measurable, then prove that (5.42) holds provided that all three integrals converge absolutely.

5.15. Prove that: (i) $\int_0^\infty (|\sin x|/x)\,dx = \infty$; and (ii) for all $a > 0$,

(5.44) $$\left| \int_0^a \frac{\sin x}{x}\,dx - \frac{\pi}{2} \right| \le \frac{2}{a}.$$

Conclude that $\int_0^\infty (\sin(x)/x)\,dx = \pi/2$. (HINT: $x^{-1} = \int_0^\infty e^{-xu}\,du$.)

5.16 (Problem 5.15, Continued). Prove that for all $a > 0$,

(5.45) $$\left| \int_0^a \left(\frac{1-\cos x}{x^2}\right) dx - \frac{\pi}{2} \right| \le \frac{3}{a}, \quad \text{and} \quad \left| \int_0^a \left(\frac{\sin x}{x}\right)^2 dx - \frac{\pi}{2} \right| \le \frac{3}{2a}.$$

Thus, $\int_0^\infty (1-\cos x)/x^2\,dx = \int_0^\infty (\sin(x)/x)^2\,dx = \pi/2$.

5.17. Use Fubini–Tonelli to compute $\sum_{n=1}^\infty (\alpha^n/n)$ and $\sum_{n=1}^\infty \alpha^n H_n$ for all $\alpha \in (0,1)$, where $H_n := \sum_{j=1}^n (1/j)$ for $n \ge 1$. (HINT: $\alpha^n/n = \int_0^\alpha x^{n-1}\,dx$.)

5.18. Use the Fubini–Tonelli theorem to compute

(5.46) $$\int_0^\infty \left(\frac{1-e^{-\alpha x}}{x}\right) e^{-\beta x}\,dx \quad \forall \alpha, \beta > 0.$$

5.19 (Hard). Consider a set-valued function $\mathbf{X}$ on some given probability space $(\Omega, \mathscr{F}, P)$. Specifically, $\mathbf{X} : \Omega \to \mathscr{P}(\mathbf{R}^d)$, where $\mathscr{P}(\mathbf{R}^d)$ denotes the power set of $\mathbf{R}^d$. We say that $\mathbf{X}$ is a *random set* if $(\omega, x) \mapsto \mathbf{1}_{\mathbf{X}(\omega)}(x)$ is product-measurable on the measure space $(\Omega \times \mathbf{R}^d, \mathscr{F} \times \mathscr{B}(\mathbf{R}^d))$. Prove:

(1) If $A \in \mathscr{B}(\mathbf{R}^d)$, then A and $\mathbf{X} \cap A$ are both random sets.

(2) If $\mathbf{X}, \mathbf{X}_1, \mathbf{X}_2, \ldots$ are random sets, then so are $\mathbf{X}^c$, $\cap_{n=1}^\infty \mathbf{X}_n$, and $\cup_{n=1}^\infty \mathbf{X}_n$.

(3) If $A \in \mathscr{B}(\mathbf{R}^d)$ satisfies $\lambda(A) < \infty$ where λ is a σ-finite measure on $(\mathbf{R}^d, \mathscr{B}(\mathbf{R}^d))$, then $\lambda(\mathbf{X} \cap A)$ is a finite random variable.

(4) For all $A \in \mathscr{B}(\mathbf{R}^d)$ such that $\lambda(A) < \infty$, and for all integers $k \ge 1$,
$$\|\lambda(\mathbf{X} \cap A)\|_k^k = \int_A \cdots \int_A P\{x_1 \in \mathbf{X}, \ldots, x_k \in \mathbf{X}\}\,\lambda(dx_1)\cdots\lambda(dx_k).$$

(5) $P\{x \in \mathbf{X}\} = 0$ for λ-almost every $x \in \mathbf{R}^d$ if and only if $\lambda(\mathbf{X}) = 0$, P-a.s.

(6) There is a non-empty random set $\mathbf{X}$ such that $P\{x \in \mathbf{X}\} = 0$ for all $x \in \mathbf{R}^d$.

Notes

(1) Example 5.11 is due to Sierpiński (1920).

Mattner (1999) has constructed a Borel set $A \subset \mathbf{R}$ and two σ-finite measures μ_1 and μ_2 on $\mathscr{B}(\mathbf{R})$ such that if we ignored measurability issues, then we would have the following:
$$\int_{-\infty}^\infty \left(\int_{-\infty}^\infty \mathbf{1}_A(x+y)\,\mu_1(dx) \right) \mu_2(dy) \ne \int_{-\infty}^\infty \left(\int_{-\infty}^\infty \mathbf{1}_A(x+y)\,\mu_2(dy) \right) \mu_1(dx).$$

Mattner's construction is interesting for at least two reasons: (i) It does not rely on the axiom of choice nor on the continuum hypothesis; and (ii) it shows that the "convolution" $y \mapsto \int f(x-y)\,\mu_1(dx)$ need not be measurable with respect to the smallest σ-algebra with respect to which all functions $\{f(\bullet - y);\ y \in \mathbf{R}\}$ are measurable.

(2) Problem 5.9 is motivated by two papers of Mukherjea (1972, 1974).

(3) The FKG inequality (Problem 5.13) is due to Fortuin, Kasteleyn, and Ginibre (1971).

Chapter 6

Independence

Nothing is too wonderful to be true.

Attributed to Michael Faraday

Our review/development of measure theory is finally complete, and we begin studying probability theory in earnest. In this chapter we introduce the all-important notion of independence, and use it to prove a precise formulation of the so-called law of large numbers. In rough terms, the latter states that the sample average of a large random sample is close to the population average. Throughout, $(\Omega, \mathscr{F}, \mathrm{P})$ is a probability space.

1. Random Variables and Distributions

For every random variable $X : \Omega \to \mathbf{R}$ we can define a set function $\mathrm{P} \circ X^{-1}$ on $(\mathbf{R}, \mathscr{B}(\mathbf{R}))$ as follows:

(6.1) $\qquad \left(\mathrm{P} \circ X^{-1} \right)(E) = \mathrm{P}\{X \in E\} \qquad \forall E \in \mathscr{B}(\mathbf{R}).$

This notation is motivated by the fact that $\{X \in E\}$ is another way to write $X^{-1}(E)$, so that $(\mathrm{P} \circ X^{-1})(E) = \mathrm{P}(X^{-1}(E))$.

Lemma 6.1. $\mathrm{P} \circ X^{-1}$ *is a probability measure on* $(\mathbf{R}, \mathscr{B}(\mathbf{R}))$.

Definition 6.2. The measure $\mathrm{P} \circ X^{-1}$ is called the *distribution* of the random variable X.

Proof of Lemma 6.1. The proof is straightforward: $(\mathrm{P} \circ X^{-1})(\varnothing) = 0$, and $\mathrm{P} \circ X^{-1}$ is countably additive on $(\mathbf{R}, \mathscr{B}(\mathbf{R}))$, since P is countably additive on $(\Omega, \mathscr{F})$ and X is a function. $\qquad \square$

Lemma 6.1 tells us that to each random variable we can associate a *real* probability space $(\mathbf{R}, \mathscr{B}(\mathbf{R}), \mathrm{P} \circ X^{-1})$. In a sense, the converse is also true: For every probability measure μ on $(\mathbf{R}, \mathscr{B}(\mathbf{R}))$, we can define $X(\omega) = \omega$ ($\omega \in \mathbf{R}$) to deduce that there exists a random variable X whose distribution is μ.

Definition 6.3. The (cumulative) *distribution function* F of a probability measure μ on $(\mathbf{R}, \mathscr{B}(\mathbf{R}))$ is defined by $F(x) = \mu((-\infty, x])$ for all $x \in \mathbf{R}$. The *distribution function* F of a random variable X is the distribution function of $\mathrm{P} \circ X^{-1}$. In other words, $F(x) := \mathrm{P}\{X \le x\}$ for all $x \in \mathbf{R}$.

Note that: (i) F is non-decreasing, right-continuous, and has left limits; (ii) $F(-\infty) := \lim_{x \to -\infty} F(x) = 0$; and (iii) $F(\infty) := \lim_{x \to \infty} F(x) = 1$. These properties characterize F; i.e.,

Theorem 6.4. *A function $F : \mathbf{R} \to [0,1]$ is the distribution function of a probability measure μ if and only if: (i) F is non-decreasing and right-continuous; and (ii) $F(-\infty) = 0$ and $F(\infty) = 1$. In addition, F and μ define one another uniquely.*

Proof. (Sketch) The necessity of (i) and (ii) has already been established. Conversely, suppose $F : \mathbf{R} \to [0,1]$ satisfies (i) and (ii). We can then *define* $\mu((a,b]) = F(b) - F(a)$ for all real numbers $a < b$. Extend the definition of μ to finite disjoint unions of intervals of type $(a_i, b_i]$ by setting

$$(6.2) \qquad \mu\left(\bigcup_{i=1}^{n}(a_i, b_i]\right) = \sum_{i=1}^{n}[F(b_i) - F(a_i)].$$

It is not difficult to check that: (a) This is a well-defined extension of μ; and (b) μ is countably additive on the algebra of all disjoint finite unions of intervals of the type $(a, b]$. These assertions are proved by adapting the proof of Lemma 3.15 on page 26 to the present case. Now we apply Carathéodory's theorem (page 27), and extend μ uniquely to a measure on all of $\mathscr{B}(\mathbf{R})$. It remains to check that this extended μ is a probability measure, but this follows from $\mu(\mathbf{R}) = \lim_{n \to \infty} \mu((-\infty, n]) = F(\infty) = 1$, thanks to the inner continuity of measures. □

Definition 6.5. If $p > 0$, then the *pth moment* of a random variable X is defined as $\mathrm{E}[X^p]$ provided that $X \ge 0$ a.s., or $X \in L^p(\mathrm{P})$.

Lemma 6.6. *If $X \ge 0$ a.s., then $\mathrm{E}[X^p] = \int_\Omega X^p \, d\mathrm{P} = \int_{-\infty}^{\infty} x^p \, \mu(dx)$, where μ denotes the distribution of X. More generally still, if $h : \mathbf{R} \to \mathbf{R}$ is Borel measurable, then*

$$(6.3) \qquad \mathrm{E}h(X) = \int_\Omega h(X) \, d\mathrm{P} = \int_{-\infty}^{\infty} h(x) \, \mu(dx),$$

provided that the integrals exist.

2. Independent Random Variables

Proof. The assertion $Eh(X) = \int_\Omega h(X)\, dP$ is a tautology. Next consider a simple function h, i.e., one of the form

(6.4) $$h(x) = \sum_{i=1}^n \alpha_i \mathbf{1}_{A_i}(x),$$

where $A_1, \ldots, A_n \in \mathscr{F}$ are disjoint and $\alpha_1, \ldots, \alpha_n \in \mathbf{R}$. It follows that $h(X)$ is a discrete random variable, and

(6.5) $$Eh(X) = \sum_{i=1}^n \alpha_i P\{X \in A_i\} = \sum_{i=1}^n \alpha_i \mu(A_i) = \int h\, d\mu.$$

For a more general non-negative function h, we can choose simple functions $h_n \uparrow h$ and appeal to the monotone convergence theorem (p. 46). The rest follows from linearity. □

Definition 6.7. The *variance* and the *standard deviation* of a random variable $X \in L^2(P)$ are respectively defined as $\operatorname{Var} X := E[(X - EX)^2]$ and $\operatorname{SD}(X) := \sqrt{\operatorname{Var} X}$. If $X, Y \in L^2(P)$, then the *covariance* and *correlation* between X and Y are respectively defined as

(6.6) $$\operatorname{Cov}(X, Y) := E[(X - EX)(Y - EY)],$$

and

(6.7) $$\rho(X, Y) := \frac{\operatorname{Cov}(X, Y)}{\operatorname{SD}(X)\operatorname{SD}(Y)},$$

where $0/0 := 1$. Two random variables X and Y are said to be *uncorrelated* if $\rho(X, Y) = 0$.

Lemma 6.8. *If $X, Y \in L^2(P)$, then $\operatorname{Var} X = \|X - EX\|_2^2 = E[X^2] - |EX|^2$, and $\operatorname{Cov}(X, Y) = E[XY] - EX\, EY$. If $X \geq 0$ a.s. then*

(6.8) $$E[X^p] = p \int_0^\infty \lambda^{p-1} P\{X \geq \lambda\}\, d\lambda \qquad \forall p > 0.$$

In particular,

(6.9) $$\sum_{n=1}^\infty P\{X \geq n\} \leq EX \leq \sum_{n=0}^\infty P\{X \geq n\}.$$

2. Independent Random Variables

Now we generalize the notion of independence that was touched on first in Chapter 1.

Definition 6.9. Events $\{E_i\}_{i=1}^n$ are *independent* if for all distinct indices $i(1), \ldots, i(l)$ in $\{1, \ldots, n\}$,

$$(6.10) \qquad \mathrm{P}\left(E_{i(1)} \cap \cdots \cap E_{i(l)}\right) = \prod_{j=1}^l \mathrm{P}\left(E_{i(j)}\right).$$

A collection $\{E_\alpha\}_{\alpha \in I}$ of events is *independent* if $E_{\alpha(1)}, \ldots, E_{\alpha(n)}$ are independent for all $\alpha(1), \ldots, \alpha(n) \in I$. For an arbitrary index set I, the σ-algebras $\{\mathscr{F}_\alpha\}_{\alpha \in I}$ are called *independent* if any finite number of events $A_{\alpha(i)} \in \mathscr{F}_{\alpha(i)}$ ($i = 1, \ldots, n$) are independent.

Definition 6.10. The random variables $X_1, \ldots, X_n : \Omega \to \mathbf{R}^d$ are *independent* if the events $\{X_i^{-1}(A_i)\}_{i=1}^n$ are independent for all $A_1, \ldots, A_n \in \mathscr{B}(\mathbf{R}^d)$. An arbitrary collection $\{X_\alpha\}_{\alpha \in I}$ is *independent* if $X_{\alpha(1)}, \ldots, X_{\alpha(n)}$ are independent for all $\alpha(1), \ldots, \alpha(n) \in I$. If $\{X_\alpha\}_{\alpha \in I}$ are independent and identically distributed random variables, then we say that the X_α's are *i.i.d.*

Equivalently, $\{X_i\}_{i=1}^n$ are independent if for all measurable $\{A_i\}_{i=1}^n$,

$$(6.11) \qquad \mathrm{P}\{X_1 \in A_1, \ldots, X_n \in A_n\} = \prod_{j=1}^n \mathrm{P}\{X_j \in A_j\}.$$

Remark 6.11. One can construct random variables X_1, X_2, X_3 such that X_i and X_j are independent whenever $i \neq j$, but X_1, X_2, X_3 are *not* independent.

Lemma 6.12. *Random variables $\{X_i\}_{i=1}^n$ are independent iff for all measurable functions $\phi_1, \ldots, \phi_n : \mathbf{R}^d \to \mathbf{R}_+$,*

$$(6.12) \qquad \mathrm{E}\left[\prod_{j=1}^n \phi_j(X_j)\right] = \prod_{j=1}^n \mathrm{E}\left[\phi_j(X_j)\right].$$

Consequently, $\{X_i\}_{i=1}^n$ are independent iff $\{h_i(X_i)\}_{i=1}^n$ are independent for all Borel-measurable functions $h_1, \ldots, h_n : \mathbf{R} \to \mathbf{R}$.

Proof. The second assertion is a ready consequence of (6.12). Therefore, we derive only the latter equation.

When the ϕ_j's are elementary functions, (6.12) is the definition of independence. By linearity (in each of the ϕ_j's), (6.12) continues to hold when the ϕ_j's are simple functions. Take limits to obtain the full result. □

Soon we will see that assuming independence places severe restrictions on the random variables in question. But first, we need a definition or two.

Definition 6.13. The σ-*algebra generated* by $\{X_i\}_{i \in I}$ is the smallest σ-algebra with respect to which all of the X_i's are measurable; it is written as $\sigma(\{X_i\}_{i \in I})$. When we say that $\{X_i\}_{i \in I}$ is independent of a σ-algebra $\mathscr{G}$, we mean that $\sigma(\{X_i\}_{i \in I})$ is independent of $\mathscr{G}$.

2. Independent Random Variables

Definition 6.14. The *tail σ-algebra* $\mathscr{T}$ of the random variables $\{X_i\}_{i=1}^{\infty}$ is the σ-algebra $\mathscr{T} = \cap_{n=1}^{\infty} \sigma(\{X_i\}_{i=n}^{\infty})$.

The following tells us that our definitions of independence are compatible. Moreover, the last portion implies that in order to prove that two real-valued random variables X and Y are independent, it is necessary as well as sufficient to prove that

(6.13) $\qquad \mathrm{P}\{X \le x, Y \le y\} = \mathrm{P}\{X \le x\}\mathrm{P}\{Y \le y\} \qquad \forall x, y \in \mathbf{R}.$

Lemma 6.15. *Let $\mathbf{A}$ and $\mathbf{B}$ denote two topological spaces.*

  (i) *For all random variables $X : \Omega \to \mathbf{A}$,*

  $$\sigma(X) = \{X^{-1}(A) : A \in \mathscr{B}(\mathbf{A})\}.$$

  (ii) *If $\{X_j\}_{j=1}^{\infty}$ are random variables all taking values in $\mathbf{A}$, then a $\mathbf{B}$-valued random variable Y is independent of $\{X_j\}_{j=1}^{\infty}$ if and only if Y is independent of $\{X_j\}_{j=1}^{n}$ for every $n \ge 1$.*

  (iii) *Let $\mathscr{B}$ and $\mathscr{A}$ be two subalgebras that respectively generate $\mathscr{B}(\mathbf{B})$ and $\mathscr{B}(\mathbf{A}^{\infty})$. If $Y^{-1}(F)$ is independent of $(X_1, X_2, \ldots)^{-1}(E)$ for all $E \in \mathscr{A}$ and all $F \in \mathscr{B}$, then Y and $\{X_j\}_{j=1}^{\infty}$ are independent.*

The proof of this is relegated to the exercises. Instead, we turn to the following consequence of independence. It confirms the assertion—made earlier—that independence is a severe restriction.

Kolmogorov's Zero-One Law. *If $\{X_i\}_{i=1}^{\infty}$ are independent random variables, then their tail σ-algebra $\mathscr{T}$ is trivial in the sense that for all $E \in \mathscr{T}$, $\mathrm{P}(E) = 0$ or 1. Consequently, any $\mathscr{T}$-measurable random variable is a constant almost surely.*

Proof. Our strategy is to prove that every $E \in \mathscr{T}$ is independent of itself, so that $\mathrm{P}(E) = \mathrm{P}(E \cap E) = \mathrm{P}(E)\mathrm{P}(E)$. Since $E \in \mathscr{T}$, it follows that E is independent of $\sigma(\{X_i\}_{i=1}^{n-1})$. Because this is true for each n, Lemma 6.15 (iii) ensures that E is independent of $\vee_{n=1}^{\infty} \sigma(\{X_i\}_{i=1}^{n-1})$, which is defined to be the smallest σ-algebra that contains $\cup_{n=1}^{\infty} \sigma(\{X_i\}_{i=1}^{n-1})$. In other words, E is independent of all the X_i's, and hence of itself.

To conclude this proof suppose Y is $\mathscr{T}$-measurable. We intend to prove that there exists a constant c such that $\mathrm{P}\{Y = c\} = 1$. For any $x \in \mathbf{R}$, the event $\{Y \le x\}$ has probability zero or one. Therefore, the distribution function F of Y is necessarily of the form $F(y) = \mathbf{1}_{[c,\infty)}(y)$, where c denotes the smallest ρ such that $\mathrm{P}\{Y \le \rho\} = 1$. This implies that $Y = c$ almost surely. $\square$

Example 6.16. Suppose $\{X_i\}_{i=1}^\infty$ are independent and define

$$A_n := \frac{X_1 + \cdots + X_n}{n}. \tag{6.14}$$

Then, $\limsup_{n\to\infty} A_n$ and $\liminf_{n\to\infty} A_n$ are almost surely constants. Furthermore, the probability that $\lim_{n\to\infty} A_n$ exists in $[-\infty, \infty]$ is zero or one. If this probability is 1, then $\lim_{n\to\infty} A_n$ is a constant almost surely.

Next we prove that independent random variables exist.

Theorem 6.17. *If $\{\mu_i\}_{i=1}^\infty$ are probability measures on $(\mathbf{R}^d, \mathscr{B}(\mathbf{R}^d))$, then there exist independent random variables $\{X_i\}_{i=1}^\infty$, all on a suitable probability space, such that the distribution of X_i is μ_i for each $i = 1, 2, \ldots$.*

Proof. For the sake of notational convenience we will assume that $d = 1$.

Let $\Omega^n := \mathbf{R}^n$, $\mathscr{F}^n := \mathscr{B}(\mathbf{R}^n)$, and $\mu^n := \mu_1 \times \cdots \times \mu_n$ for every $n \geq 1$. Clearly, $\{\mu^n\}_{n=1}^\infty$ is a consistent family of probability measures. By the Kolmogorov extension theorem (p. 60; see also Remark 5.17 therein) there exists a probability measure P on $(\mathbf{R}^\infty, \mathscr{B}(\mathbf{R}^\infty))$ that extends $\{\mu^n\}_{n=1}^\infty$. Define $X_i(\omega) = \omega_i$ for all $\omega \in \mathbf{R}^\infty$ and $i \geq 1$. Because $X_i^{-1}(E_i) = \{\omega \in \mathbf{R}^\infty : \omega_i \in E_i\}$, it follows that $\mathrm{P}\{X_i \in E_i\} = \mu_i(E_i)$ and

$$\mathrm{P}\{X_1 \in E_1, \ldots, X_n \in E_n\} = \mathrm{P}(X_1^{-1}(E_1) \cap \cdots \cap X_n^{-1}(E_n)) = \prod_{i=1}^n \mu_i(E_i). \tag{6.15}$$

Therefore, the X_i's are independent and have the asserted distributions. □

Let us conclude this section with two results of computational utility. The proofs are left to the reader.

Corollary 6.18. *If $X, Y \in L^2(\mathrm{P})$ are independent, then they are uncorrelated; i.e., $\mathrm{Cov}(X, Y) = 0$.*

The converse is false in general; see Problem 6.7.

Corollary 6.19. *If $\{X_i\}_{i=1}^n$ are uncorrelated and in $L^2(\mathrm{P})$, then*

$$\mathrm{Var}(X_1 + \cdots + X_n) = \sum_{j=1}^n \mathrm{Var} X_j. \tag{6.16}$$

In particular, this identity is valid if the X_i's are independent.

3. An Instructive Example

We now describe a class of distributions that do not fit into the classical probability models of Chapter 1.

Let $\{X_i\}_{i=1}^\infty$ denote i.i.d. random variables, all taking the values 0 and 1 with probability $\frac{1}{2}$ each, and define $Y := \sum_{i=1}^\infty 4^{-i} X_i$. If μ denotes the distribution of Y, then μ is a probability measure on $\mathscr{B}([0,1])$ that is neither discrete nor absolutely continuous. It is also not a simple combination of the latter two types of distributions. The following makes these remarks more precise.

Theorem 6.20. *The distribution μ of Y satisfies:* (i) $\mu(\{x\}) = 0$ *for all* $x \in [0,1]$; *and* (ii) *there exists a measurable set* $A \subseteq \mathbf{R}_+$ *that has zero Lebesgue measure and yet* $\mu(A) = 1$.

Proof. For all $n \geq 1$ let $Y_n := \sum_{i=1}^n 4^{-i} X_i$ and $\mathscr{Y}_n$ the set of possible values of Y_n. The cardinality of $\mathscr{Y}_n$ is 2^n because its elements are of the form $\sum_{i=1}^n 4^{-i} y_i$ where $y_i = 0$ or 1. Also, because $|Y - Y_n| \leq \sum_{i=n+1}^\infty 4^{-i} = 4^{-n}/3$, Y is a.s. within $4^{-n}/3$ of some $y \in \mathscr{Y}_n$. Therefore, $\mu(V_n) = \mathrm{P}\{Y \in V_n\} = 1$ for all $n \geq 1$, where

$$(6.17) \qquad V_n := \bigcup_{k=n}^\infty \bigcup_{y \in \mathscr{Y}_k} \left[y - \frac{4^{-k}}{3}, y + \frac{4^{-k}}{3} \right].$$

Because $V_n \supseteq V_{n+1}$, $A := \bigcap_{n=1}^\infty V_n$ has μ-measure one. On the other hand, if m denotes Lebesgue measure, then for all $n \geq 1$,

$$(6.18) \qquad m(A) \leq m(V_n) \leq 2 \sum_{k=n}^\infty \sum_{y \in \mathscr{Y}_k} \frac{4^{-k}}{3} = \frac{2}{3} \sum_{k=n}^\infty 2^{-k} = \frac{2^{2-n}}{3}.$$

This proves that $m(A) = 0$ although $\mu(A) = 1$. It remains to prove that $\mu(\{x\}) = 0$ for all $x \in [0,1]$. To this end, write $x = \sum_{i=1}^\infty 4^{-i} x_i$ where $x_i \in \{0,1,2,3\}$, and note that

$$(6.19) \qquad \mu(\{x\}) \leq \mathrm{P}\{X_1 = x_1, \ldots, X_n = x_n\} \leq 2^{-n}.$$

Let $n \to \infty$ to complete the proof. $\square$

4. Khintchine's Weak Law of Large Numbers

The weak law of Khintchine (1929) states that, with high probability, sample averages are close to population averages. We will soon see that this is a considerable improvement of the law of large numbers of Bernoulli (1713) on page 18. Although the weak law is subsumed by the forthcoming strong law of large numbers, we state and prove it first because it provides us with a good opportunity to learn more about the Markov and Chebyshev inequalities, as well as Markov's "truncation method."

Throughout this section $\{X_i\}_{i=1}^\infty$ are i.i.d. (see Definition 6.10), real-valued, and

$$(6.20) \qquad S_n := X_1 + \cdots + X_n \qquad \forall n \geq 1.$$

The Weak Law of Large Numbers. *If $\{X_i\}_{i=1}^\infty$ are in $L^1(P)$ then, as $n \to \infty$,*

$$(6.21) \qquad \frac{S_n}{n} \to EX_1 \quad \text{in } L^1(P), \text{ and hence in probability.}$$

Example 6.21. Imagine n independent Bernoulli trials where the probability of success per trial is $p \in (0, 1)$. If E_j denotes the event that the jth trial is a success, then $S_n := \sum_{j=1}^n \mathbf{1}_{E_j}$ is the total number of successes, and $S_n = \text{Bin}(n, p)$. Because S_n is a sum of n i.i.d. mean-p random variables, Khintchine's weak law of large numbers includes Bernoulli's law of large numbers (p. 18) as a special case.

Proof of the Weak Law. Thanks to Theorem 4.20 on page 43, it suffices to prove that $S_n/n \to EX_1$ in $L^1(P)$. We do this in two steps.

Step 1. The L^2-Case. If $\{X_i\}_{i=1}^\infty$ are in $L^2(P)$, then Corollary 6.19 tells us that

$$(6.22) \qquad \left\| \frac{S_n}{n} - EX_1 \right\|_2 = \text{SD}(S_n/n) = \frac{\text{SD}(X_1)}{\sqrt{n}} \to 0.$$

That is, $S_n/n \to EX_1$ in $L^2(P)$, and hence in $L^1(P)$ (Proposition 4.16, p. 42).

Step 2. The General Case. When the X_i's are assumed only to be in $L^1(P)$ we use a **truncation argument**. Choose and fix a large $\alpha > 0$, and define $X_i^\alpha := X_i \mathbf{1}_{\{|X_i| \leq \alpha\}}$ and $S_n^\alpha := X_1^\alpha + \cdots + X_n^\alpha$. By the triangle inequality, for all $n \geq 1$,

$$(6.23) \qquad \|S_n - S_n^\alpha\|_1 \leq \sum_{i=1}^n E(|X_i|; |X_i| > \alpha) = nE(|X_1|; |X_1| > \alpha).$$

Also, $|EX_1 - E[X_1^\alpha]| \leq E\{|X_1|; |X_1| > \alpha\}$. Therefore,

$$(6.24) \qquad \left\| \frac{S_n}{n} - EX_1 \right\|_1 \leq 2E(|X_1|; |X_1| > \alpha) + \left\| \frac{S_n^\alpha}{n} - E[X_1^\alpha] \right\|_1.$$

Because the X_i^α's are bounded and i.i.d., Step 1 insures that the last L^1-norm converges to zero as $n \to \infty$. Therefore,

$$(6.25) \qquad \limsup_{n \to \infty} \left\| \frac{S_n}{n} - EX_1 \right\|_1 \leq 2E(|X_1|; |X_1| \geq \alpha),$$

for all truncation levels $\alpha > 0$. Let $\alpha \to \infty$ and appeal to the dominated convergence theorem to finish. □

5. Kolmogorov's Strong Law of Large Numbers

We are ready to state and prove the law of large numbers of Kolmogorov (1933). Throughout this section $\{X_i\}_{i=1}^\infty$ are i.i.d. random variables in **R**. We will write $S_n := X_1 + \cdots + X_n$ as before.

The Strong Law of Large Numbers. *If $X_1 \in L^1(\mathrm{P})$, then*

$$(6.26) \qquad \lim_{n \to \infty} \frac{S_n}{n} = \mathrm{E}X_1 \qquad \text{a.s.}$$

Conversely, if $\limsup_{n \to \infty} |S_n/n| < \infty$ with positive probability, then the X_j's are in $L^1(\mathrm{P})$ and (6.26) holds.

Our proof hinges on two key technical results.

The Borel–Cantelli Lemma. *Let $\{A_i\}_{i=1}^\infty$ be a collection of events. If $\sum_{n=1}^\infty \mathrm{P}(A_n) < \infty$ then $\sum_{n=1}^\infty \mathbf{1}_{A_n} < \infty$ a.s. Conversely, suppose that the A_i's are pairwise independent; i.e., $\mathrm{P}(A_i \cap A_j) = \mathrm{P}(A_i)\mathrm{P}(A_j)$ whenever $i \neq j$. Then, $\sum_{n=1}^\infty \mathrm{P}(A_n) = \infty$ implies that $\sum_{n=1}^\infty \mathbf{1}_{A_n} = \infty$ a.s.*

Proof. Let $p_n := \mathrm{P}(A_n)$ for all $n \geq 1$. By the monotone convergence theorem, $\sum_{n=1}^\infty p_n = \mathrm{E}\sum_{n=1}^\infty \mathbf{1}_{A_n}$. Any non-negative $[0,\infty]$-valued random variable that is in $L^1(\mathrm{P})$ is a.s.-finite. Therefore, if $\sum_{n=1}^\infty p_n < \infty$ then $\sum_{n=1}^\infty \mathbf{1}_{A_n} < \infty$ almost surely.

The converse is more interesting: Suppose that $\sum_{n=1}^\infty p_n = \infty$ and the A_j's are pairwise independent. Let $Z_k := \sum_{n=1}^k \mathbf{1}_{A_n}$ and note that

$$(6.27) \qquad \mathrm{Var}\, Z_k = \sum_{n=1}^k p_n(1 - p_n) \leq \sum_{n=1}^k p_n = \mathrm{E}Z_k \qquad \forall k \geq 1.$$

See Corollaries 6.18 and 6.19. Chebyshev's inequality (p. 43) then yields

$$(6.28) \qquad \mathrm{P}\{|Z_k - \mathrm{E}Z_k| \geq \epsilon \mathrm{E}Z_k\} \leq \frac{1}{\epsilon^2 \mathrm{E}Z_k}.$$

Therefore, $Z_k/\mathrm{E}Z_k \xrightarrow{\mathrm{P}} 1$. It follows that $Z_k \to \infty$ in probability; i.e., $\lim_{k \to \infty} \mathrm{P}\{Z_k > \lambda\} = 1$ for all $\lambda > 0$. Because $\sum_{n=1}^\infty \mathbf{1}_{A_n} \geq Z_k$ for all k, this proves that $\sum_{n=1}^\infty \mathbf{1}_{A_n} = \infty$ almost surely. □

Before we proceed with our second technical result, let us prove the second half of the strong law.

Proof of the Strong Law (Necessity). Suppose that $E|X_1| = \infty$; we plan to prove that $\limsup_{n\to\infty} |S_n/n| = \infty$ a.s. Because $|X_n| \le |S_n|+|S_{n-1}|$,

$$(6.29) \qquad \limsup_{n\to\infty} \frac{|X_n|}{n} \le 2\limsup_{n\to\infty} \frac{|S_n|}{n}.$$

Therefore, it suffices to prove that $\limsup_{n\to\infty} |X_n|/n = \infty$ a.s.

Choose and fix an arbitrary $k > 0$, and observe that

$$(6.30) \qquad \frac{1}{k}E|X_1| \le \sum_{n=0}^{\infty} P\{|X_1| \ge kn\} = \sum_{n=0}^{\infty} P\{|X_n| \ge kn\}.$$

See Lemma 6.8. Because the left-hand side is infinite, the second half of the Borel–Cantelli lemma implies that $\limsup_{n\to\infty} |X_n|/n \ge k$ a.s. Let $k \to \infty$ to finish. $\square$

The following *maximal L^2-inequality* of Kolmogorov (1933, 1950) is the second technical result that was promised earlier.

Kolmogorov's Maximal Inequality. *Suppose $S_n = X_1 + \cdots + X_n$, where the X_j's are independent and in $L^2(P)$. Then for all $\lambda > 0$ and $n \ge 1$,*

$$(6.31) \qquad P\left\{\max_{1\le k\le n} |S_k - ES_k| \ge \lambda\right\} \le \frac{\mathrm{Var}S_n}{\lambda^2}.$$

Remark 6.22. Chebyshev's inequality (p. 43) implies the following weaker bound: $\lambda^2 P\{|S_n - ES_n| \ge \lambda\} \le \mathrm{Var}S_n$.

Proof. We may assume that $EX_i = 0$; otherwise, we can consider $X_i - EX_i$ in place of X_i. Note that $ES_n = 0$ in this case.

For all $k \ge 2$, let A_k denote the event that $|S_k| \ge \lambda$ but that $|S_i| < \lambda$ for all $i < k$. If we think of the index k as "time," then A_k denotes the event that the random process $i \mapsto S_i$ leaves $[-\lambda, \lambda]$ for the first time at time k. Because the A_k's are disjoint, $E[S_n^2] \ge \sum_{k=1}^{n} E[S_n^2; A_k]$. Because $S_n^2 \ge 2(S_n - S_k)S_k + S_k^2$, this yields

$$(6.32) \qquad E[S_n^2] \ge 2\sum_{k=1}^{n} E[S_k(S_n - S_k); A_k] + \sum_{k=1}^{n} E[S_k^2; A_k].$$

The event A_k and the random variable S_k both depend on $\{X_i\}_{i=1}^{k}$, whereas $S_n - S_k = X_{k+1} + \cdots + X_n$ is independent of $\{X_i\}_{i=1}^{k}$. Consequently, $S_n - S_k$ is independent of $S_k \mathbf{1}_{A_k}$ (Lemma 6.12). It follows that $E[(S_n - S_k)S_k; A_k] = E[S_n - S_k]E[S_k; A_k] = 0$, since $ES_n = ES_k = 0$. Whenever $\omega \in A_k$ we have $S_k^2(\omega) \ge \lambda^2$. Therefore,

$$(6.33) \qquad E[S_n^2] \ge \sum_{k=1}^{n} E[S_k^2; A_k] \ge \lambda^2 \sum_{k=1}^{n} P(A_k) = \lambda^2 P\left(\bigcup_{k=1}^{n} A_k\right).$$

5. The Strong Law

This proves the result. □

Proof of the Strong Law (Sufficiency). We suppose that $X_1 \in L^1(P)$, and strive to prove that $\lim_{n \to \infty} S_n/n = EX_1$ a.s. Throughout, we may assume, without loss of generality, that $EX_1 = 0$; otherwise, we can consider $X_i - EX_i$ in place of X_i.

The proof of the strong law simplifies considerably when we assume further that $X_1 \in L^2(P)$. This will be done in the first step. The second step uses a truncation argument to reduce the matter to the L^2 case.

Step 1. The L^2 Case. If the X_j's are in $L^2(P)$, then by the Kolmogorov maximal inequality for all $n \geq 1$ and $\epsilon > 0$,

$$(6.34) \qquad P\left\{\max_{1 \leq k \leq n} |S_k| \geq \epsilon n\right\} \leq \frac{E[S_n^2]}{n^2 \epsilon^2} = \frac{\|X_1\|_2^2}{n\epsilon^2},$$

because $E[S_n^2] = \text{Var} S_n = nE[X_1^2]$ (Corollary 6.19). Replace n by 2^n to deduce that $\sum_{n=1}^{\infty} P\{\max_{1 \leq k \leq 2^n} |S_k| \geq \epsilon 2^n\} < \infty$. By the Borel–Cantelli lemma, with probability one,

$$(6.35) \qquad \max_{1 \leq k \leq 2^n} |S_k| \leq \epsilon 2^n,$$

for all but a finite number of n's. This proves the strong law of large numbers along the subsequence 2^n.

In order to prove the strong law along n we use a **sandwich trick**. Because any integer m can be sandwiched between 2^n and 2^{n+1} for some n, it follows that $|S_m| \leq \max_{1 \leq k \leq 2^{n+1}} |S_k|$. Thanks to (6.35), with probability one, $|S_m| \leq \epsilon 2^{n+1} \leq 2\epsilon m$ for all large m. We have demonstrated that for all fixed $\epsilon > 0$ there exists a null set $N(\epsilon)$ such that

$$(6.36) \qquad \limsup_{m \to \infty} \frac{|S_m(\omega)|}{m} \leq 2\epsilon \qquad \forall \omega \notin N(\epsilon).$$

By countable subadditivity $N := \cup_{\epsilon \in \mathbf{Q}_+} N(\epsilon)$ is a null set. Because $|S_m(\omega)| = o(m)$ for all $\omega \notin N$, this proves the result in the case that $X_1 \in L^1(P)$.

Step 2. The L^1 Case. Henceforth, we assume only that $X_1 \in L^1(P)$. We follow the proof of the weak law of large numbers and truncate the X_i's. But now the truncation "levels" are chosen with more care: Define $X'_i := X_i \mathbf{1}_{\{|X_i| \leq i\}}$ for all $i \geq 1$, and let $S'_n := X'_1 + \cdots + X'_n$. Note that

$$(6.37) \qquad \sum_{i=1}^{n} P\{|X_i| > i\} = \sum_{i=1}^{n} P\{|X_1| > i\} \leq \|X_1\|_1;$$

see Lemma 6.8. Therefore, the Borel–Cantelli lemma implies that a.s., $X_n = X'_n$ for all but a finite number of integers n. In particular, $|S_n - S'_n| = o(n)$ a.s. Thus, the theorem follows once we prove that $S'_n = o(n)$ a.s.

First, we prove the simpler assertion that $\mathrm{E}S'_n = o(n)$. Since X_1 has mean zero, $\mathrm{E}S'_n = -\sum_{j=1}^n \mathrm{E}[X_1; |X_1| > j]$. Therefore, we apply the triangle inequality to deduce that $|\mathrm{E}S'_n| \le \sum_{j=1}^n \mathrm{E}(|X_1|; |X_1| > j)$. We exchange the order of summation and expectation to find that

(6.38) $$|\mathrm{E}S'_n| \le \mathrm{E}\{|X_1|\min(|X_1|,n)\}.$$

By the dominated convergence theorem, $\mathrm{E}S'_n = o(n)$, as asserted. Our next and final goal is to prove that $|S'_n - \mathrm{E}S'_n| = o(n)$ a.s. It follows from this that $|S'_n| = o(n)$ a.s., whence the proof would follow.

For all $k \ge 1$, S'_k is a sum of k independent (though not i.i.d.) random variables. According to the Kolmogorov maximal inequality,

(6.39) $$\mathrm{P}(E(n)) \le \frac{\mathrm{Var}S'_n}{n^2\epsilon^2} \quad \forall n \ge 1,\, \epsilon > 0,$$

where

(6.40) $$E(n) := \left\{\max_{1 \le k \le n} \left|S'_k - \mathrm{E}S'_k\right| \ge n\epsilon\right\}.$$

Furthermore, $\mathrm{Var}S'_n = \sum_{j=1}^n \mathrm{Var}X'_j \le \sum_{j=1}^n \mathrm{E}[(X'_j)^2]$, and the latter sum is identically equal to $\sum_{j=1}^n \mathrm{E}[X_1^2; |X_1| \le j]$. Consequently,

(6.41) $$\sum_{n=1}^\infty \mathrm{P}(E(2^n)) \le \sum_{n=1}^\infty \sum_{j=1}^{2^n} \frac{\mathrm{E}[X_1^2; |X_1| \le j]}{4^n \epsilon^2}$$
$$= \frac{1}{\epsilon^2} \sum_{j \ge 1} \mathrm{E}[X_1^2; |X_1| \le j] \sum_{n \ge \log_2(j)} 4^{-n}.$$

If $x \ge 0$, then $\sum_{n \ge x} 4^{-n}$ is at most $\sum_{n=\lfloor x \rfloor}^\infty 4^{-n} = 4^{1-\lfloor x \rfloor}/3 \le 4^{2-x}/3$, where $\lfloor\;\rfloor$ denotes the greatest-integer function. Hence,

(6.42) $$\sum_{n=1}^\infty \mathrm{P}(E(2^n)) \le \frac{16}{3\epsilon^2} \sum_{j \ge 1} \frac{\mathrm{E}[X_1^2; |X_1| \le j]}{j^2}$$
$$= \frac{16}{3\epsilon^2} \mathrm{E}\left[X_1^2 \sum_{j \in \mathbf{N}:\, j \ge |X_1|} \frac{1}{j^2}\right].$$

It is not hard to prove that $x\sum_{j \ge x} j^{-2} \le 2$ for all $x > 0$.[1] Set $x := |X_1|$ and plug this into (6.42) to find that

(6.43) $$\sum_{n=1}^\infty \mathrm{P}(E(2^n)) \le \frac{32}{3\epsilon^2} \mathrm{E}|X_1| < \infty.$$

The previous bound, and the "sandwich trick" of Step 1, together imply that $|S'_n - \mathrm{E}S'_n| = o(n)$ a.s., which verifies our goal. $\square$

---

[1] Indeed, if $x \ge 2$, then we estimate the sum with $\int_{x-1}^\infty u^{-2}\,du \le 2/x$; else, by $\sum_{j \ge 1} j^{-2} \le 2$.

6. Applications

6.1. The Weierstrass Approximation Theorem.
The Weierstrass approximation theorem states that every continuous function on $[0,1]$ can be approximated by a polynomial, uniformly to within any given $\epsilon > 0$. We now follow Bernstein (1913) and derive the Weierstrass theorem from Khintchine's weak law of large numbers.

Definition 6.23. For every continuous function $f : [0,1] \to \mathbf{R}$ define the Bernstein polynomial $\mathcal{B}_n f$ by

$$(6.44) \qquad (\mathcal{B}_n f)(x) := \sum_{j=0}^{n} \binom{n}{j} x^j (1-x)^{n-j} f\left(\frac{j}{n}\right) \qquad \forall x \in [0,1].$$

Then, $\mathcal{B}_n f$ is a polynomial of order at most n for each $n \geq 1$.

Theorem 6.24. $\lim_{n\to\infty} \mathcal{B}_n f = f$, uniformly on $[0,1]$.

Proof. Choose and fix some $p \in [0,1]$, and define $\{X_i\}_{i=1}^{\infty}$ to be independent random variables that take the values 1 and 0 with respective probabilities p and $1-p$. Recall that $S_n := X_1 + \cdots + X_n = \mathrm{Bin}(n,p)$, and note that $(\mathcal{B}_n f)(p) = \mathrm{E} f(S_n/n)$. Consider the "modulus of continuity of f":

$$(6.45) \qquad m_f(\delta) = \sup_{\substack{0 \leq r,s \leq 1:\\ |r-s| \leq \delta}} |f(r) - f(s)| \qquad \forall \delta > 0.$$

Consider the event $A := \{|(S_n/n) - p| \leq \delta\}$, and write

$$(6.46) \qquad \left|(\mathcal{B}_n f)(p) - f(p)\right| \leq T_1 + T_2,$$

where

$$(6.47) \qquad \begin{aligned} T_1 &:= \mathrm{E}\big[\,|f(S_n/n) - f(p)|\,;\,A\big], \\ T_2 &:= \mathrm{E}\big[\,|f(S_n/n) - f(p)|\,;\,A^c\big]. \end{aligned}$$

We bound T_2 first: If $K := 2\sup_x |f(x)|$, then by the Chebyshev inequality,

$$(6.48) \qquad |T_2| \leq K \mathrm{P}(A^c) \leq K \delta^{-2} \mathrm{Var}(S_n/n).$$

By Corollary 6.19, $\mathrm{Var}(S_n/n) = n^{-1} p(1-p) \leq (4n)^{-1}$. Consequently, $|T_2| \leq K/(4\delta^2 n)$ uniformly in $p \in [0,1]$. Since $|T_1| \leq m_f(\delta)$, it follows that for any $\delta > 0$,

$$(6.49) \qquad \sup_{0 \leq p \leq 1} |(\mathcal{B}_n f)(p) - f(p)| \leq m_f(\delta) + \frac{K}{4\delta^2 n}.$$

Let $n \to \infty$ and then $\delta \to 0$, in this order, to finish the proof. $\square$

Remark 6.25 (Kac's Refinement). Kac (1937) has observed that the preceding proof can yield quantitative estimates. Indeed, suppose that f is *Hölder continuous* with index $\alpha > 0$; i.e., there exists L such that $m_f(\delta) \leq L\delta^\alpha$. It follows that

$$\sup_{0 \leq p \leq 1} |(\mathcal{B}_n f)(p) - f(p)| \leq \frac{A}{n^{\alpha/2}}, \tag{6.50}$$

for a constant A that depends only on α and L. We derive (6.50) next. The only case of interest is $\alpha \in (0, 1]$. For if $\alpha > 1$ then $f' = 0$, in which case f is a constant and there is nothing to prove.

Henceforth, choose and fix $\alpha \in (0, 1]$, and define $\mathcal{D}_n := |(\mathcal{B}_n f)(p) - f(p)|$ for brevity. Because $\mathcal{D}_n = |\mathrm{E}\{f(S_n/n) - f(p)\}|$, Hölder continuity of f implies that $\mathcal{D}_n \leq L\mathrm{E}\{|(S_n/n) - p|^\alpha\}$. Since $\alpha \in (0, 1]$, Hölder's inequality asserts that $\mathrm{E}(|Z|^\alpha) \leq (\mathrm{E}\{Z^2\})^{\alpha/2}$ for all random variables Z. Therefore,

$$|(\mathcal{B}_n f)(p) - f(p)| \leq L \left[\mathrm{Var}\left(\frac{S_n}{n}\right)\right]^{\alpha/2} = L \left[\frac{p(1-p)}{n}\right]^{\alpha/2}. \tag{6.51}$$

The elementary inequality $p(1-p) \leq 1/4$ yields (6.50) with $A := L2^{-\alpha}$.

6.2. The Asymptotic Equipartition Property. Our next application of independence is one of the starting points of the work of Shannon (1948; 1949), who discovered various startling connections between the thermodynamical notion of entropy and the mathematical theory of communication. We explore one of these connections here.

First, we need some jargon from communication theory: Consider a fixed finite set $\mathbf{A} := \{\sigma_1, \ldots, \sigma_m\}$ with m distinct elements. Any element σ_i of $\mathbf{A}$ is a *letter* (or *symbol*), and $\mathbf{A}$ itself is the *alphabet*. A *word* (or *code*) $w := (w_1, \ldots, w_n)$ of length n is a vector of n letters in $\mathbf{A}$. The *relative frequency* of the letter σ_k in the word w is then

$$f_n(\sigma_k, w) = \frac{1}{n} \sum_{j=1}^n \mathbf{1}_{\{\sigma_k\}}(w_j). \tag{6.52}$$

There are a total of m^n words of length n. If we were to select one at random (all n-letter words being equally likely) and write the resulting random word as $W := (W_1, \ldots, W_n)$, then $W_1, \ldots, W_n$ are i.i.d., and $\mathrm{P}\{W_1 = \sigma_i\} = 1/m$ for all $i = 1, \ldots, m$ (check!). Therefore, by the weak law of large numbers (p. 72), for any $k = 1, \ldots, m$, $f_n(\sigma_k, W_n) \to 1/m$ in probability. That is, for a "very typical word of indefinite length," the asymptotic relative frequency of any letter is $1/m$.

What about long words with (possibly) other asymptotic frequencies? To answer this, we choose and fix a probability vector $(p_1, \ldots, p_m)$ throughout; i.e., $p_j \geq 0$ for $j = 1, \ldots, m$, and $p_1 + \cdots + p_m = 1$.

6. Applications

Definition 6.26. If $\epsilon > 0$, then an n-letter word w is said to be ϵ-*typical* if

(6.53) $$|f_n(\sigma_k, w) - p_k| \le \epsilon \qquad \forall k = 1, \ldots, m.$$

Otherwise, w is said to be ϵ-*atypical*.

The following is, in essence, Shannon's fundamental theorem of data compression (Shannon, 1948). For an improvement see Problem 6.33 below.

Theorem 6.27. *For every $n \ge 1$ and $\epsilon > 0$ define $T_n(\epsilon)$ to be the number of ϵ-typical words of length n. Then,*

(6.54) $$\left(1 - \frac{1}{n\epsilon^2}\right) 2^{n(H(p) - c\epsilon)} \le T_n(\epsilon) \le 2^{n(H(p) + c\epsilon)},$$

where $\log_2$ denotes the base-2 logarithm, $c := -\sum_{k=1}^m \log_2 p_k \ge 0$, and $H(p) := -\sum_{i=1}^m p_i \log_2 p_i$ is the entropy of the vector $p = (p_1, \ldots, p_m)$.

Suppose n tends to infinity and $\epsilon = \epsilon_n$ goes to zero so that $\epsilon^2 n \to \infty$. Then the preceding says that of the m^n words of length n, $2^{n(H(p)+o(1))}$ have more or less the property that the letters $\sigma_1, \ldots, \sigma_m$ have asymptotic relative frequencies $p_1, \ldots, p_m$, respectively. The point is that unless $p_1 = \cdots = p_m = 1/m$, $2^{n(H(p)+o(1))}$ is asymptotically far smaller than the total number of n-letter words m^n; compare with Problem 6.25. This observation is the starting point of data compression.

Proof of Theorem 6.27. Let $\{X_i\}_{i=1}^\infty$ denote i.i.d. random variables, all taking the values $\sigma_1, \ldots, \sigma_m$ with probabilities $p_1, \ldots, p_m$. Write $W_n := (X_1, \ldots, X_n)$.

For any fixed n-letter word w, $P\{W_n = w\} = \prod_{k=1}^m p_k^{nf_n(\sigma_k, w)}$. If w is ϵ-typical, then the latter probability is at least $\prod_{k=1}^m p_k^{n(p_k+\epsilon)} = 2^{-n(H(p)+c\epsilon)}$; also it is at most $2^{-n(H(p)-c\epsilon)}$. Rearrange and sum over ϵ-typical w's to find that

(6.55) $$2^{n(H(p)-c\epsilon)} \pi_n(\epsilon) \le T_n(\epsilon) \le 2^{n(H(p)+c\epsilon)} \pi_n(\epsilon),$$

where $\pi_n(\epsilon) := P\{W_n \text{ is } \epsilon\text{-typical}\}$. Because $\pi_n(\epsilon) \le 1$, we obtain the asserted upper bound for $T_n(\epsilon)$.

For the lower bound we note that

(6.56) $$1 - \pi_n(\epsilon) = P\left\{\exists k = 1, \ldots, m : |f_n(\sigma_k, W_n) - p_k| > \epsilon\right\}$$
$$\le \sum_{k=1}^m P\{|f_n(\sigma_k, W_n) - p_k| > \epsilon\}.$$

Clearly, $nf_n(\sigma_k, W_n) = \sum_{j=1}^n \mathbf{1}_{\{X_j = \sigma_k\}} = \mathrm{Bin}(n, p_k)$. In particular, it has mean np_k and variance $np_k(1-p_k) \leq np_k$. Thus, by the Chebyshev inequality,

$$(6.57) \qquad 1 - \pi_n(\epsilon) \leq \sum_{k=1}^m \frac{np_k}{n^2 \epsilon^2} = \frac{1}{n\epsilon^2}.$$

The theorem follows. □

6.3. The Glivenko–Cantelli Theorem. A histogram is a random discrete probability distribution; it depends on $\{X_i\}_{i=1}^n$, and assigns probability $p_n(x)$ to any point $x \in \mathbf{R}$, where $p_n(x)$ is the fraction of the data that is equal to x. Its cumulative distribution function is called the empirical distribution function, and defined more formally as follows.

Definition 6.28. If $\{X_i\}_{i=1}^\infty$ are i.i.d. random variables all with distribution function F, then their *empirical distribution function* F_n is

$$(6.58) \qquad F_n(x,\omega) = \frac{1}{n} \sum_{k=1}^n \mathbf{1}_{\{X_k \leq x\}}(\omega) \qquad \forall x \in \mathbf{R}, \omega \in \Omega.$$

Thus, we can view $x \mapsto F_n(x, \cdot)$ as a random (cumulative) distribution function. As we did with other sorts of random variables, we suppress the dependence on the ω variable, and write $F_n(x)$ in place of $F_n(x, \omega)$.

The following is due to Glivenko (1933) and Cantelli (1933b). In data-analytic terms, this theorem presents a uniform approximation to an unknown distribution function F based on a random sample from this distribution.

Theorem 6.29. $\lim_{n \to \infty} \sup_{x \in \mathbf{R}} |F_n(x) - F(x)| = 0$ a.s.

Proof. Since F_n and F are right-continuous,

$$(6.59) \qquad \sup_{x \in \mathbf{R}} |F_n(x) - F(x)| = \sup_{x \in \mathbf{Q}} |F_n(x) - F(x)|.$$

Thus, $\sup_{x \in \mathbf{R}} |F_n(x) - F(x)|$ is a random variable.

Fix $x \in \mathbf{R}$ and note that $nF_n(x) = \mathrm{Bin}(n, F(x))$. By the strong law of large numbers (p. 73), for any $\epsilon > 0$, $n \geq 1$, and $x \in \mathbf{R}$,

$$(6.60) \qquad |F_n(x) - F(x)| \leq \epsilon \quad \text{for all but finitely many } n\text{'s, a.s.}$$

Recall that: (i) F is non-decreasing; (ii) F is right-continuous; (iii) $F(\infty) = 1$; and (iv) $F(-\infty) = 0$. Therefore, we can find $-\infty < x_0 < \cdots < x_m < \infty$ such that: $F(x_0) \leq \epsilon$; $F(x_m) \geq 1 - \epsilon$; and

$$(6.61) \qquad \sup_{x_{j-1} \leq x < x_j} |F(x) - F(x_{j-1})| \leq \epsilon \qquad \forall j = 1, \ldots, m.$$

According to (6.60),

(6.62) $\quad \max_{0 \leq j \leq m} |F_n(x_j) - F(x_j)| \leq \epsilon \quad$ for all but finitely many n's, a.s.

Hence follows that if $x \in [x_{j-1}, x_j)$ for some $1 \leq j \leq m$, then with probability one the following holds for all sufficiently large integers n:

(6.63)
$$F_n(x) \leq F_n(x_j) \leq F(x_j) + \epsilon; \text{ and}$$
$$F(x_{j-1}) \leq F_n(x_{j-1}) + \epsilon \leq F_n(x) + \epsilon.$$

By (6.61), $F(x_j) \leq F(x_{j-1}) + \epsilon$. Since F is non-decreasing it follows that with probability one,

(6.64) $\quad \sup_{x_0 \leq x \leq x_m} |F_n(x) - F(x)| \leq 2\epsilon \quad$ for all but finitely many n's.

Choose and fix n large enough that the previous inequality is satisfied. If $x > x_m$, then $F(x) \geq F(x_m) \geq 1 - \epsilon$ and $F_n(x) \geq F(x_m) - \epsilon \geq 1 - 2\epsilon$. Therefore,

(6.65) $\quad |F_n(x) - F(x)| \leq |1 - F(x)| + |1 - F_n(x)| \leq 3\epsilon \quad \forall x > x_m.$

Similarly, if $x < x_0$, then $|F(x) - F_n(x)| \leq F(x_0) + F_n(x_0) \leq 3\epsilon$. Consequently, with probability one,

(6.66) $\quad \sup_{x \in \mathbf{R}} |F_n(x) - F(x)| \leq 3\epsilon \quad$ for all but finitely many n's.

This proves the theorem. $\square$

6.4. The Erdős Bound on Ramsey Numbers. Let us begin with a definition or two from graph theory.

Definition 6.30. The *complete graph* K_m on m vertices is a collection of m distinct vertices any two of which are connected by a unique edge. The nth (diagonal) *Ramsey number* R_n is the smallest integer N such that any bi-chromatic coloring of the edges of K_N yields a $K_n \subseteq K_N$ whose edges are all of the same color.

To understand this definition suppose $R_n = N$. Then, no matter how we color the edges of K_N using only the colors red and blue, somewhere inside K_N there exists a K_n whose edges are either all blue or all red, and N is the smallest such value. It is possible to check that $R_2 = 3$ and $R_3 = 6$, for example.

Ramsey (1930) introduced these and other Ramsey numbers to discuss ways of checking the consistency of a logical formula. See also Skolem (1933) and Erdős and Szekeres (1935).

As a key step in his proofs Ramsey proved that $R_n < \infty$ for all $n \geq 1$. Evidently, $R_n \to \infty$ as $n \to \infty$; in fact, it is obvious that $R_n \geq n$. The following theorem of Erdős (1948) presents a much better lower bound.

Theorem 6.31. *As* $n \to \infty$, $R_n \geq (c + o(1))n2^{n/2}$, *where* $1/c := e\sqrt{2}$.

Proof. We aim to prove that given any two integers $N \geq n$,

(6.67) $$\binom{N}{n} 2^{1-\binom{n}{2}} < 1 \quad \text{implies} \quad R_n > N.$$

Let us assume that this is the case, for the time being, and apply it with $N := \lfloor cn2^{n/2} \rfloor$, where $\lfloor \ \rfloor$ denote the greatest-integer function. Because $N!/(N-n)! \leq N^n$, our particular choice of N yields

(6.68) $$\binom{N}{n} \leq \frac{(cn2^{n/2})^n}{n!}.$$

Consequently, by Stirling's formula (p. 21),

(6.69) $$\binom{N}{n} 2^{1-\binom{n}{2}} \leq \frac{2c^n n^n 2^{n/2}}{n!} \sim \sqrt{\frac{2}{\pi n}},$$

which is strictly less than 1 for all large n. It remains to verify (6.67).

Consider a random coloring of the edges of K_N; i.e., if E_N denotes the set of all edges of K_N, then consider an i.i.d. collection of random variables $\{X_e\}_{e \in E_N}$ where $P\{X_e = 1\} = P\{X_e = -1\} = 1/2$. We color e red if $X_e = 1$, and blue if $X_e = -1$.

The probability that any n given vertices form a monochromatic K_n is $2^{1-\binom{n}{2}}$. Since there are $\binom{N}{n}$ choices of these n vertices, the probability that there exist n vertices that form a monochromatic K_n is less than or equal to $\binom{N}{n} 2^{1-\binom{n}{2}}$. If this is strictly less than one, then there must exist bichromatic colorings of K_N that yield no monochromatic K_n, and hence (6.67) follows. □

6.5. Percolation. Consider the d-dimensional integer lattice $\mathbf{Z}^d$ as a graph; i.e., two points $x, y \in \mathbf{Z}^d$ are connected by an edge if and only if the Euclidean distance between x and y is one. Let $\mathbf{E}^d$ denote the resulting collection of edges; $\mathbf{Z}^d$ denotes both the vertices and the graph.

Fix a number $p \in [0,1]$, and consider the resulting random graph of $\mathbf{Z}^d$ that is obtained by deleting any edge with probability $1-p$; else, the edge is kept. The decisions from edge to edge are made independently. More precisely, let $\{X_e(p)\}_{e \in \mathbf{E}^d}$ be i.i.d. with $P\{X_e(p) = 1\} = 1 - P\{X_e(p) = 0\} = p$. As always, $X_e(p) = X_e(p, \omega)$ depends on ω, but we ignore the ω. If $X_e(p) = 0$, then we think of e as a deleted edge. The resulting random subgraph can be identified with $\Gamma(p) := \{e \in \mathbf{E}^d : X_e(p) = 1\}$.

We say that *percolation* occurs if $\Gamma(p)$ has an infinite connected subgraph. By the Kolmogorov zero-one law, the probability of this event is zero or one. The basic question is to decide when the probability of percolation is one.

6. Applications

This, and some generalizations, were introduced by Broadbent and Hammersley (1957). Since then, the subject has grown to be a vast area in mathematics and physics alike; see Grimmett (1999) for a lively introduction.

Next, we prove one of the basic results of percolation. Namely, that there exists a critical probability of percolation.

Proposition 6.32. *For any $d \geq 2$, there exists a "critical probability" $p_c(\mathbf{Z}^d) \in [0,1]$ such that whenever $p < p_c(\mathbf{Z}^d)$ percolation does not occur. But if $p > p_c(\mathbf{Z}^d)$, then percolation occurs a.s.*

Remark 6.33. (1) The same is true for $d = 1$, but this is the trivial case since $p_c(\mathbf{Z}) = 1$. Indeed, if $p < 1$, then with probability one, we end up deleting infinitely many edges on both sides of the origin; see Problem 6.16. Thus, unless $p = 1$, there is no percolation on $\mathbf{Z}$.

(2) It can be shown that if $d \geq 2$, then $p_c(\mathbf{Z}^d)$ lies (strictly) in $(0,1)$. In addition, $p_c(\mathbf{Z}^2) = \frac{1}{2}$; the lower bound on $p_c(\mathbf{Z}^2)$ is due to Harris (1960), and the upper bound to Kesten (1980). To establish the weaker bound, $p_c(\mathbf{Z}^2) \geq \frac{1}{3}$, one can employ far simpler arguments. See Problem 6.28.

Proof. The trick is to appeal to a *monotonicity argument*: We can construct all of the $X_e(p)$'s, all on the same probability space, such that [for all ω]:

(6.70) $\qquad$ If $p \leq r$, then $X_e(p) \leq X_e(r)$ for all $e \in \mathbf{E}^d$.

In order to do this, we first construct, on an appropriate product space, random variables $\{U_e\}_{e \in \mathbf{E}^d}$ that are i.i.d. and distributed uniformly on $[0,1]$. Then we define $X_e(p) := \mathbf{1}_{\{U_e \in [0,p]\}}$ for all $p \in [0,1]$ and $e \in \mathbf{E}^d$. Evidently, $\{X_e(p)\}_{e \in \mathbf{E}^d}$ are i.i.d., and $\mathrm{P}\{X_e(p) = 1\} = 1 - \mathrm{P}\{X_e(p) = 0\} = p$. Moreover, (6.70) is manifest. In particular, on this probability space, if $p \leq r$, then $\Gamma(p) \subseteq \Gamma(r)$ [for all ω]. The result follows from letting $p_c(\mathbf{Z}^d)$ denote the smallest $p \in [0,1]$ such that $\Gamma(p)$ contains an infinite connected subgraph with positive probability. $\square$

6.6. Monte Carlo Integration. Suppose we were asked to find, or estimate, the value of some n-dimensional integral

(6.71) $$I(\phi) := \int_{[0,1]^n} \phi(x) \, dx.$$

Here, $\phi : \mathbf{R}^n \to \mathbf{R}$ is a Lebesgue-integrable function that is so complicated that $I(\phi)$ is not explicitly computable.

One way to proceed is to first select i.i.d. random variables $X_1, \ldots, X_N$ uniformly at random in the n-cube $[0,1]^n$. By Lemma 6.6, $\mathrm{E}\phi(X_j) = I(\phi)$

for all $j = 1, \ldots, N$. Because $\{\phi(X_i)\}_{i=1}^N$ are i.i.d. random variables with expectation $I(\phi)$, the strong law of large numbers (p. 73) implies that

$$(6.72) \qquad \lim_{N \to \infty} \frac{1}{N} \sum_{j=1}^N \phi(X_j) = I(\phi) \quad \text{a.s.}$$

The preceding suggests a way of finding numerical approximations to $I(\phi)$: First simulate N independent uniform-$[0,1]^n$ random variables $\{X_i\}_{i=1}^N$, and then average $\phi(X_1), \ldots, \phi(X_N)$. This procedure is called *Monte Carlo integration*. It was first used in the 1930s by E. Fermi in the calculation of neutron diffusion. The full power of Monte Carlo integration was discovered in the 1940s by J. von Neumann and S. Ulam. Monte Carlo integration outperforms most other numerical integration methods when N is large.

We conclude this section with a brief discussion on random-number generation. For this discussion, it might help to recall that a random "variable" is in fact a function, or a procedure.

Stated loosely, most random-number generators simulate the said procedure as follows: Start with an initial number ω_0, called a *seed*, and some predetermined function $f : \mathbf{R} \to \mathbf{R}$. Then, define iteratively $\omega_{i+1} := f(\omega_i)$. If f is sufficiently "chaotic," and if N is sufficiently large, then ω_N simulates a realization $X(\omega_0)$ of a certain random variable X. To obtain more simulations we start over with other seeds.

In order to simulate a Unif$(0,1)$ random variable X, f has to be chosen with care. The most common choice is to use a *linear congruential generator* (LCG). A linear congruential generator is described by

$$(6.73) \qquad f(x) = (ax + b) \pmod{c},$$

where a, b, and c are "well-chosen" predescribed parameters. Knuth (1981) discusses this and related methods. In particular, one finds there intuitive as well as rigorous methods for finding good choices of a, b, and c. There are also interesting non-linear examples; for a sampler see Problems 6.31 and 6.36.

Problems

6.1. Consider a continuously differentiable function $f : \mathbf{R}_+ \to \mathbf{R}_+$ that satisfies $f(0) = 0$. Suppose that either $f' \geq 0$ a.e., or f' is integrable. Then prove that for all non-negative random variables X,

$$(6.74) \qquad \mathrm{E} f(X) = \int_0^\infty f'(\lambda) \mathrm{P}\{X \geq \lambda\} \, d\lambda = \int_0^\infty f'(\lambda) \mathrm{P}\{X > \lambda\} \, d\lambda.$$

Deduce Lemma 6.8 from this.

6.2. Prove that two real-valued random variables X and Y, both defined on the same probability space, are independent if and only if for all $x, y \in \mathbf{R}$,

$$(6.75) \qquad \mathrm{P}\{X \leq x, Y \leq y\} = \mathrm{P}\{X \leq x\} \mathrm{P}\{Y \leq y\}.$$

6.3. Improve Lemma 6.8 as follows: If $X, Y \in L^2(\mathrm{P})$ then $\mathrm{Cov}(X, Y)$ is equal to

$$\int_{-\infty}^{\infty} \int_{-\infty}^{\infty} \Big(\mathrm{P}\{X > x, Y > y\} - \mathrm{P}\{X > x\}\mathrm{P}\{Y > y\} \Big) \, dx \, dy. \tag{6.76}$$

6.4. Prove that if the distribution of (X, Y) is absolutely continuous with respect to two-dimensional Lebesgue measure, then:

(1) There exists $f \in L^1(\mathbf{R}^2)$ such that $\mathrm{P}\{(X,Y) \in A\} = \iint_A f(x,y) \, dx \, dy$ for all $A \in \mathscr{B}(\mathbf{R}^2)$,

(2) There exist $f_X, f_Y \in L^1(\mathbf{R})$ such that $\mathrm{P}\{X \in B\} = \int_B f_X(x) \, dx$ and $\mathrm{P}\{Y \in B\} = \int_B f_Y(y) \, dy$ for all $B \in \mathscr{B}(\mathbf{R})$.

(3) X and Y are independent iff $f(x,y) = f_X(x) f_Y(y)$ for almost every $(x,y) \in \mathbf{R}^2$.

6.5. Verify the claim of Remark 6.11 by constructing three random variables X_1, X_2, and X_3, such that X_i and X_j are independent whenever $i \neq j$, but $\{X_1, X_2, X_3\}$ are not.

6.6. Prove Lemma 6.15.

6.7. Construct two random variables X and Y on the same probability space such that X and Y are uncorrelated but not independent.

6.8. Prove Corollaries 6.18 and 6.19. Improve the latter by proving that given any sequence $\{X_i\}_{i=1}^n$ of random variables in $L^2(\mathrm{P})$,

$$\mathrm{Var}(X_1 + \cdots + X_n) = \sum_{i=1}^n \mathrm{Var} X_i + 2 \sum\sum_{1 \le i < j \le n} \mathrm{Cov}(X_i, X_j). \tag{6.77}$$

6.9. Given a random variable X, prove that we can construct random variables $\{X_n\}_{n=1}^\infty$ with the following properties: (i) The cumulative distribution function F_n of X_n is continuous and strictly increasing; (ii) $\lim_{n \to \infty} X_n = X$ a.s.

6.10. If $\{X_n\}_{n=1}^\infty$ are i.i.d., then prove that the following are equivalent for all $p > 0$ fixed:

(1) $\mathrm{E}\{|X_1|^p\} < \infty$;

(2) $X_n = o(n^{1/p})$ a.s.; and

(3) $\max_{1 \le j \le n} |X_j| = o(n^{1/p})$ a.s.

6.11 (One-Series Theorem). Prove the theorem of Kolmogorov (1930): If $\{X_i\}_{i=1}^\infty$ are independent mean-zero random variables, and if $\sum_j \mathrm{E}[X_j^2] < \infty$, then $\sum_j X_j$ converges almost surely. Use this to prove that "the random harmonic series converges a.s." That is, if $\{\sigma_j\}_{j=1}^\infty$ are i.i.d. random variables taking the values ± 1 with probability $\frac{1}{2}$ each, then $\sum_{j=1}^\infty (\sigma_j/j)$ converges a.s.

6.12. The strong law is easier to prove when the summands have four finite absolute moments. Suppose $\{X_i\}_{i=1}^\infty$ are i.i.d. random variables with mean zero and variance one, and let $S_n = X_1 + \cdots + X_n$. Prove that, if in addition $\|X_1\|_4 < \infty$, then $\|S_n\|_4 = O(n^{1/2})$ as $n \to \infty$. Conclude the following, which is essentially due to Cantelli (1917b): If $\mathrm{E}[X_1^4]$ is finite then $S_n = o(n)$ a.s. You may not use the Kolmogorov maximal inequality in this exercise.

6.13. Let $\{X_i\}_{i=1}^\infty$ be a sequence of non-negative, i.i.d. random variables such that $\mathrm{E} X = \infty$. Prove the following complement to the strong law (p. 73):

$$\lim_{n \to \infty} \frac{X_1 + \cdots + X_n}{n} = \infty \quad \text{a.s.} \tag{6.78}$$

6.14. Let $X_1, X_2, \ldots$ be non-negative random variables in $L^2(\mathrm{P})$ that have common mean μ. Prove that if $S_n := X_1 + \cdots + X_n$ then $S_n/n \to \mu$ a.s., provided that there exists $\delta \in (0, 2)$ such that $\mathrm{Var} S_n = o(n^{2-\delta})$. Conclude that if $\{X_i\}_{i=1}^\infty$ are non-negative, identically distributed, and uncorrelated, then $S_n \sim n \mathrm{E} X_1$ a.s.

6.15. Construct independent positive random variables $\{X_i\}_{i=1}^\infty$ such that $\lim_{n \to \infty} \prod_{i=1}^n X_i$ exists almost surely but not in $L^1(\mathrm{P})$.

6.16. Suppose $\{X_j\}_{j=1}^\infty$ are i.i.d. random variables that take values in some topological space $\mathbf{X}$, and $P\{X_1 \in A\} > 0$ for some $A \in \mathscr{B}(\mathbf{X})$. Prove that with probability one infinitely many of the X_n's fall in A. Thus, if a monkey types forever and at random, then with probability one she eventually reproduces the entire works of Shakespeare.

6.17. First prove that $EX = \sum_{n=1}^\infty P\{X \geq n\}$ whenever X is a non-negative integer-valued random variable. Then use your computation to solve the following.

(1) A dresser has k distinct pairs of socks. We select, at random, one sock at a time until a pair has been drawn. Compute the expectation of the total number of draws needed.

(2) We generate, independently, Unif$(0,1)$ random variables $X_1, X_2, \ldots$ as long as the generated sequence is monotone. Prove that if X denotes the length of this random monotone sequence, then $EX = 2e - 3$ (Shultz and Leonard, 1989).

6.18. Prove that if X is independent of itself, then X is almost surely a constant.

6.19. Suppose that $\{X_i\}_{i=1}^\infty$ is a sequence of i.i.d. exponential random variables with parameter $\lambda > 0$. Verify that almost surely, $\limsup_{n\to\infty}(X_n/\ln n) = \lambda^{-1}$ and $\liminf_{n\to\infty}(\ln X_n/\ln n) = -1$.

6.20 (Paley–Zygmund Inequality). Prove that for all nonnegative $Y \in L^2(P)$ and all $\epsilon \in (0,1)$,

$$(6.79) \qquad P\{Y > \epsilon EY\} \geq (1-\epsilon)^2 \frac{\|Y\|_1^2}{\|Y\|_2^2}.$$

(Paley and Zygmund, 1932). Now suppose that $\{E_i\}_{i=1}^\infty$ are events such that $\sum_{j=1}^\infty P(E_j) = \infty$, and that

$$(6.80) \qquad \gamma = \liminf_{n\to\infty} \frac{\sum_{i=1}^n \sum_{j=1}^n P(E_i \cap E_j)}{\left(\sum_{j=1}^n P(E_j)\right)^2} < \infty.$$

Then, show that $P\{\sum_j \mathbf{1}_{E_j} = \infty\} \geq \gamma^{-1} > 0$. Verify that this improves the independence half of the Borel–Cantelli lemma (p. 73).

6.21 (Problem 6.20, Continued). Prove that if Z is a non-negative random variable in $L^2(P)$, then $P\{Z = 0\} \leq \operatorname{Var} Z / E[Z^2]$.

6.22 (Normal Numbers). Suppose X is uniformly distributed on $[0,1]$, and write its decimal expansion as $X = \sum_{j=1}^\infty 10^{-j} X_j$, where $X_j = 0, \ldots, 9$ (with some convention for terminating expansions). Prove that $\{X_i\}_{i=1}^\infty$ are i.i.d. Find their distribution. Derive the *normal-number theorem* of Borel (1909): "*Lebesgue-almost every number* $\omega \in [0,1]$ *satisfies* $\lim_{n\to\infty} N_n^\ell(\omega)/n = 0.1$ *for* $\ell = 0, \ldots, 9$." Here, $N_n^\ell(\omega) = \sum_{i=1}^n \mathbf{1}_{\{X_i(\omega)=\ell\}}$ denotes the number of times that the digit ℓ appears in the first n binary digits of ω. [If you do not find this surprising then you may wish to try to decide whether some given irrational such as $1/\sqrt{2}$, $\pi/10$, or $\ln 2$, is a normal number.]

6.23 (Problem 6.22, Continued). Choose and fix an integer $b > 1$. If X is distributed uniformly on $[0,1]$, then write its b-ary expansion $X(\omega) = \sum_{j=1}^\infty b^{-j} X_j(\omega)$. In the case that there are two ways to choose the X_j's, we always opt for the finite expansion. Prove that $\{X_j\}_{j=1}^\infty$ are i.i.d. and take the values $0, \ldots, b-1$ with equal probability. Prove:

$$(6.81) \qquad P\left\{\lim_{n\to\infty} \frac{1}{n} \sum_{i=1}^n \mathbf{1}_{\{X_i(\omega)=\ell\}} = \frac{1}{b} \;\; \forall \ell = 0, \ldots, b-1, \; \forall b \geq 2\right\} = 1.$$

6.24. Recall that if $f: \mathbf{R} \to \mathbf{R}_+$ is measurable and integrates to one, then it is a probability density function. Prove that if, in addition, f is continuous then $F(x) := \int_{-\infty}^x f(y)\,dy$ defines a distribution function with $F' = f$ almost everywhere.

6.25. Suppose $p = (p_1, \ldots, p_m)$ is a vector of probabilities; i.e., $p_i \in [0,1]$ for all $i = 1, \ldots, m$, and the p_i's add up to one. Recall the entropy $H(p)$; you may need to define $0 \ln 0 := 0$ to make sense of this in general. Prove that the (discrete) uniform distribution maximizes the entropy uniquely among all probability vectors on m fixed points. Calculate this maximum entropy. This exemplifies the method of "*the most probable distribution*" of statistical thermodynamics (Schrödinger, 1946). (HINT: For all $x \geq 0$, $-x \ln x \leq 1 - x$.)

6.26 (Information Inequality). First prove that

$$\text{(6.82)} \qquad \int_{-\infty}^{\infty} f(x)\ln g(x)\,dx \le \int_{-\infty}^{\infty} f(x)\ln f(x)\,dx,$$

for all density functions f and g on $\mathbf{R}$, where $0\ln 0 := 0$. This is called the *information inequality*. Let $H(f) := -\int_{-\infty}^{\infty} f(x)\ln f(x)\,dx$ denote the *entropy* of f. Then prove that:

(1) The Unif(a,b) density is of maximum entropy among all density functions that are supported on (a,b).

(2) The $N(\mu,\sigma^2)$ density is of maximum entropy among all densities on $\mathbf{R}$ that have mean μ and variance σ^2.

6.27. Prove that if $\{X_i\}_{i=1}^{\infty}$ are i.i.d. and in $L^2(\mathrm{P})$, and if $S_j := X_1 + \cdots + X_j$, then

$$\text{(6.83)} \qquad \mathrm{E}\left[\max_{1\le j\le n}|S_j - \mathrm{E}S_j|\right] \le 2\mathrm{SD}(X_1)\sqrt{n}$$

6.28. Let π denote a collection of edges in the graph $\mathbf{Z}^d$. We say that π is a *path of length n* if there are n edges in π and they connect neighboring points in $\mathbf{Z}^d$, starting with the origin. A path is *self-avoiding* if no 3 of its edges share a vertex.

(1) Prove that there are at most $\lambda_{n,d} := 2d(2d-1)^{n-1}$ self-avoiding paths of length n.

(2) In the context of percolation (§6.5) prove that the expected number of kept paths of length n is at most $p^n \lambda_{n,d}$. Use this to prove that $p_c(\mathbf{Z}^d)$ is at least $(2d-1)^{-1}$.

6.29. Prove that if $\{X_i\}_{i=1}^{\infty}$ are i.i.d. and in $L^1(\mathrm{P})$, then

$$\text{(6.84)} \qquad \lim_{n\to\infty}\frac{1}{\ln n}\sum_{i=1}^{n}\frac{X_i}{i} = \mathrm{E}X_1 \qquad \text{a.s.}$$

(HINT: Work along the subsequence 2^{2^n}.)

6.30. Prove that if $f:[0,1]\to\mathbf{R}$ is non-decreasing, then so is $\mathcal{B}_n f$, where $\mathcal{B}_n f$ is the Bernstein polynomial based on f. (HINT: If $U_1, U_2, \ldots$ are i.i.d. Unif$(0,1)$'s, then $S_n := \sum_{i=1}^{n}\mathbf{1}_{\{U_i\le p\}}$ is binomial.)

6.31 (Computer Problem). Newton's method is a numerical algorithm for finding the roots of a function $g:\mathbf{R}\to\mathbf{R}$. We start with a *seed* ω_0, and iteratively define

$$\text{(6.85)} \qquad \omega_{n+1} := \omega_n - \frac{g(\omega_n)}{g'(\omega_n)} \qquad \forall n\ge 0.$$

If g is a nice function, then Newton's method converges rapidly to a root of g. Consider $g(x) := x^2 + 1$, which has no roots. What happens if one computes the Newton-method iterates $\{\omega_i\}_{i=1}^{m}$ on a computer? Explore this idea for various choices of m and ω_0. Also consider the sequence defined by $\zeta_n := \frac{1}{2} + \frac{1}{\pi}\arctan\omega_n$. Convince yourself, via computer simulation, that if m is large then the distribution of ζ_m is nearly uniform.

6.32. Suppose that $\{X_j\}_{j=1}^{\infty}$ are i.i.d., and take the values ± 1 with probability one-half each. Then prove that $\lim_{n\to\infty}\sum_{i=1}^{n} X_i$ does not exist a.s.

6.33 (Problem 4.30, p. 51, Continued; Hard). Let $\{X_i\}_{i=1}^{n}$ denote a collection of mean-zero, independent, a.s.-bounded random variables such that $\mathrm{P}\{|X_i|\le c_i\}=1$ for non-random constants $\{c_i\}_{i=1}^{n}$. If $S_n := X_1 + \cdots + X_n$ and $s_n := \sum_{j=1}^{n} c_j^2$ then:

(1) Prove that $\mathrm{P}\{|S_n|\ge t\}\le 2e^{-t^2/(2s_n)}$ for all $t\ge 0$ (Hoeffding, 1963, Theorem 1). (HINT: Problem 4.24, p. 51.)

(2) Prove that if $B = \mathrm{Bin}(n,p)$ then $\mathrm{P}\{|B - np|\ge t\sqrt{n}\}\le 2e^{-t^2/2}$ for all $t\ge 0$ (Chernoff, 1952; Okamoto, 1958).

(3) Improve (6.48) to $|T_2|\le 2K\exp(-n\delta^2/2)$.

(4) Improve Theorem 6.27 by proving that
$$T_n(\epsilon) \geq \left(1 - 2me^{-n\epsilon^2/2}\right) 2^{n(H(p)-c\epsilon)}.$$

6.34 (The Erdős–Rényi Law of Large Numbers; Hard). Suppose $\{X_i\}_{i=1}^{\infty}$ are i.i.d. with the distribution $P\{X_1 = 1\} = P\{X_1 = 0\} = 1/2$. Let Y_n denote the length of the longest uninterrupted string of 1's among $X_1, \ldots, X_n$. Prove the *Erdős–Rényi law of large numbers* (1970): $Y_n \sim \log_2 n$ a.s., where $\log_2 x$ denotes the logarithm of x in base two.

6.35 (Theorem 6.29, Continued; Hard). Improve the Glivenko–Cantelli theorem (p. 80) by proving that for all fixed $\rho < 1/2$,

(6.86) $$\sup_x |F_n(x) - F(x)| = o(n^{-\rho/2}) \qquad \text{as } n \to \infty, \text{ a.s.}$$

Can $\rho = 1/2$?

6.36 (Raikov's Ergodic Theorem; Hard). Throughout this exercise we extend the domain of any function $G : [0, 1) \to \mathbf{R}$ periodically to all of $\mathbf{R}_+$ by defining $G(t) = G(t \mod 1)$ whenever $t \notin [0, 1)$.

(1) Suppose k is a fixed positive integer and $f : [0, 1) \to \mathbf{R}$ is constant on $[j2^{-k}, (j+1)2^{-k})$ for all $j = 0, \ldots, 2^k - 1$. Define $X_j(\omega) = f(2^j\omega)$ for all $\omega \in [0, 1)$ and $j \geq 1$. Prove that $\{X_{k+m}\}_{m=0}^{\infty}$ are i.i.d. What is their common distribution? (HINT: Problem 6.23.)

(2) Prove that given any continuous function $\psi : [0, 1) \to \mathbf{R}$,
$$\lim_{n \to \infty} \frac{1}{n} \sum_{j=0}^{n} \psi(2^j\omega) = \int_0^1 \psi(s)\,ds,$$
for almost all $\omega \in [0, 1)$ (Raikov, 1936).

6.37 (The Cantor–Lebesgue Function; Hard). Inspite of Problem 6.24, follow the steps below to prove that there exists a continuous distribution function F on $\mathbf{R}$ such that $F' = 0$ almost everywhere. (Why is this surprising?)

(1) Define C to be the collection of all numbers of the form $\sum_{j=1}^{\infty} 3^{-j}\zeta_j$, where $\zeta_j = 0$ or 2 for every j. This is the *middle-thirds Cantor set*. Prove that C has zero Lebesgue measure yet is uncountable.

(2) Define F to be the distribution function of $\sum_{j=1}^{\infty} 3^{-j}X_j$ where the X_j's are i.i.d., each taking the values 0 or 2 with probability $\frac{1}{2}$ each. Prove that F is continuous, and $F' = 0$ almost everywhere. Can you see a relation to Theorem 6.20?

6.38 (Problem 6.29, Continued; Harder). Suppose $\{X_i\}_{i=1}^{\infty}$ are i.i.d. and symmetric; i.e., X_1 has the same distribution as $-X_1$. Define $\ln_+ x = \ln(\max(x, e))$ for all $x > 0$, and prove the following:

(6.87) $$\text{If } \mathrm{E}\left(\frac{|X_1|}{\ln_+ |X_1|}\right) < \infty \text{ then } \lim_{n \to \infty} \frac{1}{\ln n} \sum_{i=1}^{n} \frac{X_i}{i} = 0 \qquad \text{a.s.}$$

Prove also that $\limsup_{n \to \infty} (\ln n)^{-1} \sum_{i=1}^{n}(X_i/i) = \infty$ a.s., otherwise. That is, the end result of Problem 6.29 can hold even if $X_1 \notin L^1(P)$!

6.39 (Khintchine's Inequality; Harder). Let $\{X_i\}_{i=1}^{n}$ be i.i.d. random variables with $P\{X_1 = 1\} = 1 - P\{X_1 = -1\} = 1/2$, $\{a_i\}_{i=1}^{n}$ a sequence of constants, $\sigma_n^2 := \sum_{i=1}^{n} a_i^2$, and $S_n(a) := \sum_{i=1}^{n} a_i X_i$. Prove that:

(1) $P\{|S_n(a)| \geq \lambda\} \leq 2\exp(-\lambda^2/(2\sigma_n^2))$ for all $\lambda > 0$ and $n \geq 1$.
(HINT: Problem 4.30, p. 51.)

(2) For every $p \in (1, \infty)$ there exist finite constants $A_p, B_p > 0$ such that
$$A_p \sigma_n \leq \|S_n(a)\|_p \leq B_p \sigma_n,$$
for all $n \geq 1$, regardless of the values of $a_1, \ldots, a_n$. (HINT: For the lower bound, verify first that $\|Z\|_2^2 \leq \|Z\|_p \|Z\|_q$, where $p^{-1} + q^{-1} = 1$.)

6.40 (Harder). The classical *central limit theorem* (p. 100 below) states the following: If $\{X_i\}_{i=1}^{\infty}$ are i.i.d. random variables with $EX_1 = 0$ and $\text{Var}X_1 = 1$, and if $S_n := X_1 + \cdots + X_n$, then for any real $a < b$,

(6.88) $$\lim_{n \to \infty} P(E_n) = P\{a \leq N(0,1) \leq b\},$$

where $E_n := \{a\sqrt{n} \leq S_n \leq b\sqrt{n}\}$. Define $Z_n := \sum_{j=1}^{n} \mathbf{1}_{E_j}/j$, and use the central limit theorem, without proof, to derive the following:

(6.89) $$\lim_{n \to \infty} \frac{EZ_n}{\ln n} = P\{a \leq N(0,1) \leq b\} \quad \text{and} \quad \lim_{n \to \infty} \frac{\text{SD}(Z_n)}{\ln n} = 0.$$

Conclude that $Z_n/\ln n \to P\{a \leq N(0,1) \leq b\}$ in probability as $n \to \infty$.

Notes

(1) What we call "correlation" here is more precisely known as the "correlation coefficient." Stigler (1986, pp. 297–299) discusses the history of correlation.

(2) The Kolmogorov zero-one law for random variables that are either zero or one is much older than the general form presented here (p. 69). See Cantelli (1917a) for this and its applications to the theory of inference.

(3) Truncation was invented by A. A. Markov in 1907 (Markov, 1910).

(4) The convergence half of the Borel–Cantelli Lemma (p. 73) was found by Borel (1909) in his study of normal numbers (Problem 6.22). A divergence half was devised by Cantelli (1917b) in order to prove a version of the strong law of large numbers; see Problem 6.12 for a modern version of Cantelli's strong law.

Since Borel's original reasoning contained flaws [that can nowadays be corrected], some authors refer to Cantelli's strong law of large numbers as the first strong law of its kind. More importantly, the ingenuity of Cantelli's paper is undeniable for at least two reasons: (i) The second half of the Borel–Cantelli lemma has proved to be of immense value in probability and its applications; and (ii) more significantly, Cantelli's is one of the earliest papers that explicitly states the need for a countably additive probability theory.

(5) Lindvall (1982) presents an up-to-date survey of Bernstein polynomials and their role within the theories of approximation and probability (Problem 6.30).

(6) Etemadi (1981) devised an elementary and elegant proof of the strong law of large numbers. Also, as a by-product of his proof, independence can be relaxed to pairwise independence. [This means that X_i and X_j are independent for any $i \neq j$, although $\{X_i\}_{i=1}^{\infty}$ need not be.] Our proof of the strong law is slightly less direct, but is useful in other settings.

(7) Remark 6.25 has a deeper counterpart, due to Hoeffding (1971): If f is integrable on $[0,1]$ and has bounded variation, then $\int_0^1 |(\mathcal{B}_n f)(x) - f(x)| \, dx = O(n^{-1/2})$.

(8) For a pedagogic account of Shannon's theorem (Theorem 6.27) and its impact in communication theory see Cover and Thomas (1991, pp. 53–55).

(9) The lower bound of Theorem 6.31 is the best known bound of its kind. The best known upper bound is $R_n = O(n^{-\alpha} 2^{2n})$, where α is a positive and finite constant that does not depend on n. P. Erdős has conjectured that $\log_2 R_n \sim cn$ for some $c \in (0, \infty)$; see Alon and Spencer (1991, p. 241).

(10) Proposition 6.32, and more, can be found in Hammersley (1963).

(11) Kolmogorov's one-series theorem (Problem 6.11) is an abstraction of an earlier result of Rademacher (1922); see also Steinhaus (1930). The optimal result in this area is the *three-series theorem* of Khintchine and Kolmogorov (1925).

(12) The Paley–Zygmund inequality (Problem 6.20) has been discovered and refined many times since. Some notable instances can be found within Chung and Erdős (1947, 1952), Erdős and Rényi (1959), Kochen and Stone (1964), and Rényi (1962, p. 327).

(13) Problem 6.22 invites the reader — perhaps too casually — to think about the normality of numbers such as $1/\sqrt{2}$, $\pi/10$, and $\ln 2$. In all fairness, this should come equipped with a caveat: *These are hard, old, and wide-open problems!* See Wagon (1985) for a history of $\pi/10$.

Some non-trivial numbers are known to be normal in base 10. A rather curious example is $0.12345678910111213 14\cdots$, which is obtained by enumerating the numerals (Champernowne, 1933). Another interesting example is $0.123571113\cdots$, which is obtained by enumerating all primes; see Copeland and Erdős (1946). These results have found applications in theoretical computer science; see Schnorr and Stimm (1972) and Bourke, Hitchcock, and Vinodchandran (2005).

(14) Problem 6.31 is the only computer exercise in this text. The ultimate assertion of this problem, however, is a rigorous mathematical theorem. Its proof requires a working knowledge of both theories of Markov chains and complex dynamics. Two pedagogic sources are respectively Norris (1998) and Devaney (2003).

(15) Raikov's ergodic theorem (Problem 6.36) follows from the more general "pointwise ergodic theorem" of Birkhoff (1931). Chapter 6 of Durrett (1996, pp. 341–346) contains a pedagogic account based on an elementary proof of Garsia (1965).

Erdős (1949) extended Raikov's ergodic theorem in several directions: Continuity can be reduced essentially to $f \in L^2(dx)$; and $\sum_{j=1}^{n} f(2^j \omega)/n$ can be replaced by $\sum_{j=1}^{n} f(\ell_j \omega)/n$ where $\{\ell_j\}_{j=1}^{\infty}$ satisfies the so-called "Hadamard gap condition," $\inf_{j \geq 1} \ell_{j+1}/\ell_j > 1$. The latter gap condition cannot be improved dramatically (Buczolich and Mauldin, 1999).

(16) The limiting result in Problem 6.40 can be shown to hold almost surely, and the end result is called the *almost-sure central limit theorem* (Lacey and Philipp, 1990). For a prefatory version consult Lévy (1937, p. 270). This problem and its history have been recently surveyed by Berkés (1998).

Chapter 7

The Central Limit Theorem

Experimentalists think that it is a mathematical theorem, while the mathematicians believe it to be an experimental fact.

–Gabriel Lippman, in a discussion with J. H. Poincaré about the CLT

Let S_n denote the total number of successes in n independent Bernoulli trials, where the probability of success per trial is some fixed number $p \in (0,1)$. The De Moivre–Laplace central limit theorem (p. 19) asserts that for all real numbers $a < b$,

$$(7.1) \qquad \lim_{n \to \infty} \mathrm{P}\left\{ a < \frac{S_n - np}{\sqrt{np(1-p)}} \leq b \right\} = \int_a^b \frac{e^{-x^2/2}}{\sqrt{2\pi}}\, dx.$$

We will soon see that (7.1) implies that the distribution of S_n is close to that of $N(np, \sqrt{np(1-p)})$; see Example 7.3 below. In this chapter we discuss the definitive formulation of this theorem. Its statement involves the notion of weak convergence which we discuss next.

1. Weak Convergence

Definition 7.1. Let $\mathbf{X}$ denote a topological space, and suppose $\mu, \mu_1, \mu_2, \ldots$ are probability (or more generally, finite) measures on $(\mathbf{X}, \mathscr{B}(\mathbf{X}))$. We say that μ_n *converges weakly* to μ, and write $\mu_n \Rightarrow \mu$, if

$$(7.2) \qquad \lim_{n \to \infty} \int f\, d\mu_n = \int f\, d\mu,$$

for all bounded continuous functions $f : \mathbf{X} \to \mathbf{R}$. If the respective distributions of X_n and X are μ_n and μ, and if $\mu_n \Rightarrow \mu$, then we also say that X_n *converges weakly* to X and write $X_n \Rightarrow X$. This is equivalent to saying that

$$\lim_{n\to\infty} \mathrm{E} f(X_n) = \mathrm{E} f(X), \tag{7.3}$$

for all bounded continuous functions $f : \mathbf{X} \to \mathbf{R}$.

The following result of Lévy (1937) characterizes weak convergence on **R**.

Theorem 7.2. *Let $\mu, \mu_1, \mu_2, \ldots$ denote probability measures on $(\mathbf{R}, \mathscr{B}(\mathbf{R}))$ with respective distibution functions $F, F_1, F_2, \ldots$. Then, $\mu_n \Rightarrow \mu$ if and only if*

$$\lim_{n\to\infty} F_n(x) = F(x), \tag{7.4}$$

for all $x \in \mathbf{R}$ at which F is continuous.

Equivalently, $X_n \Rightarrow X$ if and only if $\mathrm{P}\{X_n \le x\} \to \mathrm{P}\{X \le x\}$ for all x such that $\mathrm{P}\{X = x\} = 0$.

Example 7.3. Consider the De Moivre–Laplace central limit theorem, and define

$$X_n := \frac{S_n - np}{\sqrt{np(1-p)}}. \tag{7.5}$$

Let F_n denote the distribution function of X_n, and F the distribution function of $N(0,1)$. Observe that: (i) F is continuous; and (ii) (7.1) asserts that $\lim_{n\to\infty}(F_n(b) - F_n(a)) = F(b) - F(a)$. By the preceding theorem, (7.1) is saying that $X_n \Rightarrow N(0,1)$.

Theorem 7.2 cannot be improved. Indeed, it can happen that $X_n \Rightarrow X$ but F_n fails to converge to F pointwise. Next is an example of this phenomenon.

Example 7.4. First let $X = \pm 1$ with probability $\frac{1}{2}$ each. Then define

$$X_n(\omega) := \begin{cases} -1 & \text{if } X(\omega) = -1, \\ 1 + \frac{1}{n} & \text{if } X(\omega) = 1. \end{cases} \tag{7.6}$$

Then, $\lim_{n\to\infty} f(X_n) = f(X)$ for all bounded continuous functions f, whence $\mathrm{E} f(X_n) \to \mathrm{E} f(X)$. However, $F_n(1) = \mathrm{P}\{X_n \le 1\} = \frac{1}{2}$ does not converge to $F(1) = \mathrm{P}\{X \le 1\} = 1$.

In order to prove Theorem 7.2 we will need the following.

Lemma 7.5. *The set $J := \{x \in \mathbf{R} : \mathrm{P}\{X = x\} > 0\}$ is denumerable.*

1. Weak Convergence

Proof. Define

(7.7) $$J_n := \left\{ x \in \mathbf{R} : \mathrm{P}\{X = x\} \geq \frac{1}{n} \right\}.$$

Since $J = \cup_{n=1}^\infty J_n$, it suffices to prove that J_n is finite. Indeed, if J_n were infinite, then we could select a countable set $K_n \subset J_n$, and observe that

(7.8) $$1 \geq \sum_{x \in K_n} \mathrm{P}\{X = x\} \geq \frac{|K_n|}{n},$$

where $|\cdots|$ denotes cardinality. This contradicts the assumption that K_n is infinite. $\square$

Proof of Theorem 7.2. Throughout, we let X_n denote a random variable whose distribution is μ_n ($n = 1, 2, \ldots$), and X a random variable with distribution μ.

Suppose first that $X_n \Rightarrow X$. For all fixed $x \in \mathbf{R}$ and $\epsilon > 0$, we can find a bounded continuous function $f : \mathbf{R} \to \mathbf{R}$ such that

(7.9) $$f(y) \leq \mathbf{1}_{(-\infty, x]}(y) \leq f(y - \epsilon) \qquad \forall y \in \mathbf{R}.$$

[Try a piecewise-linear function f.] It follows that

(7.10) $$\mathrm{E}f(X_n) \leq F_n(x) \leq \mathrm{E}f(X_n - \epsilon).$$

Let $n \to \infty$ to obtain

(7.11) $$\mathrm{E}f(X) \leq \liminf_{n \to \infty} F_n(x) \leq \limsup_{n \to \infty} F_n(x) \leq \mathrm{E}f(X - \epsilon).$$

Equation (7.9) is equivalent to the following:

(7.12) $$\mathbf{1}_{(-\infty, x-\epsilon]}(y) \leq f(y) \quad \text{and} \quad f(y - \epsilon) \leq \mathbf{1}_{(-\infty, x+\epsilon]}(y).$$

We apply this with $y := X$ and take expectations to see that

(7.13) $$F(x - \epsilon) \leq \mathrm{E}f(X) \quad \text{and} \quad \mathrm{E}f(X - \epsilon) \leq F(x + \epsilon).$$

This and (7.11) together imply that

(7.14) $$F(x - \epsilon) \leq \liminf_{n \to \infty} F_n(x) \leq \limsup_{n \to \infty} F_n(x) \leq F(x + \epsilon).$$

Let $\epsilon \downarrow 0$ to deduce that $F_n(x) \to F(x)$ whenever F is continuous at x.

For the converse we suppose that $F_n(x) \to F(x)$ for all continuity points x of F. Our goal is to prove that $\lim_{n \to \infty} \mathrm{E}f(X_n) = \mathrm{E}f(X)$ for all bounded continuous functions $f : \mathbf{R} \to \mathbf{R}$.

In accord with Lemma 7.5, for any $\delta, N > 0$, we can find real numbers $\cdots < x_{-2} < x_{-1} < x_0 < x_1 < x_2 < \cdots$ (depending only on δ and N) such that: (i) $\max_{|i| \leq N} \sup_{y \in (x_i, x_{i+1}]} |f(y) - f(x_i)| \leq \delta$; (ii) F is continuous at x_i for all $i \in \mathbf{Z}$; and (iii) $F(x_{N+1}) \geq 1 - \delta$ and $F(x_{-N}) \leq \delta$. Let $\Lambda_N := (x_{-N}, x_{N+1}]$. By (i),

$$
\begin{aligned}
(7.15) \quad &\left| \mathrm{E}\left[f(X_n);\ X_n \in \Lambda_N\right] - \sum_{j=-N}^{N} f(x_j) \left[F_n(x_{j+1}) - F_n(x_j)\right] \right| \\
&= \left| \sum_{j=-N}^{N} \mathrm{E}\{f(X_n) - f(x_j);\ X_n \in (x_j, x_{j+1}]\} \right| \\
&\leq \sum_{j=-N}^{N} \mathrm{E}\{|f(X_n) - f(x_j)|\ ;\ X_n \in (x_j, x_{j+1}]\} \\
&\leq \delta.
\end{aligned}
$$

This remains valid if we replace X_n and F_n respectively by X and F. Note that N is held fixed, and F_n converges to F at all continuity-points of F. Therefore, as $n \to \infty$,

$$
(7.16) \quad \sum_{|j| \leq N} f(x_j) \left[F_n(x_{j+1}) - F_n(x_j)\right] \to \sum_{|j| \leq N} f(x_j) \left[F(x_{j+1}) - F(x_j)\right].
$$

By the triangle inequality,

$$
(7.17) \quad \limsup_{n \to \infty} |\mathrm{E}\{f(X_n);\ X_n \in \Lambda_N\} - \mathrm{E}\{f(X);\ X \in \Lambda_N\}| \leq 2\delta.
$$

For the remainder terms, first note that

$$
(7.18) \quad \mathrm{P}\{X_n \notin \Lambda_N\} = 1 - F_n(x_{N+1}) + F_n(x_N) \leq 2\delta.
$$

Let $N \to \infty$ to find that the same quantity bounds $\mathrm{P}\{X \notin \Lambda_N\}$. Therefore, if we let $K := \sup_{y \in \mathbf{R}} |f(y)|$, then

$$
(7.19) \quad \limsup_{n \to \infty} \mathrm{E}\{|f(X_n)|;\ X_n \notin \Lambda_N\} + \mathrm{E}\{|f(X)|;\ X \notin \Lambda_N\} \leq 4K\delta.
$$

In conjunction with (7.17), this proves that

$$
(7.20) \quad \limsup_{n \to \infty} |\mathrm{E}f(X_n) - \mathrm{E}f(X)| \leq 2\delta + 4K\delta.
$$

Let δ tend to zero to finish. $\square$

2. Weak Convergence and Compact-Support Functions

Definition 7.6. If $\mathbf{X}$ is a metric space, then $C_c(\mathbf{X})$ denotes the collection of all continuous functions $f : \mathbf{X} \to \mathbf{R}$ such that f has *compact support*; i.e., there exists a compact set K such that $f(x) = 0$ for all $x \notin K$. In addition, $C_b(\mathbf{X})$ denotes the collection of all bounded continuous functions $f : \mathbf{X} \to \mathbf{R}$.

2. Weak Convergence and Compact-Support Functions

Recall that in order to prove that $\mu_n \Rightarrow \mu$, we need to verify that $\int f \, d\mu_n \to \int f \, d\mu$ for all $f \in C_b(\mathbf{X})$. Since $C_c(\mathbf{R}^k) \subseteq C_b(\mathbf{R}^k)$, the next result simplifies our task in the case that $\mathbf{X} = \mathbf{R}^k$.

Theorem 7.7. *If $\mu, \mu_1, \mu_2, \ldots$ are probability measures on $(\mathbf{R}^k, \mathscr{B}(\mathbf{R}^k))$, then $\mu_n \Rightarrow \mu$ if and only if*

$$\lim_{n \to \infty} \int f \, d\mu_n = \int f \, d\mu \qquad \forall f \in C_c(\mathbf{R}^k). \tag{7.21}$$

Proof. We plan to prove that if $\int g \, d\mu_n \to \int g \, d\mu$ for all $g \in C_c(\mathbf{R}^k)$, then $\int f \, d\mu_n \to \int f \, d\mu$ for all $f \in C_b(\mathbf{R}^k)$. With this goal in mind, let us choose and fix such an $f \in C_b(\mathbf{R}^k)$. By considering f^+ and f^- separately, we can—and will—assume without loss of generality that $f(x) \geq 0$ for all x.

Step 1. The Lower Bound. For any $p > 0$ choose and fix a function $f_p \in C_c(\mathbf{R}^k)$ such that:

(1) For all $x \in [-p, p]^k$, $f_p(x) = f(x)$.
(2) For all $x \notin [-p-1, p+1]^k$, $f_p(x) = 0$.
(3) For all $x \in \mathbf{R}^k$, $0 \leq f_p(x) \leq f(x)$, and $f_p(x) \uparrow f(x)$ as $p \uparrow \infty$.

It follows that

$$\liminf_{n \to \infty} \int f \, d\mu_n \geq \lim_{n \to \infty} \int f_p \, d\mu_n = \int f_p \, d\mu. \tag{7.22}$$

Let $p \uparrow \infty$ and apply the dominated convergence theorem to deduce that

$$\liminf_{n \to \infty} \int f \, d\mu_n \geq \int f \, d\mu. \tag{7.23}$$

This proves half of the theorem.

Step 2. A Variant. In this step we prove that, in (7.23), f can be replaced by the indicator function of an *open* k-dimensional hypercube. More precisely, given any real numbers $a_1 < b_1, \ldots, a_k < b_k$,

$$\liminf_{n \to \infty} \mu_n\left((a_1, b_1) \times \cdots \times (a_k, b_k)\right) \geq \mu\left((a_1, b_1) \times \cdots \times (a_k, b_k)\right). \tag{7.24}$$

To prove this, we first find continuous functions $\psi_m \uparrow \mathbf{1}_{(a_1,b_1) \times \cdots \times (a_k,b_k)}$, pointwise. By definition, $\psi_m \in C_c(\mathbf{R}^k)$ for all $m \geq 1$, and

$$\liminf_{n \to \infty} \mu_n\left((a_1, b_1) \times \cdots \times (a_k, b_k)\right) \geq \lim_{n \to \infty} \int \psi_m \, d\mu_n = \int \psi_m \, d\mu. \tag{7.25}$$

Let $m \uparrow \infty$ to deduce (7.24) from the dominated convergence theorem.

Step 3. The Upper Bound. Recall f_p from Step 1 and write

$$\int f\,d\mu_n = \int_{[-p,p]^k} f\,d\mu_n + \int_{\mathbf{R}^k\setminus[-p,p]^k} f\,d\mu_n$$
(7.26)
$$\leq \int f_p\,d\mu_n + \sup_{z\in\mathbf{R}^k}|f(z)|\cdot\left[1-\mu_n\left([-p,p]^k\right)\right].$$

Now let $n\to\infty$ and appeal to (7.24) to find that

(7.27) $$\limsup_{n\to\infty}\int f\,d\mu_n \leq \int f_p\,d\mu + \sup_{z\in\mathbf{R}^k}|f(z)|\cdot\left[1-\mu\left((-p,p)^k\right)\right].$$

Let $p\uparrow\infty$ and use the monotone convergence theorem to deduce that

(7.28) $$\limsup_{n\to\infty}\int f\,d\mu_n \leq \int f\,d\mu.$$

This finishes the proof. □

3. Harmonic Analysis in Dimension One

Definition 7.8. The *Fourier transform* of a probability measure μ on $\mathbf{R}$ is

(7.29) $$\widehat{\mu}(t) := \int_{-\infty}^{\infty} e^{itx}\,\mu(dx) \qquad \forall t\in\mathbf{R},$$

where $i := \sqrt{-1}$. This definition continues to makes sense if μ is a finite measure. It also makes sense if μ is replaced by a Lebesgue-integrable function $f:\mathbf{R}\to\mathbf{R}$. In that case, we set

(7.30) $$\widehat{f}(t) := \int_{-\infty}^{\infty} e^{ixt} f(x)\,dx \qquad \forall t\in\mathbf{R}.$$

[We identify the Fourier transform of the function $f = (d\mu/dx)$ with that of the measure μ.] If X is a real-valued random variable whose distribution is some probability measure μ, then $\widehat{\mu}$ is also called the *characteristic function* of X and/or μ, and $\widehat{\mu}(t)$ is equal to $\mathrm{E}\exp(itX) = \mathrm{E}\cos(tX) + i\mathrm{E}\sin(tX)$.

Here are some of the elementary properties of characteristic functions.

Lemma 7.9. *If μ is a finite measure on $(\mathbf{R},\mathscr{B}(\mathbf{R}))$, then $\widehat{\mu}$ exists, is uniformly continuous on $\mathbf{R}$, and satisfies the following:*

(1) $\sup_{t\in\mathbf{R}}|\widehat{\mu}(t)| = \widehat{\mu}(0) = \mu(\mathbf{R})$ *and* $\widehat{\mu}(-t) = \overline{\widehat{\mu}(t)}$.
(2) $\widehat{\mu}$ *is nonnegative definite. That is,* $\sum_{j=1}^n \sum_{k=1}^n \widehat{\mu}(t_j - t_k) z_j \overline{z_k} \geq 0$ *for all $z_1,\ldots,z_n \in \mathbf{C}$ and $t_1,\ldots,t_n \in \mathbf{R}$.*

Proof. Without loss of generality, we may assume that μ is a probability measure. Otherwise we can prove the theorem for the probability measure $\nu(\cdots) = \mu(\cdots)/\mu(\mathbf{R})$, and then multiply through by $\mu(\mathbf{R})$.

Let X be a random variable whose distribution is μ; $\widehat{\mu}(t) = \mathrm{E}e^{itX}$ is always defined and bounded since $|e^{itX}| \le 1$. To prove uniform continuity, we note that for all $a, b \in \mathbf{R}$,

$$(7.31) \qquad \left|e^{ia} - e^{ib}\right| = \left|1 - e^{i(a-b)}\right| = \left|\int_0^{a-b} e^{ix}\, dx\right| \le |a-b|.$$

Consequently,

$$(7.32) \qquad \left|e^{ia} - e^{ib}\right| \le |a-b| \wedge 2.$$

It follows from this that

$$(7.33) \qquad \sup_{|s-t|\le\delta} |\widehat{\mu}(t) - \widehat{\mu}(s)| \le \sup_{|s-t|\le\delta} \mathrm{E}\left|e^{itX} - e^{isX}\right| \le \mathrm{E}\left(\delta|X| \wedge 2\right).$$

Thanks to the dominated convergence theorem, the preceding tends to 0 as δ converges down to 0. The uniform continuity of $\widehat{\mu}$ follows.

Part (1) is elementary. To prove (2) we first observe that

$$(7.34) \qquad \sum\sum_{1\le j,k\le n} \widehat{\mu}(t_j - t_k) z_j \overline{z_k} = \sum\sum_{1\le j,k\le n} \mathrm{E}e^{i(t_j - t_k)X} z_j \overline{z_k}.$$

This is the expectation of $\left|\sum_{j=1}^n e^{it_j X} z_j\right|^2$, and hence is real as well as non-negative. $\square$

Example 7.10 (§5.1, p. 11). If $X = \mathrm{Unif}(a, b)$ for some $a < b$, then $\mathrm{E}e^{itX} = (e^{itb} - e^{ita})/it(b-a)$ for all $t \in \mathbf{R}$.

Example 7.11 (Problem 1.11, p. 13). If X has the exponential distribution with some parameter $\lambda > 0$, then $\mathrm{E}e^{itX} = \lambda/(\lambda - it)$ for all $t \in \mathbf{R}$.

Example 7.12 (§5.2, p. 11). If $X = N(\mu, \sigma^2)$ for some $\mu \in \mathbf{R}$ and $\sigma \ge 0$, then $\mathrm{E}e^{itX} = \exp\left(it\mu - \tfrac{1}{2}t^2\sigma^2\right)$ for all $t \in \mathbf{R}$.

Example 7.13 (§4.1, p. 8). If $X = \mathrm{Bin}(n, p)$ for an integer $n \ge 1$ and some $p \in [0, 1]$, then $\mathrm{E}e^{itX} = (pe^{it} + 1 - p)^n$ for all $t \in \mathbf{R}$.

Example 7.14 (Problem 1.9, p. 13). If $X = \mathrm{Poiss}(\lambda)$ for some $\lambda > 0$, then $\mathrm{E}e^{itX} = \exp(-\lambda + \lambda e^{it})$ for all $t \in \mathbf{R}$.

4. The Plancherel Theorem

In this section we state and prove a variant of a result of Plancherel (1910, 1933). Roughly speaking, Plancherel's theorem shows us how to reconstruct a distribution from its characteristic function. In order to state things more precisely we need some notation.

Definition 7.15. Suppose $f, g : \mathbf{R} \to \mathbf{R}$ are measurable. Then, when defined, the *convolution* $f * g$ is the function,

$$(f * g)(x) := \int_{-\infty}^{\infty} f(x - y) g(y) \, dy. \tag{7.35}$$

Convolution is a symmetric operation; i.e., $f * g = g * f$ for all measurable $f, g : \mathbf{R} \to \mathbf{R}$. This tacitly implies that one side of the stated identity converges if and only if the other side does. Next are two less obvious properties of convolutions. Henceforth, let ϕ_ϵ denote the density function of $N(0, \epsilon^2)$; i.e.,

$$\phi_\epsilon(x) = \frac{1}{\epsilon \sqrt{2\pi}} \exp\left(-\frac{x^2}{2\epsilon^2}\right) \qquad \forall x \in \mathbf{R}. \tag{7.36}$$

The first important property of convolutions is that they provide us with smooth approximations to nice functions.

Fejér's Theorem. *If $f \in C_c(\mathbf{R})$, then $f * \phi_\epsilon$ is infinitely differentiable for all $\epsilon > 0$, and the kth derivative is $f * \phi_\epsilon^{(k)}$ for all $k \geq 1$. Moreover,*

$$\limsup_{\epsilon \to 0} {}_{x \in \mathbf{R}} |(f * \phi_\epsilon)(x) - f(x)| = 0. \tag{7.37}$$

Proof. Let $\phi_\epsilon^{(0)} := \phi_\epsilon$. Then for all $k \geq 0$ and all $\epsilon > 0$ fixed,

$$\frac{\left(f * \phi_\epsilon^{(k)}\right)(x + h) - \left(f * \phi_\epsilon^{(k)}\right)(x)}{h} \tag{7.38}$$

$$= \int_{-\infty}^{\infty} f(y) \, \frac{\phi_\epsilon^{(k)}(x + h - y) - \phi_\epsilon^{(k)}(x - y)}{h} \, dy.$$

Because $\phi_\epsilon^{(k+1)}$ is bounded and f has compact support, the bounded convergence theorem implies that $f * \phi_\epsilon^{(k)}$ is differentiable, and the derivative is $f * \phi_\epsilon^{(k+1)}$. Now we apply induction to find that the kth derivative of $f * \phi_\epsilon$ exists and is equal to $f * \phi_\epsilon^{(k)}$ for all $k \geq 1$.

Let Z denote a standard normal random variable, and note that ϕ_ϵ is the density function of ϵZ; thus, $(f * \phi_\epsilon)(x) = \mathbf{E} f(x - \epsilon Z)$. By the uniform continuity of f, $\lim_{\epsilon \to 0} \sup_{x \in \mathbf{R}} |f(x - \epsilon Z) - f(x)| = 0$ a.s. Because f is bounded, this and the bounded convergence theorem together imply the result. □

The second property of convolutions, alluded to earlier, is the Plancherel theorem.

4. The Plancherel Theorem

Plancherel's Theorem. *If μ is a finite measure on $\mathbf{R}$ and $f : \mathbf{R} \to \mathbf{R}$ is Lebesgue-integrable, then*

$$(7.39) \quad \int_{-\infty}^{\infty} (f * \phi_\epsilon)(x) \, \mu(dx) = \frac{1}{2\pi} \int_{-\infty}^{\infty} e^{-\epsilon^2 t^2/2} \widehat{f}(t) \, \overline{\widehat{\mu}(t)} \, dt \qquad \forall \epsilon > 0.$$

Consequently, if $f \in C_c(\mathbf{R})$ and $\widehat{f} \in L^1(\mathbf{R})$, then

$$(7.40) \quad \int_{-\infty}^{\infty} f \, d\mu = \frac{1}{2\pi} \int_{-\infty}^{\infty} \widehat{f}(t) \, \overline{\widehat{\mu}(t)} \, dt.$$

Proof. By the Fubini–Tonelli theorem,

$$(7.41) \quad \begin{aligned} & \frac{1}{2\pi} \int_{-\infty}^{\infty} e^{-\epsilon^2 t^2/2} \widehat{f}(t) \, \overline{\widehat{\mu}(t)} \, dt \\ &= \frac{1}{2\pi} \int_{-\infty}^{\infty} e^{-\epsilon^2 t^2/2} \left(\int_{-\infty}^{\infty} f(x) e^{itx} \, dx \right) \left(\int_{-\infty}^{\infty} e^{-ity} \mu(dy) \right) dt \\ &= \frac{1}{2\pi} \int_{-\infty}^{\infty} \int_{-\infty}^{\infty} \left(\int_{-\infty}^{\infty} e^{-\epsilon^2 t^2/2} e^{it(x-y)} \, dt \right) \mu(dy) \, f(x) \, dx. \end{aligned}$$

A direct calculation reveals that

$$(7.42) \quad \begin{aligned} \int_{-\infty}^{\infty} e^{-\epsilon^2 t^2/2} e^{it(x-y)} \, dt &= \frac{\sqrt{2\pi}}{\epsilon} \exp\left(-\frac{(x-y)^2}{2\epsilon^2} \right) \\ &= 2\pi \phi_\epsilon(x-y). \end{aligned}$$

See Example 7.12. Since f is integrable, all of the integrals in the right-hand side of (7.41) converge absolutely. Therefore, (7.39) follows from the Fubini–Tonelli theorem; (7.40) follows from (7.39) and the Fejér theorem. □

The Plancherel theorem is a deep result, and has a number of profound consequences. We state two of them.

The Uniqueness Theorem. *If μ and ν are two finite measures on $\mathbf{R}$ and $\widehat{\mu} = \widehat{\nu}$, then $\mu = \nu$.*

Proof. By the theorems of Plancherel and Fejér, $\int f \, d\mu = \int f \, d\nu$ for all $f \in C_c(\mathbf{R})$. Choose $f_k \in C_c(\mathbf{R})$ such that $f_k \downarrow 1_{[a,b]}$. The monotone convergence theorem then implies that $\mu([a,b]) = \nu([a,b])$. Thus, μ and ν agree on all finite unions of disjoint closed intervals of the form $[a,b]$. Because the said collection generates $\mathscr{B}(\mathbf{R})$, $\mu = \nu$ on $\mathscr{B}(\mathbf{R})$. □

The following convergence theorem of P. Lévy is another significant consequence of the Plancherel theorem.

The Convergence Theorem. *Suppose $\mu, \mu_1, \mu_2, \ldots$ are probability measures on $(\mathbf{R}, \mathscr{B}(\mathbf{R}))$. If $\lim_{n \to \infty} \widehat{\mu}_n = \widehat{\mu}$ pointwise, then $\mu_n \Rightarrow \mu$.*

Proof. In accord with Theorem 7.7 it suffices to prove that $\lim_{n\to\infty} \int f\,d\mu_n = \int f\,d\mu$ for all $f \in C_c(\mathbf{R})$. Thanks to the Fejér theorem, for all $\delta > 0$ we can choose $\epsilon > 0$ such that

(7.43)
$$\sup_{x\in\mathbf{R}} |(f * \phi_\epsilon)(x) - f(x)| \le \delta.$$

Apply the triangle inequality twice to see that for all $\delta > 0$,

(7.44)
$$\left|\int f\,d\mu_n - \int f\,d\mu\right| \le 2\delta + \left|\int (f*\phi_\epsilon)\,d\mu_n - \int (f*\phi_\epsilon)\,d\mu\right|$$
$$= 2\delta + \left|\int_{-\infty}^{\infty} \widehat{f}(t) e^{-\epsilon^2 t^2/2} \left(\frac{\widehat{\mu_n}(t) - \widehat{\mu}(t)}{2\pi}\right) dt\right|.$$

The last line holds by the Plancherel theorem. Since $f \in C_c(\mathbf{R})$, $\widehat{f}$ is uniformly bounded by $\int_{-\infty}^{\infty} |f(x)|\,dx < \infty$ (Lemma 7.9). Therefore, by the dominated convergence theorem,

(7.45)
$$\limsup_{n\to\infty} \left|\int f\,d\mu_n - \int f\,d\mu\right| \le 2\delta.$$

The theorem follows because $\delta > 0$ is arbitrary. $\square$

5. The 1-D Central Limit Theorem

We are ready to state and prove the main result of this chapter: The one-dimensional central limit theorem (CLT). The CLT is generally considered to be a cornerstone of classical probability theory.

The Central Limit Theorem. *Suppose $\{X_i\}_{i=1}^{\infty}$ are i.i.d., real-valued, and have two finite moments. If $S_n := X_1 + \cdots + X_n$ and $\mathrm{Var}\,X_1 \in (0,\infty)$, then*

(7.46)
$$\frac{S_n - n\mathrm{E}X_1}{\sqrt{n}} \Rightarrow N(0, \mathrm{Var}\,X_1).$$

Because $n\mathrm{E}X_1 + \sqrt{n} N(0, \mathrm{Var}\,X_1)$ and $N(n\mathrm{E}X_1, n\mathrm{Var}\,X_1)$ have the same distribution, the central limit theorem states that the distribution of S_n is close to that of $N(n\mathrm{E}X_1, n\mathrm{Var}\,X_1)$.

Proof. By considering instead $X_j^* := (X_j - \mathrm{E}X_1)/\mathrm{SD}(X_1)$ and $S_n^* := \sum_{j=1}^{n} X_j^*$ we can assume without loss of generality that the X_j's have mean zero and variance one.

We apply the Taylor expansion with remainder to deduce that for all $x \in \mathbf{R}$,

(7.47)
$$e^{ix} = 1 + ix - \frac{1}{2}x^2 + R(x),$$

where $|R(x)| \le \frac{1}{6}|x|^3 \le |x|^3$. If $|x| \le 4$, then this is a good estimate, but when $|x| > 4$, we can use $|R(x)| \le |e^{ix}| + 1 + |x| + \frac{1}{2}x^2 \le x^2$ instead. Combine terms to obtain the bound:

$$|R(x)| \le |x|^3 \wedge x^2. \tag{7.48}$$

Because the X_j's are i.i.d., Lemma 6.12 on page 68 implies that

$$\mathrm{E} e^{itS_n/\sqrt{n}} = \prod_{j=1}^{n} \mathrm{E} e^{itX_j/\sqrt{n}}. \tag{7.49}$$

This and (7.47) together imply that

$$\begin{aligned}\mathrm{E} e^{itS_n/\sqrt{n}} &= \left(1 + i\mathrm{E}\left[\frac{tX_1}{\sqrt{n}}\right] - \frac{1}{2}\mathrm{E}\left[\frac{(tX_1)^2}{n}\right] + \mathrm{E}\left[R\left(\frac{tX_1}{\sqrt{n}}\right)\right]\right)^n \\ &= \left(1 - \frac{t^2}{2n} + \mathrm{E}\left[R\left(\frac{tX_1}{\sqrt{n}}\right)\right]\right)^n.\end{aligned} \tag{7.50}$$

By (7.48) and the dominated convergence theorem,

$$n\left|\mathrm{E}\left[R\left(tX_1/\sqrt{n}\right)\right]\right| \le \mathrm{E}\left[\frac{|tX_1|^3}{\sqrt{n}} \wedge (tX_1)^2\right] = o(1) \qquad (n \to \infty). \tag{7.51}$$

By the Taylor expansion $\ln(1-z) = -z + o(|z|)$ as $|z| \to 0$, where "ln" denotes the principal branch of the logarithm. It follows that

$$\lim_{n \to \infty} \mathrm{E} e^{itS_n/\sqrt{n}} = \lim_{n \to \infty} \left(1 - \frac{t^2}{2n} + o\left(\frac{1}{n}\right)\right)^n = e^{-t^2/2}. \tag{7.52}$$

The CLT follows from the convergence theorem (p. 99) and Example 7.12 (p. 97). □

6. Complements to the CLT

6.1. The Multidimensional CLT. Now we turn our attention to the study of random variables in $\mathbf{R}^d$. Throughout, $X, X^1, X^2, \ldots$ are i.i.d. random variables that take values in $\mathbf{R}^d$, and $S_n := \sum_{i=1}^{n} X^i$. Our discussion is a little sketchy. But this should not cause too much confusion, since we encountered most of the key ideas earlier on in this chapter. Throughout this section, $\|x\|$ denotes the usual Euclidean norm of a variable $x \in \mathbf{R}^d$. That is,

$$\|x\| := \sqrt{x_1^2 + \cdots + x_d^2} \qquad \forall x \in \mathbf{R}^d. \tag{7.53}$$

Definition 7.16. The *characteristic function* of X is the function $f(t) = \mathrm{E} e^{it \cdot X}$ where $t \cdot x = \sum_{i=1}^{d} t_i x_i$ for $t \in \mathbf{R}^d$. If μ denotes the distribution of X, then f is also written as $\widehat{\mu}$.

The following is the simplest analogue of the uniqueness theorem; it is an immediate consequence of the convergence theorem (p. 99).

Theorem 7.17. *If $\mu, \mu_1, \mu_2, \ldots$ are probability measures on $(\mathbf{R}^d, \mathscr{B}(\mathbf{R}^d))$ and $\widehat{\mu_n} \to \widehat{\mu}$ pointwise, then $\mu_n \Rightarrow \mu$.*

This leads us to our next result.

Theorem 7.18. *Suppose $\{X^i\}_{i=1}^\infty$ are i.i.d. random variables in $\mathbf{R}^d$ with $\mathrm{E}X_1^i = \mu_i$, and $\mathrm{Cov}(X_1^i, X_1^j) = Q_{i,j}$ for an invertible $(d \times d)$ matrix $Q := (Q_{i,j})$. Then for all d-dimensional hypercubes $G = (a_1, b_1] \times \cdots \times (a_d, b_d]$,*

$$(7.54) \qquad \lim_{n \to \infty} \mathrm{P}\left\{ \frac{S_n - n\mu}{\sqrt{n}} \in G \right\} = \int_G \frac{e^{-\frac{1}{2}y'Q^{-1}y}}{(2\pi)^{d/2}\sqrt{\det(Q)}} \, dy.$$

That is, $(S_n - n\mu)/\sqrt{n}$ converges weakly to a multidimensional Gaussian distribution with mean vector 0 and covariance matrix Q.

The preceding theorems are the natural d-dimensional extensions of their 1-D counterparts. On the other hand, the following is inherently multi-dimensional.

The Cramér–Wold Device. $X_n \Rightarrow X$ *if and only if* $(t \cdot X_n) \Rightarrow (t \cdot X)$ *for all $t \in \mathbf{R}^d$.*

If we were to prove that X_n converges weakly, then the Cramér–Wold device boils our task down to proving the weak convergence of the one-dimensional $(t \cdot X_n)$. But this needs to be proved for all $t \in \mathbf{R}^d$.

Proof. Suppose $X_n \Rightarrow X$, and choose and fix $f \in C_b(\mathbf{R}^d)$. Because $g_t(x) = t \cdot x$ is continuous, $\mathrm{E}f(g_t(X_n))$ converges to $\mathrm{E}f(g_t(X))$ as $n \to \infty$. This is half of the theorem.

Conversely, let μ_n and μ denote the distributions of X_n and X, respectively. Then $(t \cdot X_n) \Rightarrow (t \cdot X)$ for all $t \in \mathbf{R}^d$ iff $\widehat{\mu_n}(t) \to \widehat{\mu}(t)$. The converse now follows from Theorem 7.17. □

6.2. The Projective Central Limit Theorem. The *projective CLT* describes another natural way of arriving at the standard normal distribution. In kinetic theory this CLT implies that, for an ideal gas, all normalized *Gibbs states follow the standard normal distribution*. We are concerned only with the mathematical formulation of this CLT.

Definition 7.19. Define $\mathbf{S}^{n-1} := \{x \in \mathbf{R}^n : \|x\| = 1\}$ to be the *unit sphere* in $\mathbf{R}^n$. This is topologized by the relative topology in $\mathbf{R}^n$. That is, $U \subset \mathbf{S}^{n-1}$ is *open* in $\mathbf{S}^{n-1}$ iff U is an open subset of $\mathbf{R}^n$.

Recall that an $(n \times n)$ matrix M is a *rotation* if $M'M$ is the identity.

Definition 7.20. A measure μ on $\mathscr{B}(\mathbf{S}^{n-1})$ is called the *uniform distribution* on $\mathbf{S}^{n-1}$ if: (i) $\mu(\mathbf{S}^{n-1}) = 1$; and (ii) $\mu(A) = \mu(MA)$ for all $A \in \mathscr{B}(\mathbf{S}^{n-1})$ and all $(n \times n)$ rotation matrices M. If X is a random variable whose distribution is μ, then we say that X is *distributed uniformly* on $\mathbf{S}^{n-1}$. Item (ii) states that μ is *rotation invariant*.

Theorem 7.21. *If $X^{(n)}$ is distributed uniformly on $\mathbf{S}^{n-1}$, then*

(7.55) $$\sqrt{n}\, X_1^{(n)} \Rightarrow N(0,1).$$

Remark 7.22. Without worrying too much about what this really means let X denote the first coordinate of a random variable that is distributed uniformly on the centered ball of radius $\sqrt{\infty}$ in $\mathbf{R}^\infty$. The projective CLT asserts that X is standard normal.

Before we prove Theorem 7.21 we need to demonstrate that there are, in fact, rotation-invariant probability measures on $\mathbf{S}^{n-1}$. The following is a special case of a more general result in abstract harmonic analysis.

Theorem 7.23. *For all $n \geq 1$ there exists a unique rotation-invariant probability measure on $\mathbf{S}^{n-1}$.*

Proof. Let $\{Z_i\}_{i=1}^\infty$ denote a sequence of i.i.d. standard normal random variables, and define $Z^{(n)} = (Z_1, \ldots, Z_n)$. We normalize the latter as follows:

(7.56) $$X^{(n)} := \frac{Z^{(n)}}{\|Z^{(n)}\|} \qquad \forall n \geq 1.$$

By independence, the characteristic function of $Z^{(n)}$ is $f(t) := \exp(-\|t\|^2/2)$. Because f is rotation-invariant, $Z^{(n)}$ and $MZ^{(n)}$ have the same characteristic function as long as M is an $(n \times n)$ rotation matrix. Consequently, $Z^{(n)}$ and $MZ^{(n)}$ have the same distribution for all rotations M; confer with the uniqueness theorem on page 99. It follows that the distribution of $X^{(n)}$ is rotation invariant, and hence the existence of a uniform distribution on $\mathbf{S}^{n-1}$ follows. Next we prove the more interesting uniqueness portion.

For all $\epsilon > 0$ and all sets $A \subseteq \mathbf{S}^{n-1}$ define $K_A(\epsilon)$ to be the largest number of disjoint open balls of radius ϵ that can fit inside A. By compactness, if A is closed then $K_A(\epsilon)$ is finite. The function K_A is known as *Kolmogorov ϵ-entropy*, *Kolmogorov complexity*, as well as the *packing number* of A.

Let μ and ν be two uniform probability measures on $\mathscr{B}(\mathbf{S}^{n-1})$. By the maximality condition in the definition of K_A, and by the rotational invariance of μ and ν, for all closed sets $A \subset \mathbf{S}^{n-1}$,

(7.57) $$K_A(\epsilon)\mu(B_\epsilon) \leq \mu(A) \leq (K_A(\epsilon) + 1)\mu(B_\epsilon),$$

where $B_\epsilon := \{x \in \mathbf{S}^{n-1} : \|x\| < \epsilon\}$. The preceding display remains valid if we replace μ by ν everywhere. Therefore, for all closed sets A that have positive ν-measure,

(7.58) $$\left(\frac{K_A(\epsilon)}{K_A(\epsilon)+1}\right)\frac{\mu(A)}{\nu(A)} \le \frac{\mu(B_\epsilon)}{\nu(B_\epsilon)} \le \left(\frac{K_A(\epsilon)+1}{K_A(\epsilon)}\right)\frac{\mu(A)}{\nu(A)}.$$

Consequently,

(7.59) $$\left|\frac{\mu(A)}{\nu(A)} - \frac{\mu(B_\epsilon)}{\nu(B_\epsilon)}\right| \le \frac{1}{K_A(\epsilon)}\frac{\mu(A)}{\nu(A)}.$$

We apply this with $A := \mathbf{S}^{n-1}$ to find that

(7.60) $$\left|1 - \frac{\mu(B_\epsilon)}{\nu(B_\epsilon)}\right| \le \frac{1}{K_{\mathbf{S}^{n-1}}(\epsilon)}.$$

We plug this back in (7.59) to conclude that for all closed sets A with positive ν-measure,

(7.61) $$\left|\frac{\mu(A)}{\nu(A)} - 1\right| \le \frac{1}{K_A(\epsilon)}\frac{\mu(A)}{\nu(A)} + \frac{1}{K_{\mathbf{S}^{n-1}}(\epsilon)} \qquad \forall \epsilon > 0.$$

As ϵ tends to zero, the right-hand side converges to zero. This implies that $\mu(A) = \nu(A)$ for all closed sets $A \in \mathscr{B}(\mathbf{S}^{n-1})$ that have positive ν-measure. Next, we reverse the roles of μ and ν to find that $\mu(A) = \nu(A)$ for all closed sets $A \in \mathscr{B}(\mathbf{S}^{n-1})$. Because closed sets generate all of $\mathscr{B}(\mathbf{S}^{n-1})$, the monotone class theorem (p. 30) implies that $\mu = \nu$. □

Proof of Theorem 7.21. We follow the proof of Theorem 7.23 closely, and observe that by the strong law of large numbers (p. 73), $\|Z^{(n)}\|/\sqrt{n} \to 1$ a.s. Therefore, $\sqrt{n}\,X_1^{(n)} \to Z_1$ a.s. The latter is standard normal. Since a.s.-convergence implies weak convergence, the theorem follows. □

6.3. The Replacement Method of Liapounov. There are other approaches to the CLT than the harmonic-analytic ones of the previous sections. In this section we present an alternative probabilistic method of Lindeberg (1922) who, in turn, used an ingenious "replacement method" of Liapounov (1900, pp. 362–364). This method makes clear the fact that the CLT is a *local phenomenon*. By this we mean that the structure of the CLT does not depend on the behavior of any fixed number of the increments.

In words, the method proceeds as follows: We estimate the distribution of S_n by replacing the increments, one at a time, by independent normal random variables. Then we use an idea of Lindeberg, and appeal to Taylor's theorem of calculus to keep track of the errors incurred by the replacement method.

As a nice by-product we obtain quantitative bounds on the error-rate in the CLT without further effort. To be concrete, we derive the following using the Liapounov method; the heart of the matter lies in its derivation.

Theorem 7.24. *Fix an integer $n \geq 1$, and suppose $\{X_i\}_{i=1}^n$ are independent mean-zero random variables in $L^3(\mathrm{P})$. Define $S_n := \sum_{i=1}^n X_i$ and $s_n^2 := \mathrm{Var} S_n$. Then for any three times continuously differentiable function f,*

$$(7.62) \qquad |\mathrm{E}f(S_n) - \mathrm{E}f(N(0,s_n^2))| \leq \frac{2M_f}{3\sqrt{\pi/2}} \sum_{i=1}^n \|X_i\|_3^3,$$

provided that $M_f := \sup_z |f'''(z)|$ is finite.

Proof. Let σ_i^2 denote the variance of X_i for all $i = 1, \ldots, n$, so that $s_n^2 = \sum_{i=1}^n \sigma_i^2$. By Taylor expansion,

$$(7.63) \qquad \left| f(S_n) - f(S_{n-1}) - X_n f'(S_{n-1}) - \frac{X_n^2}{2} f''(S_{n-1}) \right| \leq \frac{M_f}{6} |X_n|^3.$$

Because $\mathrm{E}X_n = 0$ and $\mathrm{E}[X_n^2] = \sigma_n^2$, the independence of the X's implies that

$$(7.64) \qquad \left| \mathrm{E}f(S_n) - \mathrm{E}f(S_{n-1}) - \frac{\sigma_n^2}{2} \mathrm{E}f''(S_{n-1}) \right| \leq \frac{M_f}{6} \|X_n\|_3^3.$$

Next consider a normal random variable Z_n that has the same mean and variance as X_n, and is independent of $X_1, \ldots, X_n$. If we apply (7.64), but replace X_n by Z_n, then we obtain

$$(7.65) \qquad \left| \mathrm{E}f(S_{n-1} + Z_n) - \mathrm{E}f(S_{n-1}) - \frac{\sigma_n^2}{2} \mathrm{E}f''(S_{n-1}) \right| \leq \frac{M_f}{6} \|Z_n\|_3^3.$$

This and (7.64) together yield

$$(7.66) \qquad |\mathrm{E}f(S_n) - \mathrm{E}f(S_{n-1} + Z_n)| \leq \frac{M_f}{6} \left(\|Z_n\|_3^3 + \|X_n\|_3^3 \right).$$

A routine computation reveals that $\|Z_n\|_3^3 = A\sigma_n^3$, where $A := \mathrm{E}\{|N(0,1)|^3\} = 2/\sqrt{\pi/2} > 1$. Since $\sigma_n^3 \leq \|X_n\|_3^3$ (Proposition 4.16, p. 42), we find that

$$(7.67) \qquad |\mathrm{E}f(S_n) - \mathrm{E}f(S_{n-1} + Z_n)| \leq \frac{2M_f}{3\sqrt{\pi/2}} \|X_n\|_3^3.$$

Now we iterate this procedure: Bring in an independent normal Z_{n-1} with the same mean and variance as X_{n-1}. Replace X_{n-1} by Z_{n-1} in (7.67) to find that

$$(7.68) \quad |\mathrm{E}f(S_n) - \mathrm{E}f(S_{n-2} + Z_{n-1} + Z_n)| \leq \frac{2M_f}{3\sqrt{\pi/2}} \left(\|X_{n-1}\|_3^3 + \|X_n\|_3^3 \right).$$

Next replace X_{n-2} by another independent normal Z_{n-2}, etc. After n steps, we arrive at

$$(7.69) \qquad |\mathrm{E}f(S_n) - \mathrm{E}f\left(\sum_{i=1}^n Z_i\right)| \leq \frac{2M_f}{3\sqrt{\pi/2}} \sum_{i=1}^n \|X_i\|_3^3.$$

The theorem follows because $\sum_{i=1}^n Z_i = N(0, s_n^2)$; see Problem 7.18. □

To understand how this can be used suppose $\{X_i\}_{i=1}^n$ are i.i.d., with mean zero and variance σ^2. We can then apply Theorem 7.24 with $f(x) := g(x\sqrt{n})$ to deduce the following.

Corollary 7.25. *If $\{X_i\}_{i=1}^n$ are i.i.d. with mean zero, variance σ^2, and three bounded moments, then for all three times continuously differentiable functions g,*

$$(7.70) \qquad \left| Eg(S_n/\sqrt{n}) - Eg(N(0,\sigma^2)) \right| \le \frac{A}{\sqrt{n}},$$

where $A := 2\sup_z |g'''(z)| \cdot \|X_1\|_3^3 / (3\sqrt{\pi/2})$.

We let $g(x) := e^{itx}$, and extend the preceding to complex-valued functions in the obvious way to obtain the central limit theorem (p. 100) under the extra condition that $X_1 \in L^3(P)$. Moreover, when $X_1 \in L^3(P)$ we find that the rate of convergence to the CLT is of the order $n^{-1/2}$.

Theorem 7.24 is not restricted to increments that are in $L^3(P)$. For the case where $X_1 \in L^{2+\rho}(P)$ for some $\rho \in (0,1)$ see Problem 7.44. Even when $X_1 \in L^2(P)$ only, Theorem 7.24 can be used to prove the CLT, viz.,

Lindeberg's Proof of the CLT. Without loss of generality, we may assume that $\mu = 0$ and $\sigma = 1$. Choose and fix $\epsilon > 0$, and define $X_i' := X_i \mathbf{1}_{\{|X_i| \le \epsilon\sqrt{n}\}}$, $S_n' := \sum_{i=1}^n X_i'$, $\mu_n := ES_n'$, and $s_n^2 := \operatorname{Var} S_n'$.

Choose and fix a function $g : \mathbf{R} \to \mathbf{R}$ such that g and its first three derivatives are bounded and continuous. According to Theorem 7.24,

$$(7.71) \qquad \begin{aligned} \left| Eg\left(\frac{S_n' - \mu_n}{\sqrt{n}}\right) - Eg\left(N\left(0, \frac{s_n^2}{n}\right)\right) \right| &\le \frac{2M_g}{3\sqrt{\pi n/2}} E\left(|X_1' - EX_1'|^3\right) \\ &\le \frac{32 M_g}{3\sqrt{\pi n/2}} \|X_1'\|_3^3. \end{aligned}$$

The last line follows from the inequality $|a+b|^3 \le 8(|a|^3 + |b|^3)$ and the fact that $\|X_1'\|_1 \le \|X_1'\|_3$ (Proposition 4.16, p. 42). Because $|X_1'|$ is bounded above by $\epsilon\sqrt{n}$,

$$(7.72) \qquad \|X_1'\|_3^3 \le \epsilon\sqrt{n}\, E\left(|X_1'|^2\right) \le \epsilon\sqrt{n}\, E\left(X_1^2\right) = \epsilon\sqrt{n}.$$

Consequently,

$$(7.73) \qquad \left| Eg\left(\frac{S_n' - \mu_n}{\sqrt{n}}\right) - Eg\left(N\left(0, \frac{s_n^2}{n}\right)\right) \right| \le \frac{32 M_g}{3\sqrt{\pi/2}} \epsilon := A\epsilon.$$

A one-term Taylor expansion simplifies the first term as follows:

$$(7.74) \quad \left| \mathrm{E}g\left(\frac{S'_n - \mu_n}{\sqrt{n}}\right) - \mathrm{E}g\left(\frac{S_n}{\sqrt{n}}\right) \right| \le \sup_z |g'(z)| \, \mathrm{E}\left|\frac{S'_n - S_n - \mu_n}{\sqrt{n}}\right|$$
$$\le \sup_z |g'(z)| \, \frac{\mathrm{SD}(S_n - S'_n)}{\sqrt{n}}.$$

Since $S_n - S'_n = \sum_{i=1}^n X_i \mathbf{1}_{\{|X_i| \ge \epsilon\sqrt{n}\}}$ is a sum of n i.i.d. random variables,

$$(7.75) \quad \mathrm{Var}(S_n - S'_n) = n\mathrm{Var}\left(X_1 \mathbf{1}_{\{|X_1| > \epsilon\sqrt{n}\}}\right) \le n\mathrm{E}\left[X_1^2; |X_1| > \epsilon\sqrt{n}\right].$$

Therefore,

$$(7.76) \quad \left| \mathrm{E}g\left(\frac{S_n}{\sqrt{n}}\right) - \mathrm{E}g\left(N\left(0, \frac{s_n^2}{n}\right)\right) \right| \le A\epsilon + \sqrt{\mathrm{E}\left[X_1^2; |X_1| > \epsilon\sqrt{n}\right]}.$$

Now, $s_n^2/n = \mathrm{Var}(S'_n)/n = \mathrm{Var}(X_1 \mathbf{1}_{\{|X_1| > \epsilon\sqrt{n}\}})$. By the dominated convergence theorem, this converges to $\mathrm{Var}\, X_1 = 1$ as $n \to \infty$. Therefore by scaling (Problem 1.14, p. 14),

$$(7.77) \quad \mathrm{E}g\left(N\left(0, \frac{s_n^2}{n}\right)\right) = \mathrm{E}g\left(N(0,1)\frac{s_n}{\sqrt{n}}\right) \to \mathrm{E}g(N(0,1)),$$

as $n \to \infty$. This, the continuity of g, and (7.76), together yield

$$(7.78) \quad \limsup_{n \to \infty} |\mathrm{E}g\left(S_n/\sqrt{n}\right) - \mathrm{E}g(N(0,1))| \le A\epsilon.$$

Because the left-hand side is independent of ϵ, it must therefore be equal to zero. It follows that $\mathrm{E}g(S_n/\sqrt{n}) \to \mathrm{E}g(N(0,1))$ if g and its first three derivatives are continuous and bounded.

Now suppose $\psi \in C_c(\mathbf{R})$ is fixed. By Fejér's theorem (p. 98), for all $\delta > 0$ we can find g such that g and its first three derivatives are bounded and continuous, and $\sup_z |g(z) - \psi(z)| \le \delta$. Because δ is arbitrary, the triangle inequality and what we have proved so far together prove that $\mathrm{E}\psi(S_n/\sqrt{n}) \to \mathrm{E}\psi(N(0,1))$. This is the desired result. □

6.4. Cramér's Theorem. In this section we use characteristic function methods to prove the following striking theorem of Cramér (1936). This section requires only a rudimentary knowledge of complex analysis.

Theorem 7.26. *Suppose X_1 and X_2 are independent real-valued random variables such that $X_1 + X_2$ is a possibly degenerate normal random variable. Then X_1 and X_2 are possibly degenerate normal random variables too.*

Remark 7.27. Cramér's theorem states that if μ_1 and μ_2 are probability measures such that $\widehat{\mu_1}(t)\widehat{\mu_2}(t) = e^{i\mu t - \sigma^2 t^2}$ ($\mu \in \mathbf{R}, \sigma \ge 0$), then μ_1 and μ_2 are Gaussian probability measures.

Remark 7.28. Cramér's theorem does not rule out the possibility that one or both of the X_i's are constants. It might help to recall our convention that $N(\mu, 0) = \mu$.

We prove Cramér's theorem by first deriving three elementary lemmas from complex analysis, and one from probability. Recall that a function $f : \mathbf{C} \to \mathbf{C}$ is *entire* function if is is analytic on $\mathbf{C}$.

Lemma 7.29 (The Liouville Theorem). *Suppose $f : \mathbf{C} \to \mathbf{C}$ is an entire function, and there exists an integer $n \geq 0$ such that*

(7.79) $$|f(z)| = O(|z|^n) \qquad \text{as } |z| \to \infty.$$

Then there exist $a_0, \ldots, a_n \in \mathbf{C}$ such that $f(z) = \sum_{j=0}^n a_j z^j$ on $\mathbf{C}$.

Remark 7.30. When $n = 0$, Lemma 7.29 asserts that *bounded entire functions are constants*. This is the more usual form of the Liouville theorem.

Proof. For any $z_0 \in \mathbf{C}$ and $\rho > 0$, define $\gamma(\theta) := z_0 + \rho e^{i\theta}$ for all $\theta \in (0, 2\pi]$. By the Cauchy integral formula on circles, for any $n \geq 0$, the nth derivative $f^{(n)}$ is analytic and satisfies

(7.80) $$\begin{aligned} f^{(n+1)}(z_0) &= \frac{(n+1)!}{2\pi i} \int_\gamma \frac{f(z)}{(z-z_0)^{n+2}} \, dz \\ &= \frac{(n+1)!}{2\pi i \rho^{n+1}} \int_0^{2\pi} \frac{f(z_0 + \rho e^{i\theta})}{e^{i(n+2)\theta}} \, d\theta. \end{aligned}$$

Since f is continuous, (7.79) tells us that there exists a constant $A > 0$ such that $|f(z_0 + \rho e^{i\theta})| \leq A\rho^n$ for all $\rho > 0$ sufficiently large and all $\theta \in [0, 2\pi)$. In particular, $|f^{(n+1)}(z_0)| \leq (n+1)! A \rho^{-1}$. Because this holds for all large $\rho > 0$, $f^{(n+1)}(z_0) = 0$ for all $z_0 \in \mathbf{C}$, whence follows the result. □

Lemma 7.31 (Schwarz). *Choose and fix $A, \rho > 0$. Suppose f is analytic on $B_\rho := \{w \in \mathbf{C} : |w| < \rho\}$, $f(0) = 0$, and $\sup_{z \subset B_\rho} |f(z)| \leq A$. Then,*

(7.81) $$|f(z)| \leq \frac{A|z|}{\rho} \qquad \text{on } B_\rho.$$

Proof. Define

(7.82) $$F(z) := \begin{cases} f(z)/z & \text{if } z \neq 0, \\ f'(0) & \text{if } z = 0. \end{cases}$$

Evidently, F is analytic on B_ρ. According to the *maximum principle*, an analytic function in a given domain attains its maximum on the boundary of the domain. Therefore, whenever $r \in (0, \rho)$, it follows that

(7.83) $$|F(z)| \leq \sup_{|w|=r} |F(w)| \leq \frac{A}{r} \qquad \forall |z| < r.$$

Let r converge upward to ρ to finish. □

The following is our final requirement from complex analysis.

Lemma 7.32 (Borel and Carathéodory). *If $f : \mathbf{C} \to \mathbf{C}$ is entire, then*

$$\text{(7.84)} \qquad \sup_{|z|\leq r/2} |f(z)| \leq 4 \sup_{|z|\leq r} |\text{Re } f(z)| + 5|f(0)| \qquad \forall r > 0.$$

Proof. Let $g(z) := f(z) - f(0)$, so that g is entire and $g(0) = 0$. Define $R(r) := \sup_{|z|\leq r} |\text{Re } g(z)|$ for all $r > 0$, and consider the function

$$\text{(7.85)} \qquad T(w) := \frac{w}{2R(r) - w} \qquad \forall |w| \leq R(r).$$

Evidently,

$$\text{(7.86)} \qquad g(z) = 2R(r) \frac{T(g(z))}{1 + T(g(z))}.$$

One can check directly that $|T(f(z))| \leq 1$ for all $z \in B_r$, and hence $T \circ g$ is analytic on B_r. Because $T(g(0)) = 0$, Lemma 7.31 implies that $|T(g(z))| \leq |z|/r$ for all $z \in B_r$. It follows that for all $z \in B_r$,

$$\text{(7.87)} \qquad |g(z)| \leq 2R(r) \frac{|z|/r}{1 - (|z|/r)}.$$

This proves that, $|g(z)| \leq 4R(r)$, uniformly for $|z| \leq r/2$, and hence,

$$\text{(7.88)} \qquad \sup_{|z|\leq r/2} |f(z) - f(0)| \leq 4 \sup_{|z|\leq r} |\text{Re } f(z) - \text{Re } f(0)|.$$

The lemma follows from this and the triangle inequality. $\square$

Finally, we need a preparatory lemma from probability.

Lemma 7.33. *If $V \geq 0$ a.s., then for any $a > 0$,*

$$\text{(7.89)} \qquad \mathrm{E}e^{aV} = 1 + a \int_0^\infty e^{ax} \mathrm{P}\{V \geq x\} \, dx.$$

In particular, suppose U is non-negative, and there exists $r \geq 1$ such that

$$\text{(7.90)} \qquad \mathrm{P}\{V \geq x\} \leq r \mathrm{P}\{U \geq x\} \qquad \forall x > 0.$$

Then, $\mathrm{E}e^{aV} \leq r \mathrm{E}e^{aU}$ for all $a > 0$.

Proof. Because $e^{aV(\omega)} = 1 + a \int_0^\infty \mathbf{1}_{\{V(\omega)\geq x\}} e^{ax} \, dx$ and the integrand is non-negative, we can take expectations and use Fubini–Tonelli to deduce (7.89). Because $r \geq 1$, the second assertion is a ready corollary of the first $\square$

Proof of Theorem 7.26. Throughout, let $Z := X_1 + X_2$; Z is normally distributed. We can assume without loss of generality that $\mathrm{E}Z = 0$; else we consider $Z - \mathrm{E}Z$ in place of Z. The proof is now carried out in two natural steps.

Step 1. Identifying the Modulus. We begin by finding the form of $\left|\mathrm{E}e^{itX_k}\right|$ for $k = 1, 2$.

Because $\mathrm{E}Z = 0$, there exists $\sigma \geq 0$ such that $\mathrm{E}\exp(zZ) = \exp(z^2\sigma^2)$ for all $z \in \mathbf{C}$. Since $|Z| \geq |X_1| - |X_2|$, if $|X_1| \geq \lambda$ and $|X_2| \leq m$ then $|Z| \geq \lambda - m$. Therefore, by independence,

$$\mathrm{P}\{|Z| \geq \lambda - m\} \geq \mathrm{P}\{|X_1| \geq \lambda\}\mathrm{P}\{|X_2| \leq m\}$$
(7.91)
$$\geq \frac{1}{4}\mathrm{P}\{|X_1| \geq \lambda\},$$

provided that we choose a sufficiently large m. Choose and fix such an m.

In accord with Lemma 7.33, $\mathrm{E}e^{c|X_1|} \leq 4e^{cm}\mathrm{E}e^{c|Z|}$ for all $c > 0$. But

(7.92) $\qquad \mathrm{E}e^{c|Z|} \leq \mathrm{E}e^{cZ} + \mathrm{E}e^{-cZ} \leq 2e^{c^2\sigma^2} \qquad \forall c > 0.$

Consequently,

(7.93) $\qquad \left|\mathrm{E}e^{zX_1}\right| \leq \mathrm{E}e^{|z|\cdot|X_1|} \leq 8\exp\left(|z|m + \sigma^2|z|^2\right) \qquad \forall z \in \mathbf{C}.$

Because $|Z| \geq |X_2| - |X_1|$, the same bound holds if we replace X_1 by X_2 everywhere. This proves that $f_k(z) := \mathrm{E}\exp(zX_k)$ exists for all $z \in \mathbf{C}$, and defines an entire function (why?).

To summarize, $\mathbf{R} \ni t \mapsto f_k(it)$ is the characteristic function of X_k, and

(7.94) $\qquad |f_k(z)| \leq 8\exp\left(|z|m + \sigma^2|z|^2\right) \qquad \forall z \in \mathbf{C}, k = 1, 2.$

Because $f_1(z)f_2(z) = \mathrm{E}\exp(zZ) = \exp(z^2\sigma^2)$, (7.94) implies that for all $z \in \mathbf{C}$ and $k = 1, 2$,

(7.95) $\qquad 8\exp\left(|z|m + \sigma^2|z|^2\right)|f_k(z)| \geq |\exp(z^2\sigma^2)| \geq \exp\left(-|z|^2\sigma^2\right).$

It follows from this and (7.94) that for all $z \in \mathbf{C}$ and $k = 1, 2$,

(7.96) $\qquad \frac{1}{8}\exp\left(-|z|m - 2\sigma^2|z|^2\right) \leq |f_k(z)| \leq 8\exp\left(|z|m + \sigma^2|z|^2\right).$

Consequently, $\ln|f_k|$ is an entire function that satisfies the growth condition (7.79) of Lemma 7.29 with $n = 2$, and hence,

(7.97) $\qquad |f_1(z)| = \exp\left(a_0 + a_1 z + a_2 z^2\right) \qquad \forall z \in \mathbf{C}.$

A similar expression holds for $|f_2(z)|$.

Step 2. Estimating the Imaginary Part. Because f_k is non-vanishing and entire, we can write

(7.98) $\qquad f_k(z) = \exp(g_k(z)),$

where g_k is entire for $k = 1, 2$. To prove this we first note that f_k'/f_k is entire, and therefore so is

(7.99) $\qquad g_k(z) := \int_0^z \frac{f_k'(w)}{f_k(w)}\,dw.$

Next we compute directly to find that $(e^{-g_k} f_k)'(z) = 0$ for all $z \in \mathbf{C}$. Because $f_k(0) = 1$ and $g_k(0) = 0$, it follows that $f_k(z) = \exp(g_k(z))$, as asserted.

It follows then that $|f_k(z)| = \exp(\operatorname{Re} g_k(z))$, and Step 1 implies that $\operatorname{Re} g_k$ is a complex quadratic polynomial for $k = 1, 2$. Thanks to this and Lemma 7.32, we can deduce that the entire function g_k satisfies (7.79) with $n = 2$. Therefore, by Liouville's theorem, $g_k(z) = \alpha_k + \beta_k z + \gamma_k z^2$ where $\alpha_1, \alpha_2, \beta_1, \beta_2, \gamma_1, \gamma_2$ are complex numbers. Consequently,

$$(7.100) \quad \mathrm{E} e^{itX_k} = f_k(it) = \exp\left(\alpha_k + it\beta_k - t^2 \gamma_k\right) \quad \forall t \in \mathbf{R}, \ k = 1, 2.$$

Plug in $t = 0$ to find that $\alpha_k = 0$. Also part (1) of Lemma 7.9 implies that $f_k(-it)$ is the complex conjugate of $f_k(it)$. We can write this out to find that

$$(7.101) \quad \exp(-it\beta_k - t^2 \gamma_k) = \exp(-it\overline{\beta_k} - t^2 \overline{\gamma_k}) \quad \forall t \in \mathbf{R}.$$

This proves that

$$(7.102) \quad it\beta_k - t^2 \gamma_k = it\overline{\beta_k} - t^2 \overline{\gamma_k} + 2\pi i N(t),$$

where $N(t)$ is integer-valued for every $t \in \mathbf{R}$. All else being continuous, this proves that N is a continuous integer-valued function. Therefore, $N(t) = N(0) = 0$, and so it follows from the preceding display that β_k and γ_k are real-valued. Because $|f_k(it)| \le 1$, we have also that $\gamma_k \ge 0$. The result follows from these calculations. $\square$

Problems

7.1. Define $C_c^\infty(\mathbf{R}^k)$ to be the collection of all infinitely differentiable functions $f : \mathbf{R}^k \to \mathbf{R}$ that have compact support. If $\mu, \mu_1, \mu_2, \ldots$ are probability measures on $(\mathbf{R}^k, \mathscr{B}(\mathbf{R}^k))$, then prove that $\mu_n \Rightarrow \mu$ iff $\int f \, d\mu_n \to \int f \, d\mu$ for all $f \in C_c^\infty(\mathbf{R}^k)$.

7.2. If $\mu, \mu_1, \mu_2, \ldots, \mu_n$ is a sequence of probability measures on $(\mathbf{R}^d, \mathscr{B}(\mathbf{R}^d))$, then show that the following are characteristic functions of probability measures:

(1) $\overline{\widehat{\mu}}$;

(2) $\operatorname{Re} \widehat{\mu}$,

(3) $|\widehat{\mu}|^2$;

(4) $\prod_{j=1}^n \widehat{\mu}_j$; and

(5) $\sum_{j=1}^n p_j \widehat{\mu}_j$, where $p_1, \ldots, p_n \ge 0$ and $\sum_{j=1}^n p_j = 1$.

Also prove that $\overline{\widehat{\mu}(\xi)} = \widehat{\mu}(-\xi)$. Consequently, if μ is a *symmetric measure* (i.e., $\mu(-A) = \mu(A)$ for all $A \in \mathscr{B}(\mathbf{R}^d)$) then $\widehat{\mu}$ is a real-valued function.

7.3. Use characteristic functions to derive Problem 1.17 on page 14. Apply this to prove that if $X = \operatorname{Unif}[-1, 1]$, then we can write it as

$$(7.103) \quad X := \sum_{j=1}^\infty \frac{X_j}{2^j},$$

where the X_j's are i.i.d., taking the values ± 1 with probability $\frac{1}{2}$ each.

7.4 (Problem 7.3, continued). Prove that

$$\text{(7.104)} \qquad \frac{\sin x}{x} = \prod_{k=1}^{\infty} \cos\left(\frac{x}{2^k}\right) \qquad \forall x \in \mathbf{R} \setminus \{0\}.$$

By continuity, this is true also for $x = 0$.

7.5. Let X and Y denote two random variables on the same probability space. Suppose that $X + Y$ and $X - Y$ are independent standard-normal random variables. Then prove that X and Y are independent normal random variables. You may not use Theorem 7.26 or its proof.

7.6. Suppose X_1 and X_2 are independent random variables. Use characteristic functions to prove that:

(1) If $X_i = \text{Bin}(n_i, p)$ for the same $p \in [0, 1]$, then $X_1 + X_2 = \text{Bin}(n_1 + n_2, p)$.
(2) If $X_i = \text{Poiss}(\lambda_i)$, then $X_1 + X_2 = \text{Poiss}(\lambda_1 + \lambda_2)$.
(3) If $X_i = N(\mu_i, \sigma_i^2)$, then $X_1 + X_2 = N(\mu_1 + \mu_2, \sigma_1^2 + \sigma_2^2)$.

7.7. Let X have the gamma distribution with parameters (α, λ). Compute, carefully, the characteristic function of X. Use it to prove that if $X_1, X_2, \ldots$ are i.i.d. exponential random variables with parameter λ each, then $S_n := X_1 + \cdots + X_n$ has a gamma distribution. Identify the latter distribution's parameters.

7.8. Let f be a symmetric and bounded probability density function on $\mathbf{R}$. Suppose there exists $C > 0$ and $\alpha \in (0, 1]$ such that

$$\text{(7.105)} \qquad f(x) \sim C|x|^{-(1+\alpha)} \qquad \text{as } |x| \to \infty.$$

Prove that

$$\text{(7.106)} \qquad \widehat{f}(t) = 1 - D|t|^\alpha + o(|t|^\alpha) \qquad \text{as } |t| \to 0,$$

and compute D. Check also that $D < \infty$. What happens if $\alpha > 1$?

7.9 (Lévy's Concentration Inequality). Prove that if μ is a probability measure on the line, then

$$\text{(7.107)} \qquad \mu\left(\left\{x : |x| > \frac{1}{\epsilon}\right\}\right) \leq \frac{7}{\epsilon} \int_0^\epsilon (1 - \text{Re}\,\widehat{\mu}(t))\, dt \qquad \forall \epsilon > 0.$$

(HINT: Start with the right-hand side.)

7.10 (Fourier Series). Suppose X is a random variable that takes values in $\mathbf{Z}^d$ and has mass function $p(x) = P\{X = x\}$. Define $\widehat{p}(t) = Ee^{it \cdot X}$, and derive the following *inversion formula*:

$$\text{(7.108)} \qquad p(x) = \frac{1}{(2\pi)^d} \int_{[-\pi, \pi]^d} \exp(-it \cdot x)\, \widehat{p}(t)\, dt \qquad \forall x \in \mathbf{Z}^d.$$

Is the latter identity valid for all $x \in \mathbf{R}^d$?

7.11. Derive the following variant of Plancherel's theorem (p. 99): For any $a < b$ and all probability measures μ on $(\mathbf{R}, \mathscr{B}(\mathbf{R}))$,

$$\text{(7.109)} \qquad \lim_{\epsilon \downarrow 0} \frac{1}{2\pi i} \int_{-\infty}^{\infty} e^{-\epsilon^2 t^2/2} \left(\frac{e^{-ita} - e^{-itb}}{t}\right) \widehat{\mu}(t)\, dt = \mu((a, b)) + \frac{\mu(\{a\}) + \mu(\{b\})}{2}.$$

7.12 (Inversion Theorem). Derive the *inversion theorem*: If μ is a probability measure on $\mathscr{B}(\mathbf{R}^k)$ such that $\widehat{\mu}$ is integrable $[dx]$, then μ is absolutely continuous with respect to the Lebesgue measure on $\mathbf{R}^k$. Moreover, then μ has a uniformly continuous density function f, and

$$\text{(7.110)} \qquad f(x) = \frac{1}{(2\pi)^k} \int_{\mathbf{R}^k} e^{-it \cdot x} \widehat{f}(t)\, dt \qquad \forall x \in \mathbf{R}^k.$$

7.13 (The Triangular Distribution). Consider the density function $f(x) := (1 - |x|)^+$ for $x \in \mathbf{R}$. If the density function of X is f, then compute the characteristic function of X. Prove that f itself is the characteristic function of a probability measure. (HINT: Problem 7.12.)

7.14. Suppose f is a probability density function on $\mathbf{R}$; i.e., $f \geq 0$ a.e. and $\int_{-\infty}^{\infty} f(x)\, dx = 1$.

(1) We say that f is of *positive type* if $\widehat{f}$ is non-negative and integrable. Prove that if f is of positive type, then $f(x) \le f(0)$ for all $x \in \mathbf{R}$.

(2) Prove that if f is of positive type, then $g(x) := \widehat{f}(x)/(2\pi f(0))$ is a density function, and $\widehat{g}(t) = f(t)/f(0)$. (HINT: Problem 7.12.)

(3) Compute the characteristic function of $g(x) = \frac{1}{2}\exp(-|x|)$. Use this to conclude that $f(x) := \pi^{-1}(1+x^2)^{-1}$ is a probability density function whose characteristic function is $\widehat{f}(t) = \exp(-|t|)$. The function f defines the so-called *Cauchy density function*. [Alternatively, you may use contour integration to arrive at the end result.]

7.15 (Riemann–Lebesgue lemma). Prove that $\lim_{|t|\to\infty} \mathrm{E}e^{it\cdot X} = 0$ for all k-dimensional absolutely continuous random variables X. Can the absolute-continuity condition be removed altogether? (HINT: Consider first a nice X.)

7.16. Suppose X and Y are two independent random variables; X is absolutely continuous with density function f, and the distribution of Y is μ. Prove that $X+Y$ is absolutely continuous with density function

(7.111) $$(f * \mu)(x) := \int f(x-y)\,\mu(dy).$$

Prove also that if Y is absolutely continuous with density function g, then the density function of $X+Y$ is $f*g$.

7.17. Prove that the CLT (p. 100) continues to hold when $\sigma = 0$.

7.18. A probability measure μ on $(\mathbf{R}, \mathscr{B}(\mathbf{R}))$ is said to be *infinitely divisible* if for any $n \ge 1$ there exists a probability measure ν such that $\widehat{\mu} = (\widehat{\nu})^n$. Prove that the normal and the Poisson distributions are infinitely divisible. So is the probability density

(7.112) $$f(x) := \frac{1}{\pi(1+x^2)} \qquad \forall x \in \mathbf{R}.$$

This is called the *Cauchy distribution*. (HINT: Problem 7.14.)

7.19. Prove that if $\{X_i\}_{i=1}^\infty$ are i.i.d. uniform-$[0,1]$ random variables, then

(7.113) $$\frac{4\sum_{i=1}^n iX_i - n^2}{n^{3/2}} \quad \text{converges weakly.}$$

Identify the limiting distribution.

7.20 (Extreme Values). If $\{X_i\}_{i=1}^\infty$ are i.i.d. standard normal random variables, then find non-random sequences $a_n, b_n \to \infty$ such that $a_n \max_{1\le i\le n} X_i - b_n$ converges weakly. Identify the limiting distribution. Replace "standard normal" by "mean-λ exponential," where $\lambda > 0$ is a fixed number, and repeat the exercise.

7.21. Let $\{X_i\}_{i=1}^\infty$ denote independent random variables such that

(7.114) $$X_j = \begin{cases} \pm j & \text{each with probability } (4j^2)^{-1}, \\ \pm 1 & \text{with probability } \frac{1}{2} - (4j^2)^{-1}. \end{cases}$$

Prove that

(7.115) $$\frac{S_n}{\mathrm{SD}(S_n)} \Rightarrow N(0,\sigma^2),$$

and compute σ.

7.22 (An abelian CLT). Suppose that $\{X_i\}_{i=1}^\infty$ are i.i.d. with $\mathrm{E}X_1 = 0$ and $\mathrm{E}[X_1^2] = \sigma^2 < \infty$. First establish that $\sum_{i=1}^\infty r^i X_i$ converges almost surely for all $r \in (0,1)$. Then, prove that

(7.116) $$\sqrt{1-r}\sum_{i=0}^\infty r^i X_i \Rightarrow N(0,\gamma^2) \qquad \text{as } n \to \infty,$$

and compute γ (Bovier and Picco, 1996).

7.23. State and prove a variant of Theorem 7.18 that does not assume Q to be non-singular.

7.24 (Liapounov Condition). In the notation of Problem 7.38 below assume there exists $\delta > 0$ such that

(7.117) $$\lim_{n \to \infty} \frac{1}{s_n^{2+\delta}} \sum_{j=1}^{n} \mathrm{E}\left[|X_j - \mu_j|^{2+\delta}\right] = 0.$$

Prove the theorem of Liapounov (1900, 1922):

(7.118) $$\frac{S_n - \sum_{j=1}^{n} \mu_j}{s_n} \Rightarrow N(0,1).$$

Check that the variables of Problem 7.21 do not satisfy (7.129).

7.25. Compute

(7.119) $$\lim_{n \to \infty} e^{-n} \left(1 + n + \frac{n^2}{2} + \frac{n^3}{3!} + \frac{n^4}{4!} + \cdots + \frac{n^n}{n!}\right).$$

7.26 (The Simple Walk). Let $\mathbf{e}_1, \ldots, \mathbf{e}_d$ denote the usual basis vectors of $\mathbf{R}^d$; i.e., $\mathbf{e}_1 = (1, 0, \ldots, 0)$, $\mathbf{e}_2 = (0, 1, 0, \ldots, 0)$, etc. Consider i.i.d. random variables $\{X_i\}_{i=1}^{\infty}$ such that

(7.120) $$\mathrm{P}\{X_1 = \pm \mathbf{e}_j\} = \frac{1}{2d}.$$

Then the random process $S_n = X_1 + \cdots + X_n$ with $S_0 = 0$ is the *simple walk* on $\mathbf{Z}^d$. It starts at zero and moves to each of the neighboring sites in $\mathbf{Z}^d$ with equal probability, and the process continues in this way ad infinitum. Find vectors a_n and constants b_n such that $(S_n - a_n)/b_n$ converges weakly to a nontrivial limit distribution. Compute the latter distribution.

7.27 (Problem 7.26, continued). Consider a collection of points $\Pi = \{\pi_i\}_{i=0}^{n}$ in $\mathbf{Z}^d$. We say that Π is a *lattice path of length* n if $\pi_0 = 0$, and for all $i = 2, \ldots, n-1$ the distance between π_i and π_{i+1} is one. Prove that all lattice paths Π of length n are equally likely for the first n steps in a simple walk.

7.28 (Problem 7.27, continued). Let $N_n(d)$ denote the number of length-n lattice-paths $\{\pi_i\}_{i=0}^{n}$ such that $\pi_n = 0$. Then prove that

(7.121) $$N_n(d) = \frac{1}{(2\pi)^d} \int_{[-\pi,\pi]^d} \left[2 \sum_{j=1}^{d} \cos t_j\right]^n dt$$

if $n \geq 2$ is even; else $N_n(d) = 0$. Conclude the 1655 *Wallis formula*:

(7.122) $$\int_{-\pi}^{\pi} (\cos t)^n \, dt = \binom{n}{n/2} \frac{\pi}{2^{n-1}},$$

valid for all even $n \geq 2$. (HINT: Problem 7.10.)

7.29. Suppose $\{X_i\}_{i=1}^{\infty}$ are i.i.d., mean-zero, and in $L^2(\mathrm{P})$. Prove that there exists a positive constant c such that

(7.123) $$\mathrm{E}\left(\max_{1 \leq j \leq n} |S_j|\right) \geq c \, \mathrm{SD}(X_1) \sqrt{n} \qquad \forall n \geq 1.$$

Compare to Problem 6.27 on page 87.

7.30. Suppose that $X_n \Rightarrow X$ and $Y_n \Rightarrow Y$ as $n \to \infty$, where $X_n, Y_n, X,$ and Y are real-valued.

(1) Prove that if Y is non-random, then $Y_n \to Y$ in probability. Conclude from this that $(X_n, Y_n) \Rightarrow (X, Y)$.

(2) Prove that if $\{X_n\}_{n=1}^{\infty}$ and $\{Y_n\}_{n=1}^{\infty}$ are independent from one another, then (X_n, Y_n) converges weakly to (X, Y).

(3) Find an example where $X_n \Rightarrow X$, $Y_n \Rightarrow Y$, and $(X_n, Y_n) \not\Rightarrow (X, Y)$.

7.31 (Variance-Stabilizing Transformations). Suppose $g : \mathbf{R} \to \mathbf{R}$ has at least three bounded continuous derivatives, and let $X_1, X_2, \ldots$ be i.i.d. and in $L^2(\mathrm{P})$. Prove that

$$\sqrt{n}\,[g(\bar{X}_n) - g(\mu)] \Rightarrow N(0, \sigma^2), \tag{7.124}$$

where $\bar{X}_n := n^{-1} \sum_{i=1}^n X_i$, $\mu := \mathrm{E} X_1$, and $\sigma := \mathrm{SD}(X_1) g'(\mu)$. Also prove that

$$\mathrm{E} g(\bar{X}_n) - g(\mu) = \frac{\sigma^2 g''(\mu)}{2n} + o\left(\frac{1}{n}\right) \quad \text{as } n \to \infty. \tag{7.125}$$

7.32 (Microcanonical Distributions). Prove that if $X^{(n)}$ is distributed uniformly on $\mathbf{S}^{n-1}$, then $(X_1^{(n)}, \ldots, X_k^{(n)}) \Rightarrow Z$ for any fixed $k \geq 1$, where $Z = (Z_1, \ldots, Z_k)$ and the Z_i's are i.i.d. standard normals.

7.33. Choose and fix an integer $n \geq 1$ and let $X_1, X_2, \ldots$ be i.i.d. with common distribution given by $\mathrm{P}\{X_1 = k\} = 1/n$ for $k = 1, \ldots, n$. Let T_n denote the smallest integer $l \geq 1$ such that $X_1 + \cdots + X_l > n$, and compute $\lim_{n \to \infty} \mathrm{P}\{T_n = k\}$ for all k.

7.34 (Uniform Integrability). Suppose $X, X_1, X_2, \ldots$ are real-valued random variables such that: (i) $X_n \Rightarrow X$; and (ii) $\sup_n \|X_n\|_p < \infty$ for some $p > 1$. Then prove that $\lim_{n \to \infty} \mathrm{E} X_n = \mathrm{E} X$. (HINT: See Problem 4.28 on page 51.) Use this to prove the following: Fix some $p_0 \in (0, 1)$, and define $f(t) = |t - p_0|$ ($t \in [0, 1]$). Then prove that there exists a constant $c > 0$ such that the Bernstein polynomial $\mathcal{B}_n f$ satisfies

$$|(\mathcal{B}_n f)(p_0) - f(p_0)| \geq \frac{c}{\sqrt{n}} \quad \forall n \geq 1. \tag{7.126}$$

Thus, (6.50) on page 78 is sharp (Kac, 1937).

7.35 (Hard). Define the *Fourier* map $\mathcal{F} f = \hat{f}$ for $f \in L^1(\mathbf{R}^k)$. Prove that

$$\|f\|_{L^2(\mathbf{R}^k)} = \frac{1}{(2\pi)^{k/2}} \|\mathcal{F} f\|_{L^2(\mathbf{R}^k)} \quad \forall f \in L^1(\mathbf{R}^k) \cap L^2(\mathbf{R}^k). \tag{7.127}$$

This is sometimes known as the *Plancherel theorem*. Use it to extend $\mathcal{F}$ to a homeomorphism from $L^2(\mathbf{R}^k)$ onto itself. Conclude from this that if μ is a finite measure on $\mathscr{B}(\mathbf{R}^k)$ such that $\int_{\mathbf{R}^k} |\hat{\mu}(t)|^2 \, dt < \infty$, then μ is absolutely continuous with respect to the Lebesgue measure on $\mathbf{R}^k$. WARNING: The formula $(\mathcal{F} f)(t) = \int_{\mathbf{R}^k} f(x) e^{it \cdot x} \, dx$ is valid only when $f \in L^1(\mathbf{R}^k)$.

7.36 (An Uncertainty Principle; Hard). Prove that if $f : \mathbf{R} \to \mathbf{R}$ is a probability density function that is zero outside $[-\pi, \pi]$, then there exists $t \notin [-1/2, 1/2]$ such that $\hat{f}(t) \neq 0$ (Donoho and Stark, 1989). (Hint: View f as a function on $[-\pi, \pi]$, and develop it as a Fourier series. Then study the Fourier coefficients.)

7.37 (Hard). Choose and fix $\lambda_1, \ldots, \lambda_m > 0$ and $a_1, \ldots, a_m \in \mathbf{R}$. Then prove that if $m < \infty$, then f_m defines the characteristic function of a probability measure, where

$$f_m(t) := \exp\left(-\sum_{j=1}^m \lambda_j \left(1 - \cos(a_j t)\right)\right) \quad \forall t \in \mathbf{R},\ 1 \leq m \leq \infty. \tag{7.128}$$

Prove that f_∞ is a characteristic function provided that $\sum_j (a_j^2 \wedge |a_j|) \lambda_j < \infty$. (HINT: Consult Example 7.14 on page 97.)

7.38 (Lindeberg CLT; Hard). Let $\{X_i\}_{i=1}^\infty$ be independent $L^2(\mathrm{P})$-random variables in $\mathbf{R}$, and for all n define $s_n^2 = \sum_{j=1}^n \mathrm{Var} X_j$ and $\mu_n = \mathrm{E} X_n$. In addition, suppose that $s_n \to \infty$, and

$$\lim_{n \to \infty} \frac{1}{s_n^2} \sum_{j=1}^n \mathrm{E}\left[(X_j - \mu_j)^2;\, |X_j - \mu_j| > \epsilon s_n\right] = 0 \quad \forall \epsilon > 0. \tag{7.129}$$

Prove the *Lindeberg CLT* (1922):

$$\frac{S_n - \sum_{j=1}^n \mu_j}{s_n} \Rightarrow N(0, 1). \tag{7.130}$$

Check that the variables of Problem 7.21 do not satisfy (7.129).

7.39 (Hard). Let (X,Y) be a random vector in $\mathbf{R}^2$ and for all $\theta \in (0, 2\pi]$ define

(7.131) $$X_\theta := \cos(\theta)X + \sin(\theta)Y \quad \text{and} \quad Y_\theta := \sin(\theta)X - \cos(\theta)Y.$$

Prove that if X_θ and Y_θ are independent for all $\theta \in (0, 2\pi]$, then X and Y are independent normal variables. (HINT: Use Cramér's theorem to reduce the problem to the case that X and Y are symmetric; or you can consult the original paper of Kac (1939).)

7.40 (Skorohod's Theorem; Hard). Weak convergence does not imply a.s. convergence. To wit, $X_n \Rightarrow X$ does not even imply that any of the random variables $\{X_n\}_{n=1}^\infty$ and/or X live on the same probability space. The converse, however, is always true; check that $X_n \Rightarrow X$ whenever $X_n \to X$ almost surely. On the other hand, if you are willing to work on *some* probability space, then weak convergence is equivalent to a.s. convergence as we now work to prove.

(1) If F is a distribution function on $\mathbf{R}$ that has a continuous inverse, and if U is uniformly distributed on $(0, 1)$, then find the distribution function of $F^{-1}(U)$.

(2) Suppose $F_n \Rightarrow F$: All are distribution functions; each has a continuous inverse. Then prove that $\lim_{n\to\infty} F_n^{-1}(U) = F^{-1}(U)$ a.s.

(3) Use this to prove that whenever $X_n \Rightarrow X_\infty$, we can find, on a suitable probability space, random variables X'_n and X' such that: (i) For every $1 \le n \le \infty$, X'_n has the same distribution as X_n; and (ii) $\lim_n X'_n = X'$ almost surely Skorohod (1961, 1965).

(HINT: Problem 6.9.)

7.41 (Ville's CLT; Hard). Let Ω denote the collection of all permutations of $1,\ldots,n$, and let P be the probability measure that puts mass $(n!)^{-1}$ on each of the $n!$ elements of Ω. For each $\omega \in \Omega$ define $X_1(\omega) = 0$, and for all $k = 2,\ldots,n$ let $X_k(\omega)$ denote the number of *inversions* of k in the permutation ω; i.e., the number of times $1,\ldots,k-1$ precede k in the permutation ω. [For instance, suppose $n=4$. If $\omega = \{3,1,4,2\}$, then $X_2(\omega) = 1$, $X_3(\omega) = 0$, and $X_4(\omega) = 2$.]

Prove that $\{X_i\}_{i=1}^n$ are independent. Compute their distribution, and prove that the total number of inversions $S_n := \sum_{i=1}^n X_i$ in a *random permutation* satisfies

(7.132) $$\frac{S_n - (n^2/4)}{n^{3/2}} \Rightarrow N(0, 1/36).$$

(HINT: Problem 7.38.)

7.42 (A Poincaré Inequality; Hard). Suppose X and Y are independent standard normal random variables.

(1) Prove that for all twice continuously differentiable functions $f, g : \mathbf{R} \to \mathbf{R}$ that have bounded derivatives,

$$\mathrm{Cov}(f(X), g(X)) = \int_0^1 \mathrm{E}\left[f'(X)g'\left(sX + \sqrt{1-s^2}\,Y\right)\right] ds.$$

(HINT: Check it first for $f(x) := \exp(itx)$ and $g(x) := \exp(i\tau x)$.)

(2) Conclude the "Poincaré inequality" of Nash (1958):

$$\mathrm{Var} f(X) \le \|f'(X)\|_2^2.$$

7.43 (Problem 7.18, continued; Harder). Prove that the uniform distribution on $(0,1)$ is not infinitely divisible. (HINT: $\widehat{\mu} \ne (\widehat{\nu})^3$. Simpler derivations exist, but depend on more advanced Fourier-analytic methods.)

7.44 (Harder). Suppose $\{X_i\}_{i=1}^n$ are i.i.d. mean-zero variance-σ^2 random variables such that $\mathrm{E}\{|X_1|^{2+\rho}\} < \infty$ for some $\rho \in (0,1)$. Then prove that there exists a constant A, independent of n, such that

(7.133) $$|\mathrm{E}g(S_n/\sqrt{n}) - \mathrm{E}g(N(0, \sigma^2))| \le \frac{A}{n^{\rho/2}},$$

provided that g has three bounded and continuous derivatives.

Notes

(1) The term "central limit theorem" seems to be due to Pólya (1920). Our treatment covers only the beginning of a rich and well-developed theory (Lévy, 1937; Feller, 1966; Gnedenko and Kolmogorov, 1968).

(2) The present form of the CLT is due to Lindeberg (1922). See also Problem 7.38 on page 115. Zabell (1995) discusses the independent discovery of the Lindeberg CLT (1922) by the nineteen-year-old Turing (1934). See also Note (8) below.

(3) Fejér's Theorem (p. 98) appeared in 1900. Tandori (1983) discusses the fascinating history of the problem, as well as the life of Fejér.

(4) Equation (7.40) is sometimes referred to as the *Parseval identity*, named after M.-A. Parseval des Chénes for his 1801 discovery of a discrete version of (7.40) in the context of Fourier series.

(5) For an amusing consequence of Problem 7.4 plug in $x = \pi/2$ and solve to obtain the 1593 *Viéte formula* for computing π:

$$\pi = 2 \left[\frac{\sqrt{2}}{2} \frac{\sqrt{2+\sqrt{2}}}{2} \frac{\sqrt{2+\sqrt{2+\sqrt{2}}}}{2} \frac{\sqrt{2+\sqrt{2+\sqrt{2+\sqrt{2}}}}}{2} \cdots \right]^{-1}.$$

(6) Lévy (1925, p. 195) has found the following stronger version of the convergence theorem: "*If $L(t) = \lim_n \widehat{\mu_n}(t)$ exists and is continuous in a neighborhood of $t = 0$, then there exists a probability measure μ such that $L = \widehat{\mu}$ and $\mu_n \Rightarrow \mu$.*" Lévy's argument was simplified by Glivenko (1936).

(7) The term "projective CLT" is non-standard. Kac (1956, p. 182, fn. 7) states that this result "is due to Maxwell but is often ascribed to Borel." See also Kac (1939, p. 728), as well as Problem 7.39 above. The mentioned attribution of Kac seems to agree with that of Borel (1925, p. 92). For a historical survey see the final section of Diaconis and Freedman (1987), as well as Stroock and Zeitouni (1991, Introduction).

(8) The term "Liapounov replacement method" is non-standard. Many authors ascribe this method incorrectly to Lindeberg (1922). Lindeberg used the replacement method in order to deduce the modern-day statement of the CLT.

Trotter (1959) devised a fixed-point proof of the Lindeberg CLT. His proof can be viewed as a translation—into the langauge of analysis—of the replacement method of Liapounov. In this regard see also Hamedani and Walter (1984).

(9) Cramér's theorem (p. 107) is intimately connected to general central limit theory (Gnedenko and Kolmogorov, 1968; Lévy, 1937). The original proof of Cramér's theorem uses hard analytic-function theory. The ascription in Lemma 7.32 comes from Veech (1967, Lemma 7.1, p. 183).

(10) Problem 7.5 goes at least as far back as 1941; see the collected works of Bernšteĭn (1964, pp. 314–315).

(11) Problem 7.41 is borrowed from Ville (1943).

(12) Problem 7.42 is due to Nash (1958), and plays a key role in his estimate for the solution to the Dirichlet problem. The elegant method outlined here is due to Houdré, Pérez-Abreu, and Surgailis (1998).

Chapter 8

Martingales

One can measure the importance of a scientific work by the number of earlier publications rendered superfluous by it.

–David Hilbert

If X has two finite moments and $\mu = \mathrm{E}X$, then we can check easily that $\mathrm{E}[(X-x)^2] = \mathrm{Var}X + (\mu - x)^2$. Consequently, $\inf_{x \in \mathbf{R}} \mathrm{E}[(X-x)^2] = \mathrm{Var}X$, and the infimum is achieved uniquely at $x = \mu$. In other words, μ is the best predictor of the value of X, where "best" is meant in the sense of $L^2(\mathrm{P})$.

For example, imagine a two-person game wherein Player 1 tosses a fair coin 100 times independently. Player 2, who is unaware of the outcome of the coin tosses, is supposed to guess the total number X of heads. Because $X = \mathrm{Bin}(100, 1/2)$, $\mu = \mathrm{E}X = 50$. The preceding paragraph then asserts that Player 2 should predict X to be its expectation; i.e., "$\mu = 50$ heads."

Now suppose that Player 2 is aware of the outcome of the first 10 tosses (say). Let Y denote the total number of heads in the first 10 tosses. Then clearly Player 2's best possible prediction of X is "$Y + 45$ heads." This is the conditional expectation of X given the value of Y, and "conditional expectations" is the topic with which we begin the chapter.

1. Conditional Expectations

1.1. Abstract Conditioning. The entire subject of conditioning emanates from the following abstract result. Throughout, let $(\Omega, \mathscr{F}, \mathrm{P})$ be a probability space, and $\mathscr{G}$ a sub-σ-algebra of $\mathscr{F}$ (i.e., $\mathscr{G} \subseteq \mathscr{F}$).

Proposition 8.1. *If $X \in L^1(P)$, then there exists an a.s.-unique random variable $\Pi_{\mathscr{G}} X$ that: (i) is integrable; (ii) is $\mathscr{G}$-measurable; and (iii) satisfies $E[\xi X] = E[\xi \Pi_{\mathscr{G}} X]$ for all bounded, $\mathscr{G}$-measurable random variables ξ.*

We should pay close attention to the seemingly innocuous property (ii). For instance, note that if X is $\mathscr{G}$-measurable, then (ii) implies that $X = \Pi_{\mathscr{G}} X$ a.s. (why?). In particular, X is always equal to $\Pi_{\mathscr{F}} X$ a.s.

Definition 8.2. Hereafter, the random variable $\Pi_{\mathscr{G}} X$ of Proposition 8.1 is written as $E[X \mid \mathscr{G}]$, and called the *conditional expectation of X given $\mathscr{G}$*. For all random variables $Y_1, \ldots, Y_m$ define $\sigma(\{Y_i\}_{i=1}^m) = \sigma(Y_1, \ldots, Y_m)$ to be the σ-algebra generated by $Y_1, \ldots, Y_m$. Frequently, we write $E[X \mid Y_1, \ldots, Y_m]$ in place of $E[X \mid \sigma(Y_1, \ldots, Y_m)]$.

Proof. Let $\nu(A) := E[X; A] = \int_A X \, dP$ for all $A \in \mathscr{F}$. This defines a finite measure on $\mathscr{F}$ that is absolutely continuous with respect to P. We claim that for all bounded and $\mathscr{F}$-measurable random variables Z,

$$(8.1) \qquad \int Z \, d\nu = \int ZX \, dP.$$

It suffices to prove this for $Z \geq 0$; else, consider Z^+ and Z^- separately. Now (8.1) holds tautologically if $Z = \mathbf{1}_A$ for some $A \in \mathscr{G}$. So it holds also for a simple $\mathscr{G}$-measurable function Z. The rest of the claim follows from the monotone convergence theorem.

Since $\mathscr{G} \subseteq \mathscr{F}$, both ν and P are finite measures on $\mathscr{G}$. Because $\nu \ll P$, the Radon–Nikodým theorem (p. 47) assures us that there exists a $\Pi_{\mathscr{G}} X \in L^1(\Omega, \mathscr{G}, P)$ such that $\int \xi \, d\nu = \int \xi \Pi_{\mathscr{G}} X \, dP$ for all bounded, $\mathscr{G}$-measurable ξ. By the very definition of L^1-spaces, $\Pi_{\mathscr{G}} X$ is $\mathscr{G}$-measurable; it is also unique a.s. [P]. Thanks to (8.1), $\int \xi X \, dP = \int \xi \Pi_{\mathscr{G}} X \, dP$ for all bounded $\mathscr{G}$-measurable random variables ξ, and Proposition 8.1 follows. □

Abstract conditioning will become clearer the more we use conditional expectations. The following theorem and its proof together form a first step in this direction.

Theorem 8.3. *Conditional expectations have the following properties:*

(1) *If $X \geq 0$ a.s. then $E[X \mid \mathscr{G}] \geq 0$ a.s. Also, $E[E(X \mid \mathscr{G})] = EX$, $E[X \mid \mathscr{F}] = X$, and $E(X \mid \{\varnothing, \Omega\}) = EX$ a.s.*

(2) *If $X_1, X_2, \ldots, X_n \in L^1(P)$ and $a_1, a_2, \ldots, a_n \in \mathbf{R}$, then a.s.,*

$$E\left[\sum_{j=1}^n a_j X_j \,\bigg|\, \mathscr{G}\right] = \sum_{j=1}^n a_j E[X_j \mid \mathscr{G}].$$

1. Conditional Expectations

(3) If Z is $\mathscr{G}$-measurable and $ZX \in L^1(\mathrm{P})$, then $\mathrm{E}[ZX \,|\, \mathscr{G}]$ a.s. exists and is equal to $Z\mathrm{E}[X \,|\, \mathscr{G}]$.

(4) (Conditional Jensen) If $\psi : \mathbf{R} \to \mathbf{R}$ is convex and $\psi(X) \in L^1(\mathrm{P})$, then with probability one,
$$\mathrm{E}[\psi(X) \,|\, \mathscr{G}] \geq \psi(\mathrm{E}[X \,|\, \mathscr{G}]).$$

(5) (Conditional Fatou) If $\{X_i\}_{i=1}^{\infty}$ are integrable and non-negative, then with probability one,
$$\mathrm{E}\left[\liminf_{n \to \infty} X_n \,\bigg|\, \mathscr{G}\right] \leq \liminf_{n \to \infty} \mathrm{E}[X_n \,|\, \mathscr{G}].$$

(6) (Conditional Bounded Convergence) If $\{X_i\}_{i=1}^{\infty}$ are bounded and a.s.-convergent, then with probability one,
$$\mathrm{E}\left[\lim_{n \to \infty} X_n \,\bigg|\, \mathscr{G}\right] = \lim_{n \to \infty} \mathrm{E}[X_n \,|\, \mathscr{G}].$$

(7) (Conditional Monotone Convergence) If $X_1 \leq X_2 \leq X_3 \leq \cdots$ are all in $L^1(\mathrm{P})$, then with probability one,
$$\mathrm{E}[X_n \,|\, \mathscr{G}] \nearrow \mathrm{E}\left[\lim_{n \to \infty} X_n \,\bigg|\, \mathscr{G}\right] \quad \text{as } n \to \infty.$$

(8) (Conditional Dominated Convergence) If $\mathrm{E}\{\sup_n |X_n|\} < \infty$ and $\lim_{n \to \infty} X_n$ exists a.s., then with probability one,
$$\mathrm{E}\left[\lim_{n \to \infty} X_n \,\bigg|\, \mathscr{G}\right] = \lim_{n \to \infty} \mathrm{E}[X_n \,|\, \mathscr{G}].$$

(9) (Conditional Hölder) Suppose $X \in L^p(\mathrm{P})$ for some $p > 1$ and $Y \in L^q(\mathrm{P})$ where $p^{-1} + q^{-1} = 1$. Then with probability one,
$$\left|\mathrm{E}\left[XY \,|\, \mathscr{G}\right]\right| \leq (\mathrm{E}\{|X|^p \,|\, \mathscr{G}\})^{1/p} (\mathrm{E}\{|Y|^q \,|\, \mathscr{G}\})^{1/q}.$$

(10) (Conditional Minkowski) If $X, Y \in L^p(\mathrm{P})$ for some $p \geq 1$, then with probability one,
$$(\mathrm{E}\{|X+Y|^p \,|\, \mathscr{G}\})^{1/p} \leq (\mathrm{E}\{|X|^p \,|\, \mathscr{G}\})^{1/p} + (\mathrm{E}\{|Y|^p \,|\, \mathscr{G}\})^{1/p}.$$

Proof. In order to save space we write $\Pi_{\mathscr{H}} X$ in place of $\mathrm{E}[X \,|\, \mathscr{H}]$ for any σ-algebra $\mathscr{H}$.

Step 1. Verification of Part (1). First suppose $X \geq 0$, and for all $n \geq 0$ define $A_n := \{\Pi_{\mathscr{G}} X \leq -1/n\}$. Because A_n is $\mathscr{G}$-measurable,

(8.2) $$\mathrm{E}[X; A_n] = \mathrm{E}\left[\Pi_{\mathscr{G}} X; \, A_n\right].$$

The left-hand side is non-negative whereas the right-hand side is $\leq -\mathrm{P}(A_n)/n$. It follows that $\mathrm{P}(A_n) = 0$ for all $n \geq 1$. By monotonicity, $\mathrm{P}(\cup_{n \geq 1} A_n) = \lim_{n \to \infty} \mathrm{P}(A_n) = 0$. Because $\cup_{n \geq 1} A_n = \{\Pi_{\mathscr{G}} X < 0\}$, this proves that if $\mathrm{P}\{X \geq 0\} = 1$ then $\Pi_{\mathscr{G}} X \geq 0$ a.s.

Note that $EX = E[X;\Omega] = E[\Pi_{\mathscr{G}}X;\Omega] = E[\Pi_{\mathscr{G}}X]$, and $E[\xi X] = E[\xi\Pi_{\mathscr{G}}X]$ for all bounded random variables ξ. Because $\Pi_{\mathscr{G}}X$ and X are both $\mathscr{F}$-measurable, the uniqueness portion of Proposition 8.1 implies that $X = \Pi_{\mathscr{G}}X$ a.s. [This fact came up earlier, just after Definition 8.2.]

Finally, $\Pi_{\{\varnothing,\Omega\}}X$ is a non-random constant a.s. because it is measurable with respect to the trivial σ-algebra $\{\varnothing,\Omega\}$. It follows that $\Pi_{\{\varnothing,\Omega\}}X = E(\Pi_{\{\varnothing,\Omega\}}X) = EX$ a.s. This proves part (1).

Step 2. Verification of Part (2). By induction it suffices to assume that $n = 2$. For all bounded $\mathscr{G}$-measurable random variables ξ,

$$
\begin{aligned}
E\left[\xi\Pi_{\mathscr{G}}\left(a_1X_1 + a_2X_2\right)\right] &= E\left[\xi\left(a_1X_1 + a_2X_2\right)\right] \\
&= a_1 E[\xi X_1] + b_1 E[\xi X_2] \\
&= a_1 E\left[\xi\Pi_{\mathscr{G}}X_1\right] + a_2 E\left[\xi\Pi_{\mathscr{G}}X_2\right] \\
&= E\left[\xi\left(a_1\Pi_{\mathscr{G}}X_1 + a_2\Pi_{\mathscr{G}}X_2\right)\right].
\end{aligned}
\tag{8.3}
$$

Since $\Pi_{\mathscr{G}}(a_1X_1+a_2X_2)$ and $a_1\Pi_{\mathscr{G}}X_1+a_2\Pi_{\mathscr{G}}X_2$ are both $\mathscr{G}$-measurable, they must be equal a.s.

Step 3. Verification of Part (3). Thanks to part (2) we can assume, without loss of generality, that $Z, X \geq 0$. Else, we can consider $X^{\pm}$ and $Z^{\pm}$.

Our goal is to prove that $E[Z\xi X] = E[Z\xi\Pi_{\mathscr{G}}X]$ for all bounded $\mathscr{G}$-measurable ξ. If Z is bounded then $Z\xi$ is a bounded $\mathscr{G}$-measurable random variable and we are done. For the general case let $Z_n := \min(Z,n)$. Then each Z_n is bounded and $\mathscr{G}$-measurable, and hence $E[Z_n\xi X] = E[Z_n\xi\Pi_{\mathscr{G}}X]$. Because X and $\Pi_{\mathscr{G}}X$ are both non-negative (part (1)) we can write $\xi = \xi^+ + \xi^-$, take $n \to \infty$, and appeal to the monotone convergence theorem to finish.

Step 4. Conclusion. Parts (1) through (3) show that conditional expectations have all the salient features of (ordinary, unconditional) expectations. The rest of the proof follows by mimicking integration theory arguments. □

1.2. Conditioning and Prediction. Let X denote the outcome of an experiment which has not yet been performed. Intuitively speaking, $E[X \,|\, \mathscr{G}]$ denotes our "best prediction" of the as-yet-unseen value of X, given that we know the values of all $\mathscr{G}$-measurable random variables. By using this "prediction interpretation" we can give heuristic justifications for the various properties of conditional expectations. For example:

- $E[X \,|\, Y]$ is our best prediction of X if we know the value of Y.
- $E[X \,|\, \mathscr{F}] = X$ a.s. because if we know all $\mathscr{F}$-measurable random variables, then we certainly know the value of X.
- $E[X \,|\, \{\varnothing,\Omega\}]$ is our best prediction of X given that we know nothing. This is because there are no truly random (i.e., non-constant)

1. Conditional Expectations

random variables that are measurable with respect to $\{\varnothing, \Omega\}$. In the first paragraph of this chapter we saw already that $\mathrm{E}X$ is our best prediction of X given no knowledge. Therefore, $\mathrm{E}X = \mathrm{E}[X \mid \{\varnothing, \Omega\}]$ a.s.

- If ξ is $\mathscr{G}$-measurable, then the knowledge of $\mathscr{G}$ renders ξ a constant. Therefore, $\mathrm{E}[\xi X \mid \mathscr{G}] = \xi \mathrm{E}[X \mid \mathscr{G}]$ a.s.

And so on. Here is why the prediction interpretation works.

Proposition 8.4. *If* $\mathrm{E}[X^2] < \infty$, *then for all $\mathscr{G}$-measurable* $\xi \in L^2(\mathrm{P})$,

$$(8.4) \qquad \mathrm{E}\left[(X - \mathrm{E}[X \mid \mathscr{G}])^2\right] \leq \mathrm{E}\left[(X - \xi)^2\right].$$

The inequality is an equality if and only if $\xi = \mathrm{E}[X \mid \mathscr{G}]$ *a.s.*

Proof. Thanks to conditional Jensen's inequality (part (4) of Theorem 8.3) $\mathrm{E}[X \mid \mathscr{G}]$ has two finite moments. In particular, the expectations here are all finite.

Because $\mathrm{E}[X \mid \mathscr{G}] - \xi$ is $\mathscr{G}$-measurable, we can "pull it out" of the conditional expectation to find that

$$(8.5) \qquad \begin{aligned}&\mathrm{E}\left[(X - \mathrm{E}[X \mid \mathscr{G}])\left(\mathrm{E}[X \mid \mathscr{G}] - \xi\right) \,\Big|\, \mathscr{G}\right] \\ &= (\mathrm{E}[X \mid \mathscr{G}] - \xi)\, \mathrm{E}\left[(X - \mathrm{E}[X \mid \mathscr{G}]) \,\Big|\, \mathscr{G}\right] \qquad \text{a.s.,}\end{aligned}$$

and this is zero. Consequently, $\mathrm{E}[(X - \xi)^2 \mid \mathscr{G}]$ is a.s. equal to $\mathrm{E}[(X - \mathrm{E}\{X \mid \mathscr{G}\})^2 \mid \mathscr{G}] + \mathrm{E}[(\mathrm{E}\{X \mid \mathscr{G}\} - \xi)^2 \mid \mathscr{G}]$. Take expectations and appeal to part (1) of Theorem 8.3 to derive the following "parallelogram law":

$$(8.6) \qquad \mathrm{E}\left[(X - \xi)^2\right] = \mathrm{E}\left[(X - \mathrm{E}[X \mid \mathscr{G}])^2\right] + \mathrm{E}\left[(\xi - \mathrm{E}[X \mid \mathscr{G}])^2\right].$$

This proves the proposition. $\square$

Our newly found intuition about conditional expectations might suggest that if X and $\mathscr{G}$ are independent, then $\mathrm{E}[X \mid \mathscr{G}] = \mathrm{E}X$ a.s. [After all, knowing $\mathscr{G}$ does not teach us anything about X.] This is indeed the case.

Theorem 8.5. *Suppose* $X \in L^1(\mathrm{P})$ *and* $\mathscr{G}, \mathscr{G}_1, \mathscr{G}_2$ *are three sub-σ-algebras of* $\mathscr{F}$. *If X is independent of $\mathscr{G}$, then* $\mathrm{E}[X \mid \mathscr{G}] = \mathrm{E}X$ *a.s. If $\mathscr{G}_1 \subseteq \mathscr{G}_2$, then*

$$(8.7) \qquad \mathrm{E}\left[\mathrm{E}(X \mid \mathscr{G}_2) \,\Big|\, \mathscr{G}_1\right] = \mathrm{E}\left[\mathrm{E}(X \mid \mathscr{G}_1) \,\Big|\, \mathscr{G}_2\right] = \mathrm{E}\left[X \mid \mathscr{G}_1\right] \qquad a.s.$$

Equation (8.7) is known as the *towering property of conditional expectations*. Intuitively speaking, it asserts that we can use only the lesser amount of available information to predict X. We emphasize that the condition $\mathscr{G}_1 \subseteq \mathscr{G}_2$ cannot be dropped in general.

Proof. All bounded, $\mathscr{G}$-measurable random variables ξ are independent of X by default. Therefore, $\mathrm{E}[\xi X] = \mathrm{E}\xi\,\mathrm{E}X = \mathrm{E}[\xi\mathrm{E}(X)]$. Because $W = \mathrm{E}X$ is $\mathscr{G}$-measurable, the uniqueness portion of Proposition 8.1 implies that $\mathrm{E}X = \mathrm{E}[X\,|\,\mathscr{G}]$ a.s.

The second part of the theorem contains two assertions. For the first note that $\mathrm{E}[X\,|\,\mathscr{G}_1]$ is $\mathscr{G}_1$- and hence $\mathscr{G}_2$-measurable. Therefore, by Theorem 8.3, $\mathrm{E}[\mathrm{E}(X\,|\,\mathscr{G}_1)\,|\,\mathscr{G}_2] = \mathrm{E}[X\,|\,\mathscr{G}_1]$ a.s.

For the remainder of the proof let ξ be a bounded $\mathscr{G}_1$-measurable random variable. Because $\mathscr{G}_1 \subseteq \mathscr{G}_2$, ξ is $\mathscr{G}_2$-measurable too. Therefore,

$$(8.8) \qquad \mathrm{E}\left[\xi\mathrm{E}\left(X\,|\,\mathscr{G}_2\right)\right] = \mathrm{E}\left[\xi X\right] = \mathrm{E}\left[\xi\mathrm{E}(X\,|\,\mathscr{G}_1)\right].$$

This has the desired effect. □

Let us return briefly to the preamble to this chapter and analyze the two-person game mentioned there. Recall that Player 1 tosses a fair coin 100 times independently. Player 2 knows the value of Y, which is the number of heads in the first 10 tosses. If X denotes the total number of heads tossed by Player 1, then $X - Y$ and Y are independent. Therefore, $\mathrm{E}[X\,|\,Y] = \mathrm{E}[X - Y\,|\,Y] + \mathrm{E}[Y\,|\,Y] = \mathrm{E}[X - Y] - Y$. Since $X - Y = \mathrm{Bin}(90, 1/2)$, $\mathrm{E}[X - Y] = 45$, whence $\mathrm{E}[X\,|\,Y] = 45 + Y$ a.s. This calculation was arrived at earlier, but the argument lacked rigorous justification.

1.3. From Classical to Abstract Conditional Expectation.

Definition 8.6. If $X \in L^1(\mathrm{P})$ and B is an event of positive probability, then *the (classical) conditional expectation of X given B is defined as*

$$(8.9) \qquad \mathrm{E}[X\,|\,B] := \frac{\mathrm{E}[X;B]}{\mathrm{P}(B)} = \frac{1}{\mathrm{P}(B)}\int_B X\,d\mathrm{P}.$$

In general, abstract conditional expectations are genuine random variables, whereas classical conditional expectations are constants. But the two notions are related, as we will see next.

Proposition 8.7. *Choose and fix $X \in L^1(\mathrm{P})$, and let $B_1, \ldots, B_n$ be disjoint events of positive probability such that $\cup_{i=1}^n B_i = \Omega$. Recall that $\sigma(\{B_i\}_{i=1}^n)$ denotes the σ-algebra generated by the B_i's. Then,*

$$(8.10) \qquad \mathrm{E}\left[X\,\Big|\,\sigma\left(\{B_i\}_{i=1}^n\right)\right](\omega) = \sum_{i=1}^n \mathbf{1}_{B_i}(\omega)\mathrm{E}\left[X\,|\,B_i\right] \qquad a.s.$$

Proof. It suffices to consider the case where X is bounded (Theorem 8.3).

The σ-algebra $\sigma(\{B_i\}_{i=1}^n)$ is generated by the random variables $\{\mathbf{1}_{B_i}\}_{i=1}^n$. Equivalently, $\sigma(\{B_i\}_{i=1}^n)$ is generated by all random variables of the form

$\sum_{i=1}^{n} c_i \mathbf{1}_{B_i}$, where $c_1, \ldots, c_n \in \mathbf{R}$ are non-random constants. Define

$$(8.11) \qquad \Psi(c) := \mathrm{E}\left[\left(X - \sum_{i=1}^{n} c_i \mathbf{1}_{B_i}\right)^2\right] \qquad \forall c := (c_1, \ldots, c_n) \in \mathbf{R}^n.$$

It is possible to check that Ψ attains its unique minimum when $c_i = \mathrm{E}[X \,|\, B_i]$ ($i = 1, \ldots, n$). Therefore, Proposition 8.4 finishes the proof. □

We can apply the preceding to compute certain conditional expectations. Indeed, suppose Y is a discrete, possibly abstract-valued, random variable that takes the values $y_1, \ldots, y_n$ with positive probability each, and no other values are possible. Apply Proposition 8.7 with $B_i := \{Y = y_i\}$ ($i = 1, \ldots, n$) to deduce the following formula; see also Problem 8.8.

Corollary 8.8. *For all $X \in L^1(\mathrm{P})$, and for Y as above,*

$$(8.12) \qquad \mathrm{E}\left[X \,|\, Y\right] = \sum_{i=1}^{n} \mathbf{1}_{\{Y=y_i\}} \mathrm{E}\left[X \,|\, Y = y_i\right] \qquad a.s.$$

Equivalently, for all $i \le n$, $\mathrm{E}[X \,|\, Y] = \mathrm{E}[X \,|\, Y = y_i]$ a.s. on $\{Y = y_i\}$. Alternatively, if we define $\phi(y) := \mathrm{E}[Y \,|\, Y = y]$, then $\mathrm{E}[X \,|\, Y] = \phi(Y)$ a.s. Note that we are careful to not write $\mathrm{E}[X \,|\, Y] = \mathrm{E}[X \,|\, Y = Y]$. [In fact, $\mathrm{E}[X \,|\, Y = Y] = \mathrm{E}X$. Why?]

1.4. Conditional Probabilities. Conditional probabilities follow readily from conditional expectations via the assignment

$$(8.13) \qquad \mathrm{P}(A \,|\, \mathscr{G}) = \mathrm{E}[\mathbf{1}_A \,|\, \mathscr{G}] \qquad \forall A \in \mathscr{F}.$$

Their salient properties are not hard to derive, and are listed in the following:

Proposition 8.9. *For every sub-σ-algebra $\mathscr{G} \subseteq \mathscr{F}$:*

(i) $\mathrm{P}(\varnothing \,|\, \mathscr{G}) = 0$ a.s.

(ii) $\mathrm{P}(A \,|\, \mathscr{G}) = 1 - \mathrm{P}(A^c \,|\, \mathscr{G})$ a.s. for all $A \in \mathscr{F}$.

(iii) *For all disjoint measurable sets $A_1, A_2, \ldots$ there exists a null set outside which $\mathrm{P}(\cup_{i=1}^{\infty} A_i \,|\, \mathscr{G}) = \sum_{i=1}^{\infty} \mathrm{P}(A_i \,|\, \mathscr{G})$.*

Example 8.10. Consider an event A, and an event B such that $0 < \mathrm{P}(B) < 1$. Then, according to Corollary 8.8, the random variable $\mathrm{P}(A \,|\, \sigma(B))$ is equal to $\mathrm{P}(A \,|\, B)$ on B and $\mathrm{P}(A \,|\, B^c)$ on B^c. In this way, we can relate abstract conditional probabilities to the classical ones of Chapter 1 (§2, p. 4).

2. Filtrations and Semi-Martingales

Definition 8.11. A *stochastic process* (or a random process, or a process) is a collection of random variables. If $\mathscr{F}_1 \subseteq \mathscr{F}_2 \subseteq \cdots$ are sub-σ-algebras of $\mathscr{F}$, then $\{\mathscr{F}_n\}_{n=1}^\infty$ is a *filtration*. A process $\{X_n\}_{n=1}^\infty$ is *adapted* to a filtration $\{\mathscr{F}_n\}_{n=1}^\infty$ if X_n is $\mathscr{F}_n$-measurable for every $n \geq 1$.

Suppose we start with a random process of our choice, call it $\{X_i\}_{i=1}^\infty$, and *define* $\mathscr{F}_n := \sigma(\{X_i\}_{i=1}^n)$. Clearly, $\mathscr{F}_n \subseteq \mathscr{F}_{n+1}$, and by definition $\{X_n\}_{n=1}^\infty$ is adapted to $\{\mathscr{F}_n\}_{n=1}^\infty$. Suppose, in addition, that $X_n \in L^1(\mathrm{P})$ for all $n \geq 1$. We can think of $\{X_i\}_{i=1}^\infty$ as a random process that evolves in (discrete) time. Then, a sensible prediction of the value of the process at time $n+1$, given the values of the process by time n, is $\mathrm{E}[X_{n+1} \,|\, \mathscr{F}_n]$. We say that $X = \{X_i\}_{i=1}^\infty$ is a martingale if this predicted value is X_n. In this way, you should convince yourself that fair games are martingales, and in a sense, the converse is also true.

Definition 8.12. A stochastic process $X = \{X_n\}_{n=1}^\infty$ is a *submartingale* with respect to a filtration $\mathscr{F} = \{\mathscr{F}_n\}_{n=1}^\infty$ if:

(i) X is adapted to $\mathscr{F}$.

(ii) $X_n \in L^1(\mathrm{P})$ for all $n \geq 1$.

(iii) For each $n \geq 1$, $\mathrm{E}[X_{n+1} \,|\, \mathscr{F}_n] \geq X_n$ a.s.

The process X is a *supermartingale* if $-X$ is a submartingale. It is a *martingale* if it is both a sub- and a supermartingale; it is a *semi-martingale* if it can be written as $X_n = Y_n + Z_n$ where $\{Y_i\}_{i=1}^\infty$ is a martingale and $\{Z_i\}_{i=1}^\infty$ is a *bounded-variation process*; i.e., $Z_n = U_n - V_n$ where $U_1 \leq U_2 \leq \cdots$ and $V_1 \leq V_2 \leq \cdots$ are integrable adapted processes.

Occasionally, we call a finite sequence $\{X_i\}_{i=1}^n$ a submartingale if: (i) X_i is $\mathscr{F}_i$-measurable for all $1 \leq i \leq n$; (ii) $X_i \in L^1(\mathrm{P})$ for $1 \leq i \leq n$; and (iii) $\mathrm{E}[X_{i+1} \,|\, \mathscr{F}_i] \geq X_i$ for all $1 \leq i < n$. A similar remark applies to super- and semi-martingales.

Here are a few examples of martingales.

Example 8.13 (Independent Sums). Suppose that a fair game is played repeatedly. Every game is independent of all others, and results in ± 1 dollar for the gambler. We can model this by letting $\{X_i\}_{i=1}^\infty$ be i.i.d. random variables with the values ± 1 with probability one-half each. We think of X_i as the gambler's win from game i, where negative win means loss. In this way, the gambler's cumulative fortune after k games is $S_k := X_1 + \cdots X_k$. By independence, $\mathrm{E}[X_k \,|\, X_1, \ldots, X_{k-1}] = \mathrm{E}X_k = 0$. Therefore, S is a martingale with respect to the filtration $\{\mathscr{F}_i\}_{i=1}^\infty$ where $\mathscr{F}_n := \sigma(\{X_i\}_{i=1}^n)$. More generally still, if $S_n = X_1 + \cdots + X_n$, where the X_j's are independent

2. Filtrations and Semi-Martingales

(not necessarily i.i.d.) and have mean zero, then S is a martingale with respect to $\mathscr{F}$. Check that $\mathscr{F}_n$ is also equal to $\sigma(\{S_i\}_{i=1}^n)$.

For our second class of examples, we need a definition.

Definition 8.14. A stochastic process $\{A_i\}_{i=1}^\infty$ is *previsible* with respect to a given filtration $\{\mathscr{F}_i\}_{i=1}^\infty$ if A_n is $\mathscr{F}_{n-1}$-measurable for every $n \geq 1$, where $\mathscr{F}_0$ always denotes the minimal σ-algebra $\{\varnothing, \Omega\}$.

Example 8.15 (Martingale Transforms). Let $\{S_n\}_{n=1}^\infty$ be a martingale with respect to a filtration $\{\mathscr{F}_n\}_{n=1}^\infty$. Define $S_0 := 0$, y_0 a constant, and A a previsible process with respect to $\mathscr{F}$. Now consider the process Y defined by

$$(8.14) \qquad Y_n := y_0 + \sum_{j=1}^n A_j(S_j - S_{j-1}) \qquad \forall n \geq 0.$$

The process Y is called the *martingale transform* of S. It is a straightforward task to prove that Y is a martingale. Here is an example of how Y arises naturally: Suppose we play a fair game repeatedly and independently every time. Let X_i denote the amount of win/loss for the ith play of the game, so that $S_n := X_1 + \cdots + X_n$ denotes the total win/loss by the nth play. Of course, S is a martingale (Example 8.13). Now suppose we can bet A_n dollars on the nth play; we are allowed to choose A_n based on what we have seen so far. That is, at time n, we are privy to the values of $X_1, \ldots, X_{n-1}$. Then, the martingale transform $Y_n = \sum_{i=1}^n A_i X_i = \sum_{i=1}^n A_i(S_i - S_{i-1})$ describes the win/loss after the nth play. Note that the martingale transform of (8.14) has the equivalent definition, $Y_{n+1} - Y_n = A_{n+1}(S_{n+1} - S_n)$ for $n \geq 0$, where $Y_0 = y_0$. We may think of this, informally, as the discrete analogue of the "stochastic differential identity," $dY = A\,dS$.

Example 8.16 (Doob Martingales). Let $\mathscr{F}$ be a filtration and $Y \in L^1(\mathrm{P})$. Then martingales of the form $X_n := \mathrm{E}[Y \,|\, \mathscr{F}_n]$ are called *Doob martingales*.

Lemma 8.17. *If X is a submartingale with respect to a filtration $\mathscr{F}$, then it is also a submartingale with respect to the filtration generated by X itself. That is, for all n, $\mathrm{E}[X_{n+1} \,|\, X_1, \ldots, X_n] \geq X_n$ a.s.*

Proof. For all n and all $A \in \mathscr{B}(\mathbf{R})$, $X_n^{-1}(A) \in \mathscr{F}_n$. Because $\sigma(X_1, \ldots, X_n)$ is the smallest σ-algebra that contains $X_n^{-1}(A)$ for all $A \in \mathscr{B}(\mathbf{R})$, it follows that $\sigma(X_1, \ldots, X_n) \subseteq \mathscr{F}_n$ for all n. Consequently, by the towering property of conditional expectations (Theorem 8.5), with probability one,

$$(8.15) \qquad \begin{aligned} \mathrm{E}[X_{n+1} \,|\, X_1, \ldots, X_n] &= \mathrm{E}\Big[\mathrm{E}(X_{n+1} \,|\, \mathscr{F}_n) \,\Big|\, X_1, \ldots, X_n\Big] \\ &\geq \mathrm{E}[X_n \,|\, X_1, \ldots, X_n] = X_n. \end{aligned}$$

The last equality is a consequence of Theorem 8.3. $\square$

Lemma 8.18. *If X is a martingale and ψ is convex, then $\psi(X)$ is a submartingale, provided that $\psi(X_n) \in L^1(P)$ for all n. If X is a submartingale and ψ is a nondecreasing convex function, then $\psi(X)$ is a submartingale, provided that $\psi(X_n) \in L^1(\mathrm{P})$ for all n.*

Proof. Thanks to the conditional form of Jensen's inequality (Theorem 8.3), $\mathrm{E}[\psi(X_{n+1}) \mid \mathscr{F}_n] \geq \psi(\mathrm{E}[X_{n+1} \mid \mathscr{F}_n])$ a.s. This holds for any process X and any convex function ψ as long as $\psi(X_n) \in L^1(\mathrm{P})$.

If X is a martingale, then $\psi(\mathrm{E}[X_{n+1} \mid \mathscr{F}_n]) = \psi(X_n)$ a.s., whence follows the result.

If, in addition, ψ is nondecreasing but X is a submartingale, then $\psi(\mathrm{E}[X_{n+1} \mid \mathscr{F}_n]) \geq \psi(X_n)$ a.s., which has the desired result. □

Remark 8.19. If X is a martingale, then X^+, $|X|^p$, and e^X are submartingales, provided that they are integrable at each time n. If X is a submartingale then X^+ and e^X are also submartingales as long as they are integrable. However, one can construct a submartingale whose absolute value is not a submartingale; e.g., consider $X_k := -1/k$.

The definition of semi-martingales is motivated by the following, whose proof is as interesting as the fact itself:

Doob's Decomposition. *Any submartingale X can be written as $X_n = Y_n + Z_n$, where Y is a martingale, and Z is a non-negative previsible a.s.-increasing process with $Z_n \in L^1(\mathrm{P})$ for all n. In particular, sub- and supermartingales are semi-martingales, and any semi-martingale can be written as the difference of a sub- and a supermartingale.*

Proof. Define $X_0 := 0$ and $d_j := X_j - X_{j-1}$ ($j = 1, 2, \ldots$), so that $X_n := \sum_{j=1}^n d_j$. Let $Z_n := \sum_{j=1}^n \mathrm{E}[d_j \mid \mathscr{F}_{j-1}]$ and $Y_n := \sum_{j=1}^n (d_j - \mathrm{E}[d_j \mid \mathscr{F}_{j-1}])$. A direct computation reveals that this yields the promised decomposition. □

The preceding is one among many decomposition theorems for semi-martingales. Next is the decomposition theorem of Krickeberg (1963, Satz 33, p. 131). See also Krickeberg (1965, Theorem 33, p. 144). Before introducing it however, we need a brief definition.

Definition 8.20. $\{X_i\}_{i=1}^\infty$ *is bounded in $L^p(\mathrm{P})$ if $\sup_n \|X_n\|_p < \infty$.*

Krickeberg's Decomposition. *Suppose X is a submartingale which is bounded in $L^1(\mathrm{P})$. Then we can write $X_n = Y_n - Z_n$, where Y is a martingale and Z is a non-negative supermartingale.*

Proof. By the submartingale property, $Y_n = \lim_{m \to \infty} \mathrm{E}[X_m \mid \mathscr{F}_n]$ exists a.s. as an increasing limit. Note that Y is an adapted process, and $Y_n \geq$

X_n. Moreover, by the monotone convergence theorem, $\mathrm{E}Y_n = \lim_m \mathrm{E}X_m = \sup_m \mathrm{E}X_m$, which is finite since X is bounded in $L^1(\mathrm{P})$. Finally, we appeal to the towering property of conditional expectations (Theorem 8.5) and the conditional form of the monotone convergence theorem (Theorem 8.3) to find that $\mathrm{E}[Y_{n+1} \,|\, \mathscr{F}_n] = \lim_{m \to \infty} \mathrm{E}[X_m \,|\, \mathscr{F}_n] = Y_n$ a.s. This proves that Y is a martingale and $Z_n = Y_n - X_n \geq 0$. Also, because Y is a martingale and X is a submartingale,

$$(8.16) \quad \mathrm{E}[Z_{n+1} \,|\, \mathscr{F}_n] = \mathrm{E}[Y_{n+1} \,|\, \mathscr{F}_n] - \mathrm{E}[X_{n+1} \,|\, \mathscr{F}_n] \leq Y_n - X_n = Z_n,$$

almost surely. This completes our proof. □

One of the implications of Doob's decomposition is that any submartingale X is bounded below by some martingale. The Krickeberg decomposition implies a powerful converse to this: *Every L^1-bounded submartingale is also bounded above by a martingale.*

Remark 8.21. The preceding processes Y and Z are bounded in $L^1(\mathrm{P})$. Here is a proof: $\mathrm{E}|Y_n| \leq \sup_k \mathrm{E}|X_k| + \mathrm{E}Z_n$; thus it suffices to show that $\mathrm{E}Z_n$ is bounded in n. But the martingale property of Y implies that $\mathrm{E}Z_n = \mathrm{E}Y_1 - \mathrm{E}X_n$, whence we have $|\mathrm{E}Z_n| \leq |\mathrm{E}Y_1| + \sup_k \mathrm{E}|X_k|$. Also, we may remark that $Y \geq 0$ whenever $X \geq 0$.

3. Stopping Times and Optional Stopping

Definition 8.22. A *stopping time* (with respect to a filtration $\mathscr{F}$) is a random variable $T : \Omega \to \mathbf{N} \cup \{\infty\}$ such that $\{T = k\} \in \mathscr{F}_k$ for all $k \in \mathbf{N}$. This is equivalent to saying that $\{T \leq k\} \in \mathscr{F}_k$ for every $k \in \mathbf{N}$.

You should think of $\mathscr{F}_k$ as the total amount of information available by time k. For example, if we know $\mathscr{F}_k$, then we know whether or not $A \in \mathscr{F}_k$ has occurred by time k. With this in mind, the above can be interpreted as saying that T is a stopping time if and only if we only need to know the state of things by time k to decide measurably whether or not $T \leq k$.

Example 8.23. Non-random times are stopping times (check!). Next suppose $\{X_i\}_{i=1}^{\infty}$ is a stochastic process that is adapted to a filtration $\mathscr{F}$. If $A \in \mathscr{B}(\mathbf{R})$, then $T(\omega) := \inf\{n \geq 1 : X_n(\omega) \in A\}$ is a stopping time provided that we define $\inf \varnothing := \infty$. Indeed, $\{T = k\} = \cap_{j=1}^{k-1} \{X_j \notin A\} \cap \{X_k \in A\}$ for every $k \geq 2$. Because $\{T = 1\} = \{X_1 \in A\}$, we find that $\{T = k\} \in \mathscr{F}_k$ for every $k \geq 1$. The random variable T is the first time the process X enters the set A. Likewise, one shows that the kth time that X enters A is a stopping time for all $k \geq 1$. This example is generic in the following sense: If T is a stopping time with respect to a filtration, then there exists an adapted process X such that $T := \inf\{j \geq 1 : X_j = 1\}$, where $\inf \varnothing := \infty$. A simple recipe for X is $X_j(\omega) := \mathbf{1}_{\{T=j\}}(\omega)$.

Remark 8.24. The previous example shows that the kth time a process enters a Borel set is a stopping time for any $k \geq 1$. Although this produces a large collection of stopping times, not all random times are stopping times. For instance, consider

$$(8.17) \qquad L(\omega) := \sup\{n \geq 1 : X_n(\omega) \in A\},$$

where $\sup \varnothing := 0$ and A is a Borel set. Thus, L is the last time X enters A, and $\{L = k\} = \cap_{j=k+1}^{\infty} \{X_j \notin A\} \cap \{X_k \in A\}$. This is in $\mathscr{F}_k$ if and only if $X_k, X_{k+1}, \ldots$ are all Borel functions of $X_1, \ldots, X_k$; a property that does not generally hold. (For example, consider the case when the X_n's are independent.)

Lemma 8.25. *If $\{T_i\}_{i=1}^n$ are stopping times, then so too are $T_1 + \cdots + T_n$, $\min_{1 \leq j \leq n} T_j$, and $\max_{1 \leq j \leq n} T_j$.*

Consider a finite (or an a.s.-finite) stopping time T (with respect to a given underlying filtration of course). Define

$$(8.18) \qquad \mathscr{F}_T := \left\{ A \in \mathscr{F} : A \cap \{T \leq k\} \in \mathscr{F}_k \quad \forall k \geq 1 \right\}.$$

The subscript "T" of $\mathscr{F}_T$ is there only to remind us of the relation of the collection $\mathscr{F}_T$ to the stopping time T. It is not the case that $\mathscr{F}_T$ is a function of $T(\omega)$, for instance.

Lemma 8.26. *If T is an a.s.-finite stopping time, then $\mathscr{F}_T$ is a σ-algebra. If S and T are a.s.-finite stopping times such that $S \leq T$ a.s., then $\mathscr{F}_S \subseteq \mathscr{F}_T$. Furthermore, $\mathrm{E}[Y \mid \mathscr{F}_T]\mathbf{1}_{\{T=n\}} = \mathrm{E}[Y \mid \mathscr{F}_n]\mathbf{1}_{\{T=n\}}$ a.s. for all $Y \in L^1(\mathrm{P})$ and $n \geq 1$. Finally, if X is adapted to $\mathscr{F}$ then X_T is $\mathscr{F}_T$-measurable, where $X_T(\omega) := X_{T(\omega)}(\omega)$ for all $\omega \in \Omega$.*

The following theorem is due to Doob (1953, 1971); see also Hunt (1966). It is our first important result on semi-martingales.

The Optional Stopping Theorem. *Suppose S and T are a.s.-bounded stopping times such that $\mathrm{P}\{S \leq T\} = 1$. If X is a submartingale, then $\mathrm{E}[X_T \mid \mathscr{F}_S] \geq X_S$ a.s. If X is a supermartingale, then $\mathrm{E}[X_T \mid \mathscr{F}_S] \leq X_S$ a.s. If X is martingale, then $\mathrm{E}[X_T \mid \mathscr{F}_S] = X_S$ a.s.*

This result has the following interpretation in terms of a fair game. Suppose $\{X_i\}_{i=1}^{\infty}$ are i.i.d. mean-zero random variables so that $S_n := X_1 + \cdots + X_n$ can be thought of as the reward—or loss—at time n in a fair game. Because $\mathrm{E}S_n = 0$ for all $n \geq 1$, we do not expect to win with certitude at non-random times n. The optional stopping theorem states that the same fact holds for bounded stopping times n. In other words, when playing a fair game, there is no free lunch unless you are clairvoyant.

4. Applications to Random Walks

Proof. It suffices to consider the submartingale case.

We can find a non-random $K > 0$ such that with probability one, $S \le T \le K$. Now the trick is to write things in terms of $d_n := X_n - X_{n-1}$, where $X_0 := 0$. Equivalently, $X_n = \sum_{j=1}^n d_j$, and hence $X_T = \sum_{j=1}^K d_j \mathbf{1}_{\{j \le T\}}$ a.s. A similar expression holds for X_S. Therefore, for all $A \in \mathscr{F}_S$,

$$\begin{aligned}
\mathrm{E}[X_T - X_S; A] &= \sum_{j=1}^K \mathrm{E}\left[d_j \mathbf{1}_{\{S < j \le T\} \cap A}\right] \\
&= \sum_{j=1}^K \mathrm{E}\left[\mathrm{E}\left(d_j \mathbf{1}_{\{S < j \le T\} \cap A} \,\middle|\, \mathscr{F}_{j-1}\right)\right] \\
&= \sum_{j=1}^K \mathrm{E}\left[\mathrm{E}[d_j \,|\, \mathscr{F}_{j-1}] \mathbf{1}_{\{S < j \le T\} \cap A}\right].
\end{aligned}$$ (8.19)

We have used the facts that: (a) $\{T \ge j\} = \{T \le j-1\}^c \in \mathscr{F}_{j-1}$; and (b) $A \cap \{S < j \le T\} \in \mathscr{F}_{j-1}$. [Fact (b) is a consequence of the identity, $\{S < j\} \cap A = \{S \le j-1\} \cap A \in \mathscr{F}_{j-1}$.] By the definition of a submartingale, $\mathrm{E}[d_j \,|\, \mathscr{F}_{j-1}] \ge 0$ almost surely. This implies that $\mathrm{E}[X_T; A] \ge \mathrm{E}[X_S; A]$, which is equivalent to the desired result. □

Our next result follows immediately from the preceding one. But it is important, and deserves special mention.

Corollary 8.27. *Suppose T is a stopping time with respect to a filtration $\mathscr{F}$ and X is a submartingale (respectively, supermartingale or martingale) with respect to $\mathscr{F}$. Then, $\{X_{T \wedge n}\}_{n=1}^\infty$ is a submartingale (respectively, supermartingale or martingale) with respect to $\{\mathscr{F}_{T \wedge n}\}_{n=1}^\infty$.*

4. Applications to Random Walks

Definition 8.28. If $\{X_i\}_{i=1}^\infty$ are i.i.d. random variables in $\mathbf{R}^m$, then $\{S_n\}_{n=1}^\infty$ is called a *random walk*, where $S_n := X_1 + \cdots + X_n$.

Henceforth, consider the case that $m = 1$. It follows, after centering, that every L^1-random walk is a martingale.

Lemma 8.29. *If $S_n = X_1 + \cdots + X_n$ defines a random walk in one dimension and $X_1 \in L^1(\mathrm{P})$, then $\{S_n - n\mathrm{E}X_1\}_{n=1}^\infty$ is a mean-zero martingale. Suppose, in addition, that $X_1 \in L^2(\mathrm{P})$ and $\mathrm{E}X_1 = 0$. Then $\{S_n^2 - n\mathrm{Var}X_1\}_{n=1}^\infty$ is a mean-zero martingale.*

4.1. Wald's Identity. If $\{S_n\}_{n=1}^\infty$ is a random walk whose increments $\{X_i\}_{i=1}^\infty$ have a finite mean μ, then Lemma 8.29 implies immediately that $\mathrm{E}S_n = n\mu$. This identity generalizes to stopping times, as we prove next.

Theorem 8.30. *Consider a random walk defined by $S_n = \sum_{i=1}^n X_i$ where $X_1 \in L^1(\mathrm{P})$. Let $\mathscr{F}_n$ denote the σ-algebra generated by $\{X_i\}_{i=1}^n$ ($n = 1, 2, \ldots$) with respect to which T is a stopping time with $\mathrm{E}T < \infty$. Then $\mathrm{E}S_T = \mathrm{E}X_1 \, \mathrm{E}T$.*

Proof. Combine Corollary 8.27 with Lemma 8.29 to find that for all $n \geq 1$, $\mathrm{E}[S_{T \wedge n}] = \mathrm{E}X_1 \, \mathrm{E}[T \wedge n]$. As n tends to infinity, $\mathrm{E}[T \wedge n] \nearrow \mathrm{E}T$. Thus, it suffices to prove that

$$\text{(8.20)} \qquad \mathrm{E} \sup_n |S_{T \wedge n}| \leq \mathrm{E}|X_1| \, \mathrm{E}T,$$

whence it is finite. But for all $j \geq 1$, $|S_{T \wedge n}| = |\sum_{k=1}^\infty \mathbf{1}_{\{T \wedge n \geq k\}} X_k| \leq \sum_{k=1}^\infty \mathbf{1}_{\{T \geq k\}} |X_k|$. Take expectations to find that $\mathrm{E} \sup_n |S_{T \wedge n}|$ is at most $\sum_{k=1}^\infty \mathrm{E}\{|X_k|; T \geq k\}$. (Why can we interchange the infinite sum with the expectation integral?) Because $\{T \geq k\} = \{T \leq k-1\}^c \in \mathscr{F}_{k-1}$ is independent of X_k, (8.20) follows from Lemma 6.12 on page 68. $\square$

4.2. Gambler's Ruin Problem.

Definition 8.31. A random walk $S_n = X_1 + \cdots + X_n$ is called a *nearest-neighborhood walk* if $X_1 \in \{-1, +1\}$ a.s.; i.e., if at all times $n = 1, 2, \ldots$, we have $S_n = S_{n-1} \pm 1$ almost surely.

In other words, S_n is a nearest-neighborhood walk if there exists $p \in [0, 1]$ such that $\mathrm{P}\{X_1 = 1\} = p = 1 - \mathrm{P}\{X_1 = -1\}$. The case $p = \frac{1}{2}$ is particularly special and has its own name.

Definition 8.32. If $\mathrm{P}\{X_1 = 1\} = \mathrm{P}\{X_1 = -1\} = \frac{1}{2}$, then $\{S_n\}_{n=1}^\infty$ is called the *simple walk*.

We can think of a nearest-neighborhood walk S_n as the amount of money won (lost if negative) in n independent plays of a game, where in each play one wins or loses a dollar with probabilities p and $1 - p$ respectively. Then the simple walk corresponds to the fortune process of the gambler in the case that the game is fair.

Suppose that the gambler is playing against the house, there is a maximum house limit of h dollars, and the gambler's resources amount to a total of g dollars. Then consider the first time that either the house or the gambler is forced to stop playing. That is,

$$\text{(8.21)} \qquad T := \inf \{j \geq 1 : S_j = -g \text{ or } h\},$$

where $T(\omega) = \inf \varnothing := \infty$ amounts to the statement that the particular realization ω of the game is played indefinitely.

Lemma 8.33. *With probability one, $T < \infty$.*

4. Applications to Random Walks

Proof. The proof appeals to the following "continuity property" of S: If $S_n > h$, then there exists $i < n$ such that $S_i = h$.

Define $m = g + h$, and consider the events

(8.22) $\qquad E_k := \{X_{k+1} = \cdots = X_{k+m} = 1\} \qquad \forall k \geq 0.$

Evidently, $\{E_{im}\}_{i=0}^\infty$ are independent, have the same probability of occurring, and $\mathrm{P}(E_{im}^c) = 1 - p^m > 0$ for all $i \geq 0$. By the Borel–Cantelli lemma (p. 73) infinitely many of the E_{im}'s occur a.s. In particular, there a.s. exists a random finite integer τ such that E_τ occurs. [More precisely, $E_k \cap \{\tau = k\} \neq \varnothing$ for all $k \geq 1$.] On $\{S_\tau \notin (-g, h)\}$, we have $T \leq \tau < \infty$. Also, on $\{S_\tau \in (-g, h)\}$ we have $S_{\tau+m} = S_\tau + X_{\tau+1} + \cdots + X_{\tau+m} > -g + X_{\tau+1} + \cdots + X_{\tau+m} = h$. Therefore, $T < \tau + m$ is finite on $\{S_\tau(\omega) \in (-g, h)\}$ too. $\qquad\square$

The "gambler's ruin" problem is to compute, in terms of the parameter p, the probability that the gambler is ruined. In symbols, we seek to know

(8.23) $\qquad \mathfrak{Ruin}(p) := \mathrm{P}\{S_T = -g\}.$

The following *gambler's ruin formula* is our first important application of the optional stopping theorem (Corollary 8.27).

Theorem 8.34. *For all $p \in (0, 1)$ define $\zeta := (1 - p)/p$. Then,*

(8.24) $\qquad \mathfrak{Ruin}(p) = \begin{cases} \dfrac{h}{g + h} & \text{if } p = 1/2, \\[2mm] \dfrac{\zeta^{h+g} - \zeta^h}{\zeta^{h+g} - 1} & \text{if } p \neq 1/2. \end{cases}$

Note that $\mathfrak{Ruin}$ is continuous at $p = 1/2$.

Proof. Let us begin with the case $p = 1/2$: By the optional stopping theorem, and by Lemma 8.29, $\{S_{T \wedge n}\}_{n=1}^\infty$ is a mean-zero bounded martingale. Boundedness follows from $\sup_n |S_{T \wedge n}| \leq \max(g, h) < \infty$. Therefore, $\mathrm{E}S_{T \wedge n} = 0$ for all n. The dominated convergence theorem and the fact that a.s. $T < \infty$ (Lemma 8.33) together imply that $\mathrm{E}S_T = \mathrm{E}[\lim_n S_{T \wedge n}] = \lim_n \mathrm{E}S_{T \wedge n} = 0$. Because $\omega \mapsto S_T(\omega)$ is a simple function,

(8.25) $\qquad 0 = \mathrm{E}S_T = -g\mathfrak{Ruin}(1/2) + h(1 - \mathfrak{Ruin}(1/2)).$

Solve to obtain the expression for the ruin probability in the case $p = 1/2$.

When $p \neq \frac{1}{2}$, we have to only find a suitable bounded martingale and then follow the preceding argument. But it is not hard to see that $\{\zeta^{S_{T \wedge n}}\}_{n=1}^\infty$ is a bounded mean-one martingale. A similar reasoning as the one used in the $p = \frac{1}{2}$ case shows that $\mathrm{E}\zeta^{S_T} = 1$, whence follows the result. $\qquad\square$

5. Inequalities and Convergence

We now use the optional stopping theorem to prove the second set of maximal inequalities of this book (Doob, 1940).

Doob's Maximal Inequalities. *If X is a submartingale, then for all $\lambda > 0$ and $n \geq 1$,*

(8.26)
$$\lambda P\left\{\max_{1\leq j\leq n} X_j \geq \lambda\right\} \leq E\left[X_n; \max_{1\leq j\leq n} X_j \geq \lambda\right] \leq E[X_n^+],$$
$$\lambda P\left\{\min_{1\leq j\leq n} X_j \leq -\lambda\right\} \leq E[X_n^+] - EX_1.$$

We can add the two parts of (8.26) to find that $\lambda P\{\max_{j\leq n}|X_j| \geq \lambda\} \leq 2E[X_n^+] - E[X_1]$. This readily yields our next result.

Corollary 8.35. *If X is a submartingale, then for all $\lambda > 0$ and $n \geq 1$,*

(8.27) $$\lambda P\left\{\max_{1\leq j\leq n} |X_j| \geq \lambda\right\} \leq 2E|X_n| - EX_1.$$

Proof of Doob's Inequalities. If we let $\inf \varnothing := \infty$, then T is a stopping time, where $T := \inf\{j \geq 1 : X_j \geq \lambda\}$, and $\{T \leq n\} = \{\max_{j\leq n} X_j \geq \lambda\}$. Because $E[X_n^+] \geq E[X_n; T \leq n] = \sum_{j=1}^n E[X_n; T = j]$, and since $\{T = j\} \in \mathscr{F}_j$ for all $j \geq 1$, the submartingale property implies that $E[X_n^+] \geq \sum_{j=1}^n E[X_j; T = j] \geq \lambda P\{T \leq n\}$. This proves the first Doob inequality.

For the second portion of (8.26) define

(8.28) $$\tau := \inf\{1 \leq j \leq n : X_j \leq -\lambda\},$$

where $\inf \varnothing := \infty$. By the optional stopping theorem, $EX_1 \leq EX_{\tau \wedge n}$. Since $X_\tau \leq -\lambda$ on $\{\tau < \infty\}$,

(8.29) $$EX_{\tau \wedge n} = E[X_\tau; \tau \leq n] + E[X_n; \tau > n] \leq -\lambda P\{\tau \leq n\} + E[X_n^+].$$

This completes the proof. □

The following convergence theorem of Doob (1940) is a consequence of the Doob inequalities.

The Martingale Convergence Theorem. *Let X be a submartingale. Suppose either: (i) X is bounded in $L^1(P)$; or (ii) X is non-positive a.s. Then, $\lim_{n\to\infty} X_n$ exists and is finite a.s.*

Proof. We follow the general outline of the proof of the strong law of large numbers (p. 73): We first prove things in the L^2-case; then truncate down to $L^1(P)$. This is achieved in four easy steps.

Step 1. The Non-negative L^2-Bounded Case. If X is non-negative and bounded in $L^2(P)$, then for all $n, k \geq 1$,

$$\begin{aligned}
\|X_{n+k} - X_n\|_2^2 &= \|X_{n+k}\|_2^2 + \|X_n\|_2^2 - 2E[X_{n+k}X_n] \\
&= \|X_{n+k}\|_2^2 + \|X_n\|_2^2 - 2E\left[E(X_{n+k}\mid\mathscr{F}_n)X_n\right] \\
&\leq \|X_{n+k}\|_2^2 - \|X_n\|_2^2.
\end{aligned} \tag{8.30}$$

According to Lemma 8.18, X^2 is a submartingale since $X_n \geq 0$ for all n. Therefore, $\|X_n\|_2 \nearrow \sup_m \|X_m\|_2$ as $n \nearrow \infty$. It follows that $\{X_n\}_{n=1}^\infty$ is a Cauchy sequence in $L^2(P)$, and so it converges in $L^2(P)$. Let X_∞ be the $L^2(P)$-limit of X_n, and find $n_k \uparrow \infty$ such that $\|X_\infty - X_{n_k}\|_2 \leq 2^{-k}$. By Chebyshev's inequality,

$$\sum_{k=1}^\infty P\{|X_\infty - X_{n_k}| \geq \epsilon\} \leq \frac{1}{\epsilon^2}\sum_{k=1}^\infty 4^{-k} < \infty \qquad \forall \epsilon > 0. \tag{8.31}$$

Thus, by the Borel–Cantelli lemma (p. 73), $\lim_{k\to\infty} X_{n_k} = X_\infty$ a.s. On the other hand,

$$\begin{aligned}
\|X_{n_{k+1}} - X_{n_k}\|_1 &\leq \|X_{n_{k+1}} - X_{n_k}\|_2 \\
&\leq \|X_\infty - X_{n_k}\|_2 + \|X_\infty - X_{n_{k+1}}\|_2 \\
&\leq 2^{-k} + 2^{-k+1} \\
&= 3 \cdot 2^{-k}.
\end{aligned} \tag{8.32}$$

Therefore, Corollary 8.35 shows us that for all $\epsilon > 0$,

$$\begin{aligned}
\sum_{k=1}^\infty P\left\{\max_{n_k \leq j \leq n_{k+1}} |X_j - X_{n_k}| \geq \epsilon\right\} &\leq \frac{2}{\epsilon}\sum_{k=1}^\infty \|X_{n_{k+1}} - X_{n_k}\|_1 \\
&\leq \frac{6}{\epsilon}\sum_{k=1}^\infty 2^{-k} < \infty.
\end{aligned} \tag{8.33}$$

We have used the fact that $\{X_{n+j} - X_n\}_{j=0}^\infty$ is a submartingale for each fixed n with respect to the filtration $\{\mathscr{F}_{j+n}\}_{j=0}^\infty$, and that this submartingale starts at 0. It follows from the Borel–Cantelli lemma that

$$\lim_{k\to\infty} \max_{n_k \leq j \leq n_{k+1}} |X_j - X_{n_k}| = 0 \qquad \text{a.s.} \tag{8.34}$$

Because $X_{n_k} \to X_\infty$ a.s., we have proved that $\lim_{m\to\infty} X_m = X_\infty$ a.s. As $X_\infty \in L^2(P)$ it follows that X_∞ is a.s. finite.

Step 2. The Non-positive Case. If $X_n \leq 0$ is a submartingale, then $\exp(X_n)$ is a bounded non-negative submartingale (Lemma 8.18). Thanks to Step 1, $\lim_{n\to\infty}\exp(X_n)$ exists and is finite a.s.

Step 3. The Non-negative L^1-Bounded Case. If X_n is a non-negative submartingale that is bounded in $L^1(P)$, then thanks to the Krickeberg decomposition (p. 128), we can write $X_n = Y_n - Z_n$ where Y is a non-negative martingale and Z is a non-negative supermartingale; see also Remark 8.21. By Step 2, $\lim_{n \to \infty} Y_n$ and $\lim_{n \to \infty} Z_n$ exist and are finite a.s. Consequently, $\lim_{n \to \infty} X_n$ exists and is finite a.s.

Step 4. The L^1-Bounded Case. If X is an L^1-bounded submartingale, then we can write $X_n = Y_n^+ - Y_n^- - Z_n$ where Y is an L^1-bounded martingale and Z is an L^1-bounded non-negative supermartingale. Because Y^+ and Y^- are non-negative submartingales (Lemma 8.18), Step 3 implies that $\lim_{n \to \infty} Y_n^+$ and $\lim_{n \to \infty} Y_n^-$ exist a.s. Similarly, Z is a non-negative supermartingale and $\lim_{n \to \infty} Z_n$ exists almost surely; see Step 2. This completes the proof. □

6. Further Applications

Martingale theory provides us with a powerful set of analytical tools and, as such, it is not surprising that it has made an impact on a large number of diverse mathematical problems. This section contains a few applications of this theory. More examples can be found in the exercises.

6.1. Kolmogorov's Zero–One Law. Let $\{X_i\}_{i=1}^{\infty}$ denote independent random variables, and define $\mathscr{T}$ to be the corresponding tail σ-algebra; see Definition 6.14. Recall that the Kolmogorov zero-one law (p. 69) asserts that $\mathscr{T}$ is trivial. Here is a martingale proof: Let $\mathscr{F}_n := \sigma(\{X_i\}_{i=1}^n)$, and consider any event $A \in \mathscr{T}$. Because A is independent of $\mathscr{F}_n$, we have the a.s. identity $\mathrm{P}(A \,|\, \mathscr{F}_n) = \mathrm{P}(A)$. But $\mathrm{P}(A \,|\, \mathscr{F}_n)$ defines a Doob martingale. Therefore, in accord with the martingale convergence theorem, $L = \lim_{n \to \infty} \mathrm{P}(A \,|\, \mathscr{F}_n)$ exists a.s. and in $L^1(\mathrm{P})$. We claim that $L = \mathbf{1}_A$ a.s. This would then prove that $\mathrm{P}(A) = \mathbf{1}_A$ a.s., which is Kolmogorov's zero-one law.

Fix an integer $k \geq 1$ and note that $\mathrm{E}[L; B] = \lim_{n \to \infty} \mathrm{E}[\mathrm{P}(A \,|\, \mathscr{F}_n); B] = \mathrm{P}(A \cap B)$ for all $B \in \mathscr{F}_k$. By the monotone class theorem (p. 30), $\mathrm{E}[L; B] = \mathrm{P}(A \cap B)$ for all $B \in \mathscr{F}_\infty$, where $\mathscr{F}_\infty$ denotes the smallest σ-algebra that contains $\cup_{n=1}^\infty \mathscr{F}_n$. Because L is $\mathscr{F}_\infty$-measurable and $A \in \mathscr{F}_\infty$, this ensures that $L = \mathrm{P}(A \,|\, \mathscr{F}_\infty) = \mathbf{1}_A$ a.s. The result follows.

6.2. Lévy's Borel–Cantelli Lemma. In this section we describe an optimal improvement to the Borel–Cantelli lemma (p. 73; see also Problem 6.20, p. 86). This improvement is due to Lévy (1937, Corollary 68, p. 249). First, we need a definition.

Definition 8.36. If E and F are events, then we say that $E = F$ almost surely when $\mathbf{1}_E(\omega) = \mathbf{1}_F(\omega)$ for almost every ω.

Theorem 8.37. *If $\{\mathscr{F}_n\}_{n=1}^\infty$ is a filtration and $E_1, E_2, \ldots$ are events such that $E_n \in \mathscr{F}_n$ for all $n \geq 1$, then*

$$(8.35) \quad \{E_n \text{ occurs infinitely often}\} = \left\{\sum_{n=2}^\infty \mathrm{P}(E_n \,|\, \mathscr{F}_{n-1}) = \infty\right\} \quad a.s.$$

Consequently, the two events, $F_1 := \{\sum_{n=2}^\infty \mathrm{P}(E_n \,|\, \mathscr{F}_{n-1}) = \infty\}$ and $F_2 := \{E_n \text{ occurs infinitely often}\}$ have the same probability.

The proof of Lévy's Borel–Cantelli lemma rests on a general result about martingales with bounded increments.

Theorem 8.38. *Suppose $\{X_n\}_{n=1}^\infty$ is a martingale such that $|X_n - X_{n-1}| \leq \alpha$ a.s. for all $n \geq 1$, where α is a positive non-random constant. Consider the events $L_1 := \{\sup_n X_n < \infty\}$, $L_2 := \{\inf_n X_n > -\infty\}$, and $L_3 := \{\lim_{n \to \infty} X_n \text{ exists and is finite}\}$. Then $L_1 = L_2 = L_3$ a.s.*

Proof. For any $\lambda > 0$ define

$$(8.36) \quad T_\lambda := \inf\{n \geq 1 : X_n > \lambda\},$$

where $\inf \varnothing := \infty$. By the optional stopping theorem (Corollary 8.27), $\{X_{n \wedge T_\lambda}\}_{n=1}^\infty$ is a martingale. Moreover, the fact that the increments of X are at most α implies that $X_{T_\lambda} \leq \alpha + \lambda$. Therefore, $\alpha + \lambda - X_{n \wedge T_\lambda}$ defines a non-negative martingale. This must converge a.s. Consequently, for any $\lambda > 0$ there exists a null set off which $\lim_{n \to \infty} X_{n \wedge T_\lambda}$ exists and is finite. Take the union of these null sets, as λ ranges over all positive rationals, to deduce that outside one null-set N, $\lim_{n \to \infty} X_{n \wedge T_\lambda}$ exists and is finite for all rational $\lambda > 0$.

If $\omega \in L_1$ then $T_\lambda(\omega)$ is infinite for all rational $\lambda > \sup_n X_n(\omega)$. Therefore, $L_1 \cap N^c \subset L_3$. By considering the martingale $-X$ we find also that $L_2 \cap N^c \subset L_3$. This proves that $(L_1 \cup L_2) \cap N^c \subset L_3$, whence $\mathbf{1}_{L_1 \cup L_2} \leq \mathbf{1}_{L_3}$ a.s. Since $L_3 \subseteq (L_1 \cap L_2)$, the result follows. □

Proof of Theorem 8.37. The variables $X_n = \sum_{i=2}^n \{\mathbf{1}_{E_i} - \mathrm{P}(E_i \,|\, \mathscr{F}_{i-1})\}$ ($n \geq 1$) define a martingale with bounded increments. In the notation of Theorem 8.38, $L_1 = \{\sum_i \mathbf{1}_{E_i} < \infty\}$ and $L_2 = \{\sum_i \mathrm{P}(E_i \,|\, \mathscr{F}_{i-1}) < \infty\}$. [This merits a moment's thought.] The proof follows. □

6.3. Khintchine's LIL. Suppose $\{X_i\}_{i=1}^\infty$ are i.i.d. random variables taking the values ± 1 with probability $1/2$ each, and define $S_n := X_1 + \cdots + X_n$. By the strong law of large numbers (p. 73), $S_n/n \to 0$ a.s. This particular form of the strong law first appeared in the context of the normal number theorem of Borel (1909). See Problem 6.22 on page 86.

One would like to know how fast S_n/n converges to zero. The central limit theorem suggests that S_n/n cannot tend to zero much faster than

$n^{-1/2}$. The correct asymptotic size of S_n/n was found in a series of successive improvements by Hausdorff in 1913 (see 1949, pp. 420–421), Hardy and Littlewood (1914), Steinhaus (1922), and Khintchine (1923). The definitive result, along these lines, is the *law of the iterated logarithm* (LIL) of Khintchine (1924):

$$(8.37) \quad \limsup_{n \to \infty} \frac{S_n}{(2n \ln \ln n)^{1/2}} = -\liminf_{n \to \infty} \frac{S_n}{(2n \ln \ln n)^{1/2}} = 1 \quad \text{a.s.}$$

Khintchine's LIL has a remarkable extension that is valid for sums of general i.i.d. random variables with finite variance.

The Law of the Iterated Logarithm. *If $\{X_i\}_{i=1}^\infty$ are i.i.d. random variables in $L^2(\mathrm{P})$ and $S_n := X_1 + \cdots + X_n$, then*

$$(8.38) \quad \limsup_{n \to \infty} \frac{S_n - n\mathrm{E}X_1}{(2n \ln \ln n)^{1/2}} = \mathrm{SD}(X_1) \quad \text{a.s.}$$

When the X_i's are bounded this was proved by Kolmogorov (1929). Cantelli (1933a) improved Kolmogorov's theorem to the case that $X_1 \in L^{2+\delta}(\mathrm{P})$ for some $\delta > 0$. Then the LIL for general mean-zero finite-variance increments remained elusive for nearly a decade, until Hartman and Wintner (1941) devised an ingenious truncation method which reduced the general LIL to that of Kolmogorov.

We will derive the LIL only in the case that the X's are normal. The theorem, in its full generality, is much more difficult to prove.

Proof of the LIL for Normal Increments. Without loss of generality, we may assume that the X_i's are standard normal random variables. Define

$$(8.39) \quad \Lambda := \limsup_{m \to \infty} \frac{S_m}{(2m \ln \ln m)^{1/2}}.$$

According to the Kolmogorov 0-1 law (p. 69), Λ is almost surely a constant. Our task is to prove that $\Lambda = 1$. We do this in three steps.

Step 1. A Large-Deviations Estimate. Fix a $t > 0$ and define $M_n := \exp\left(tS_n - \frac{1}{2}t^2 n\right)$. Let $\mathscr{F}_n := \sigma(\{X_i\}_{i=1}^n)$, and verify that M is a non-negative mean-one martingale. Moreover,

$$(8.40) \quad \left\{ \max_{1 \le j \le n} S_j \ge nt \right\} \subseteq \left\{ \max_{1 \le j \le n} M_j \ge \exp(nt^2/2) \right\}.$$

According to Doob's maximal inequality (p. 134), for all integers $n \ge 1$ and all real numbers $t > 0$,

$$(8.41) \quad \mathrm{P}\left\{ \max_{1 \le j \le n} S_j \ge nt \right\} \le \exp\left(-nt^2/2\right).$$

6. Further Applications

Step 2. The Upper Bound. Choose and fix $c > \theta > 1$, and define $\theta_k := \lfloor \theta^k \rfloor$ ($k = 1, 2, \ldots$). We apply (8.41) to deduce that as $k \to \infty$,

$$(8.42) \quad P\left\{\max_{1 \le j \le \theta_k} S_j \ge (2c\theta_{k-1} \ln \ln \theta_{k-1})^{1/2}\right\} \le \exp\left(-\frac{c\theta_{k-1} \ln \ln \theta_{k-1}}{\theta_k}\right)$$
$$= k^{-(c/\theta) + o(1)}.$$

Thus, the left-hand side is summable in k. By the Borel–Cantelli lemma (p. 73), with probability one there exists a random variable k_0 such that for all $k \ge k_0$, $\max_{1 \le j \le \theta_k} S_j < (2c\theta_{k-1} \ln \ln \theta_{k-1})^{1/2}$. For all $m > \theta_{k_0}$ we can find $k > k_0$ such that $\theta_{k-1} \le m \le \theta_k$, and hence,

$$(8.43) \quad S_m \le \max_{j \le \theta_k} S_j \le (2c\theta_{k-1} \ln \ln \theta_{k-1})^{1/2} \le (2cm \ln \ln m)^{1/2}.$$

This proves that $\Lambda \le c^{1/2}$. Because the latter holds for all $c > \theta > 1$, it follows that $\Lambda \le 1$. This is fully one-half of the LIL in the case of standard normal increments.

We can also apply the preceding to the process $-S$ to obtain the following:

$$(8.44) \quad \limsup_{m \to \infty} \frac{|S_m|}{(2m \ln \ln m)^{1/2}} \le 1 \quad \text{a.s.}$$

Step 3. A Lower Estimate. There exists a constant $A > 0$ such that

$$(8.45) \quad P\{X_1 \ge \lambda\} \ge \frac{Ae^{-\lambda^2/2}}{\lambda} \quad \forall \lambda > 1.$$

See Problem 1.16, page 14 for a hint. Choose and fix $\theta > 1$, and define $\theta_k := \lfloor \theta^k \rfloor$. Consider the events

$$(8.46) \quad E_k := \left\{S_{\theta_{k+1}} - S_{\theta_k} \ge (2\sigma_{k+1}^2 \ln_+ \ln_+ \theta_k)^{1/2}\right\},$$

where $\ln_+ x := \ln(x \vee e)$, and

$$(8.47) \quad \sigma_{k+1}^2 := \text{Var}(S_{\theta_{k+1}} - S_{\theta_k}) = \theta_{k+1} - \theta_k.$$

The E_k's are independent events, and because of (8.45), for all k large,

$$(8.48) \quad P(E_k) = P\left\{N(0,1) \ge (2 \ln \ln \theta_k)^{1/2}\right\} \ge \frac{A}{\ln \theta_k (2 \ln \ln \theta_k)^{1/2}}.$$

Consequently, $\sum_k P(E_k) = \infty$, and hence by the independence part of the Borel–Cantelli Lemma,

$$(8.49) \quad \limsup_{k \to \infty} \frac{S_{\theta_{k+1}} - S_{\theta_k}}{(2(\theta_{k+1} - \theta_k) \ln \ln \theta_k)^{1/2}} \ge 1 \quad \text{a.s.}$$

Since $\theta_{k+1} - \theta_k \sim \theta_{k+1}(1 - \theta^{-1})$ as $k \to \infty$,

$$\text{(8.50)} \qquad \limsup_{k \to \infty} \frac{S_{\theta_{k+1}} - S_{\theta_k}}{(2\theta_{k+1} \ln \ln \theta_k)^{1/2}} \geq \left(1 - \frac{1}{\theta}\right)^{1/2} \quad \text{a.s.}$$

Thanks to this, (8.44), and the fact that $\theta_{k+1} \sim \theta \cdot \theta_k$ as $k \to \infty$,

$$\text{(8.51)} \qquad \begin{aligned} \Lambda &\geq \limsup_{k \to \infty} \frac{S_{\theta_{k+1}}}{(2\theta_{k+1} \ln \ln \theta_k)^{1/2}} \\ &\geq \limsup_{k \to \infty} \frac{S_{\theta_{k+1}} - S_{\theta_k}}{(2\theta_{k+1} \ln \ln \theta_k)^{1/2}} - \limsup_{k \to \infty} \frac{|S_{\theta_k}|}{(2\theta_{k+1} \ln \ln \theta_k)^{1/2}} \\ &\geq \left(1 - \frac{1}{\theta}\right)^{1/2} - \frac{1}{\theta^{1/2}} \quad \text{a.s.} \end{aligned}$$

Let $\theta \uparrow \infty$ to find that $\Lambda \geq 1$. □

6.4. Lebesgue's Differentiation Theorem. The fundamental theorem of calculus asserts that if $f : \mathbf{R} \to \mathbf{R}$ is continuous then $F(\omega) := \int_0^\omega f(x)\, dx$ is differentiable and $F' = f$. In fact, we have the stronger result that

$$\text{(8.52)} \qquad \lim_{\delta \downarrow 0} \frac{1}{\delta} \int_\omega^{\omega+\delta} f(y)\, dy = f(\omega),$$

uniformly for all ω in a given compact set. Here is why: For all $\omega \in \mathbf{R}$ and $\delta > 0$,

$$\text{(8.53)} \qquad \left| \frac{1}{\delta} \int_\omega^{\omega+\delta} f(y)\, dy - f(\omega) \right| \leq \frac{1}{\delta} \int_\omega^{\omega+\delta} |f(y) - f(\omega)|\, dy.$$

Therefore, (8.52) follows from the uniform continuity of f on compact sets. There is a surprising extension of this, due to H. Lebesgue, that holds for all integrable functions f. The following is the celebrated differentiation theorem of Lebesgue.

Theorem 8.39. *If $\int_0^1 |f(x)|\, dx < \infty$, then (8.52) holds for almost every $\omega \in [0,1]$.*

Consider the Steinhaus probability space $([0,1], \mathscr{B}([0,1]), \mathrm{P})$. In probabilistic language, Theorem 8.39 states that (8.52) holds almost surely provided that $f \in L^1(\mathrm{P})$. In order to derive this formulation we need a maximal inequality for the following function Mf that is known as the *Hardy–Littlewood maximal function* (1930, Theorem 17). First we extend f to a function on $\mathbf{R}$ by setting $f(\omega) := f(0)$ if $\omega < 0$ and $f(\omega) := f(1)$ if $\omega > 1$. Then, we define:

$$\text{(8.54)} \quad (Mf)(\omega) = M(f)(\omega) = \sup_{\delta \in [0, 1-\omega]} \frac{1}{\delta} \int_\omega^{\omega+\delta} |f(y)|\, dy \qquad \forall \omega \in [0,1],$$

where $0/0 := 0$ to ensure that $(Mf)(1) = 0$.

Theorem 8.40. *For all $\lambda > 0$, $p \geq 1$, and $f \in L^p(P)$,*

(8.55) $$P\{Mf \geq \lambda\} \leq \frac{8^p}{\lambda^p}\|f\|_p^p.$$

Let us first prove Theorem 8.39 assuming the preceding maximal-function inequality. Theorem 8.40 is proved subsequently.

Proof of Theorem 8.39. For notational convenience, define the "averaging operators" A_δ as follows:

(8.56) $\quad A_\delta(f)(\omega) := (A_\delta f)(\omega) := \dfrac{1}{\delta}\displaystyle\int_\omega^{\omega+\delta} f(y)\,dy \quad \forall \omega \in [0,1],\ f \in L^1(P).$

Thus, we have the pointwise equality, $(Mf)(\omega) = \sup_{\delta \in [0, 1-\omega]} A_\delta(|f|)(\omega)$.

Throughout this proof we tacitly extend the domain of *all* continuous functions $g : [0,1] \to \mathbf{R}$ to $\mathbf{R}$ by setting $g(\omega) := g(0)$ for $\omega < 0$ and $g(\omega) := g(1)$ for $\omega > 1$.

Because continuous functions are dense in $L^1(P)$ (Problem 4.18, p. 50), for every $n \geq 1$ we can find a continuous function g_n such that $\|g_n - f\|_1 \leq n^{-1}$. Let $\mathscr{L} := \limsup_{\delta \downarrow 0} |A_\delta f - f|$ to find that

(8.57)
$$\begin{aligned}\mathscr{L} &\leq \lim_{\delta \downarrow 0} |A_\delta g_n - g_n| + \limsup_{\delta \downarrow 0} |A_\delta g_n - A_\delta f| + |g_n - f|\\ &= \limsup_{\delta \downarrow 0} |A_\delta g_n - A_\delta f| + |g_n - f|\\ &\leq M(|g_n - f|) + |g_n - f|.\end{aligned}$$

If $\mathscr{L} \geq \lambda$, then by the triangle inequality one of the two terms on the right-most side must be at least $\lambda/2$. Therefore, we can write

(8.58) $$P\{\mathscr{L} \geq \lambda\} \leq T_1 + T_2,$$

where

(8.59) $\quad T_1 := P\left\{M(|g_n - f|) \geq \dfrac{\lambda}{2}\right\} \quad \text{and} \quad T_2 := P\left\{|g_n - f| \geq \dfrac{\lambda}{2}\right\}.$

We estimate T_1 and T_2 separately.

On one hand, we can apply Theorem 8.40, with $p=1$, to deduce that

(8.60) $$T_1 \leq \frac{16}{\lambda}\|g_n - f\|_1 \leq \frac{16}{\lambda n}.$$

On the other hand, we appeal to Chebyshev's inequality (p. 43) to find that

(8.61) $$T_2 \leq \frac{2}{\lambda}\|g_n - f\|_1 \leq \frac{2}{\lambda n}.$$

Consequently, $P\{\mathscr{L} \geq \lambda\} \leq 18/(\lambda n)$ for all $n \geq 1$. Let $n \to \infty$ and $\lambda \downarrow 0$, in this order, to deduce that $\mathscr{L} = 0$ a.s. This proves the theorem. $\square$

Proof of Theorem 8.40. Because f can be replaced with $|f|$, we can assume without any loss in generality that $f \geq 0$. Also, we will extend the domain of the definition of f by setting $f(\omega) := 0$ for all $\omega \in \mathbf{R} \setminus [0,1]$.

Define $\mathscr{F}_n^0$ to be the collection of all *dyadic intervals* in $(0,1]$. That is, $I \in \mathscr{F}_n^0$ if and only if $I = (j2^{-n}, (j+1)2^{-n}]$ where $j \in \{0, \ldots, 2^n - 1\}$ and $n \geq 0$. Define $\mathscr{F}_n$ to be the σ-algebra generated by $\mathscr{F}_n^0$. Since every element of $\mathscr{F}_n^0$ is a union of two of the elements of $\mathscr{F}_{n+1}^0$, it follows that

$$(8.62) \qquad \mathscr{F}_n \subseteq \mathscr{F}_{n+1} \qquad \forall n \geq 0.$$

That is, $\{\mathscr{F}_n\}_{n=0}^\infty$ is a filtration; it is known as the *dyadic filtration*.

We can view the function f as a random variable, and compute $M_n := \mathrm{E}[f \mid \mathscr{F}_n]$ using Corollary 8.8:

$$(8.63) \qquad M_n(\omega) = \sum_{Q \in \mathscr{F}_n^0} \mathbf{1}_Q(\omega) 2^n \int_Q f(y)\, dy \qquad \text{for almost all } \omega \in [0,1].$$

It should be recognized that the preceding sum consists of one term only.

Next define $\mathscr{G}_n^0$ to be the collection of all *shifted dyadic intervals* of the form $J = (j2^{-n} + 2^{-n-1}, (j+1)2^{-n} + 2^{-n-1}]$, where $j \in \mathbf{Z}$ and $n \geq 0$. Let $\mathscr{G}_n$ denote the σ-algebra generated by the intervals in $\mathscr{G}_n^0$, and define $N_n := \mathrm{E}[f \mid \mathscr{G}_n]$. Because f vanishes outside $[0,1]$,

$$(8.64) \qquad N_n(\omega) = \sum_{Q \in \mathscr{G}_n^0} \mathbf{1}_Q(\omega) 2^n \int_Q f(y)\, dy \qquad \text{for almost all } \omega \in [0,1].$$

Consider $\omega \in (0,1)$ and $\delta \in (0, 1-\omega)$. There exists $n = n(\omega) \geq 0$ such that $2^{-n-1} \leq \delta \leq 2^{-n}$. We can find $I(\omega) \in \mathscr{F}_n^0$ and $J(\omega) \in \mathscr{G}_n^0$—both containing ω—such that $(\omega, \omega+\delta) \subset I(\omega) \cup J(\omega)$. Because $f \geq 0$, $f \equiv 0$ off $[0,1]$, and $\delta \geq 2^{-n-1}$, this implies that

$$(8.65) \qquad \frac{1}{\delta} \int_\omega^{\omega+\delta} f(y)\, dy \leq 2^{n+1} \left(\int_{I(\omega)} f(y)\, dy + \int_{J(\omega)} f(y)\, dy \right)$$
$$= 2 \left(M_n(\omega) + N_n(\omega) \right).$$

Optimize over all δ to find that $Mf \leq 2 \sup_n M_n + 2 \sup_n N_n$. Therefore, for all $\lambda > 0$,

$$(8.66) \qquad \mathrm{P}\{Mf \geq \lambda\} \leq \mathrm{P}\left\{ \sup_{n \geq 0} M_n \geq \frac{\lambda}{4} \right\} + \mathrm{P}\left\{ \sup_{n \geq 0} N_n \geq \frac{\lambda}{4} \right\}.$$

Note that $M_n + N_n$ is not a martingale because M and N are adapted to different filtrations. However, M and N *are* martingales in their respective filtrations. We apply the first maximal inequality of Doob (p. 134) to the submartingale defined by $|M_n|^p$ to find that

(8.67) $$P\left\{\sup_{n\geq 0} M_n \geq \frac{\lambda}{4}\right\} \leq \frac{4^p}{\lambda^p} \sup_{n\geq 0} E\left(|M_n|^p\right) \leq \frac{4^p}{\lambda^p} \|f\|_p^p.$$

[The last inequality follows from the conditional form of Jensen's inequality.] A similar inequality holds for N. We can combine our bounds to obtain,

(8.68) $$P\{Mf \geq \lambda\} \leq \frac{2 \cdot 4^p}{\lambda^p} \|f\|_p^p.$$

The theorem follows because $2 \cdot 4^p \leq 8^p$. □

The following corollary of Theorem 8.40, essentially due to Hardy and Littlewood (1930, Theorem 17), is noteworthy as it has a number of interesting consequences in real and harmonic analysis.

Corollary 8.41. *If $p > 1$ and $f \in L^p(P)$ then*

(8.69) $$\int_0^1 |(Mf)(t)|^p \, dt \leq \left(\frac{8p}{p-1}\right)^p \int_0^1 |f(t)|^p \, dt.$$

6.5. Option-Pricing in Discrete Time. We now take a look at an application of martingale theory to the mathematics of securities in finance.

In this example we consider a simplified case where there is only one type of stock whose value changes at times $n = 1, 2, 3, \ldots, N$. We start with y_0 dollars at time 0. During the time period $(n, n+1)$ we look at the performance of this stock up to time n. Based on this information, we may decide to buy A_{n+1}-many shares.

Negative investments are also allowed in the marketplace: If $A_n(\omega) \leq 0$ for some n and ω, then we are *selling short* for that ω. This means that we sell $A_n(\omega)$ stocks that we do not own, hoping that when the clock strikes N, we will earn enough to pay our debts.

Let S_n denote the value of the stock at time n. We simplify the model further by assuming that $|S_{n+1} - S_n| = 1$. That is, the stock value fluctuates by exactly one unit at each time step, and the stock value is updated precisely at time n for every $n = 1, 2, \ldots$. The only unexplained variable is the ending time N; this is the so-called *time to maturity* and will be explained later. Now we can place things in a more precise framework.

Let Ω denote the collection of all possible $\omega = (\omega_1, \ldots, \omega_N)$ where every ω_j takes the values ± 1. Intuitively, $\omega_j = 1$ if and only if the value of our stock went up by 1 dollar at time j. Thus, $\omega_j = -1$ means that the stock went down by a dollar, and Ω is the collection of all stock movements that are theoretically possible.

Define the functions $S_1, \ldots, S_N$ by $S_0(\omega) := 0$, and

(8.70) $$S_n(\omega_1, \ldots, \omega_n) := \omega_1 + \cdots + \omega_n \qquad \forall n = 1, \ldots, N.$$

We may abuse the notation slightly and write $S_n(\omega)$ in place of $S_n(\omega_1,\ldots,\omega_n)$. In this way, $S_n(\omega)$ represents the value of the stock at time n, and corresponds to the stock movements $\omega_1,\ldots,\omega_n$. During the time interval $(n, n+1)$, we may look at $\omega_1,\ldots,\omega_n$, choose a number $A_{n+1}(\omega) = A_{n+1}(\omega_1,\ldots,\omega_n)$, which might depend on $\omega_1,\ldots,\omega_n$, and buy $A_{n+1}(\omega)$-many shares. If our starting fortune at time 0 is y_0, then our fortune at time n depends on $\{A_i(\omega)\}_{i=1}^{n-1}$, and is given by

$$(8.71) \quad Y_n(\omega) = Y_n(\omega_1,\ldots,\omega_n) = y_0 + \sum_{j=1}^{n} A_j(\omega)\left[S_j(\omega) - S_{j-1}(\omega)\right],$$

as n ranges from 1 to N. The sequence $\{A_i(\omega)\}_{i=1}^{N}$ is our investment *strategy*. Recall that it depends on the stock movements $\{\omega_i\}_{i=1}^{N}$ in a "previsible manner"; i.e., for each $n \geq 2$, $A_n(\omega)$ depends only on $\omega_1,\ldots,\omega_{n-1}$. [$A_1$ does not depend on ω.]

A *European call option* is a gamble wherein we purchase the right to buy the stock at a given price C—the *strike* or *exercise price*—at time N.

Suppose we have the option to call at C dollars. If it happens that $S_N(\omega) > C$, then we have gained $(S_N(\omega) - C)$ dollars. This is because we can buy the stock at C dollars and then instantaneously sell the stock at $S_N(\omega)$. On the other hand, if $S_N(\omega) \leq C$ then it is not wise to buy at C. Therefore, no matter what happens, the value of our option at time N is $(S_N(\omega) - C)^+$. An important question that needs to be settled is this:

(8.72) *What is the fair price for a call at C?*

This was answered by Black and Scholes (1973) and Merton (1973) for a related, but slightly different, model. The connections to probability were discovered later by Harrison and Kreps (1979) and Harrison and Pliska (1981). The present model, the so-called "binomial pricing model," is due to Cox, Ross, and Rubenstein (1979). In order to explain their solution to (8.72) we need a brief definition from finance.

Definition 8.42. A strategy A is a *hedging strategy* if:

(i) Using A does not lead us to bankruptcy; i.e., $Y_n(\omega) \geq 0$ for all $n = 1,\ldots,N$.

(ii) Y attains the value of the stock at time N; i.e.,
$$Y_N(\omega) = (S_N(\omega) - C)^+.$$

Of course any strategy A is also previsible.

Let us posit that there are no "arbitrage opportunities," where arbitrage is synonymous to "free lunch." That is, we assume there are no risk-free investments. Then, in terms of our model, y_0 is the "fair price of a given option" if, starting with y_0 dollars, we can find a hedging/investment strategy that yields the value of the said option at time N, *no matter how the stock values behave.*

The solution of Black and Scholes (1973), transcribed to the present simplified setting, depends on first making $(\Omega, \mathscr{P}(\Omega))$ into a probability space. Here, $\mathscr{P}(\Omega)$ denotes the power set of Ω. Define the probability measure P so that $X_j(\omega) = \omega_j$ are i.i.d. taking the values ± 1 with probability $\frac{1}{2}$ each. In words, under the measure P, the stock values fluctuate at random but in a fair manner. Another, yet equivalent, way to define P is as the product measure:

$$(8.73) \qquad P(d\omega) = Q(d\omega_1) \cdots Q(d\omega_N) \qquad \forall \omega \in \Omega,$$

where $Q(\{1\}) = Q(\{-1\}) = 1/2$. Using this probability space $(\Omega, \mathscr{P}(\Omega), P)$, $\{A_i\}_{i=1}^\infty$, $\{S_i\}_{i=1}^\infty$, and $\{Y_i\}_{i=1}^\infty$ are stochastic processes, and we can present the so-called Black–Scholes formula for the fair price y_0 of a European option.

The Black–Scholes Formula. *A hedging strategy exists iff*

$$(8.74) \qquad y_0 = E\left[(S_N - C)^+\right].$$

Proof (Necessity). We first prove Theorem 6.5 assuming that a hedging strategy A exists. If so then the process Y_n defined in (8.71) is a martingale; see Example 8.15. Moreover, by the definition of a hedging strategy, $Y_n \geq 0$ for all n, and $Y_N = (S_N - C)^+$ a.s. (in fact for all ω). On the other hand, martingales have a constant mean; i.e., $EY_N = EY_1 = y_0$, thanks to (8.71). Therefore, we have shown that $y_0 = E[(S_N - C)^+]$ as desired. $\square$

In order to prove the second half we need the following.

The Martingale Representation Theorem. *In $(\Omega, \mathscr{P}(\Omega), P)$, the process S is a mean-zero martingale. Any other martingale M is a martingale transform of S; i.e., there exists a previsible process H such that*

$$(8.75) \qquad M_n = EM_1 + \sum_{j=1}^n H_j(S_j - S_{j-1}) \qquad \forall n = 1, \ldots, N.$$

Proof. Because Lemma 8.29 proves that S is a mean-zero martingale, we can concentrate on proving that M is a martingale transform.

Since M is adapted, M_n is a function of $\omega_1, \ldots, \omega_n$ only. We abuse the notation slightly, and write

$$(8.76) \qquad M_n(\omega) = M_n(\omega_1, \ldots, \omega_n) \qquad \forall \omega \in \Omega.$$

The martingale property states that $E[M_{n+1} | \mathscr{F}_n] = M_n$ a.s. Now suppose $\phi_1, \ldots, \phi_n$ are bounded and ϕ_j is a function of ω_j only. Then, thanks to the independence of the ω_j's,

$$\mathrm{E}\left[\prod_{j=1}^{n}\phi_j \cdot M_{n+1}\right]$$

(8.77)
$$= \frac{1}{2}\int_\Omega \prod_{j=1}^{n}\phi_j(\omega_j)M_{n+1}(\omega_1,\ldots,\omega_n,-1)\,\mathrm{Q}(d\omega_1)\cdots\mathrm{Q}(d\omega_n)$$
$$+ \frac{1}{2}\int_\Omega \prod_{j=1}^{n}\phi_j(\omega_j)M_{n+1}(\omega_1,\ldots,\omega_n,1)\,\mathrm{Q}(d\omega_1)\cdots\mathrm{Q}(d\omega_n).$$

That is, we can write

(8.78) $$\mathrm{E}\left[\prod_{j=1}^{n}\phi_j \cdot M_{n+1}\right] = \mathrm{E}\left[\prod_{j=1}^{n}\phi_j \cdot N_n\right],$$

where

(8.79) $$N_n(\omega) := \frac{1}{2}M_{n+1}(\omega_1,\ldots,\omega_n,1) + \frac{1}{2}M_{n+1}(\omega_1,\ldots,\omega_n,-1).$$

Note that N_n is $\mathscr{F}_n$-measurable for every $n = 1, \ldots, N$. Therefore, (8.77) and the martingale property of M together show that $M = N$ a.s. This leads us to the formula

(8.80) $$M_n(\omega) = \frac{1}{2}M_{n+1}(\omega_1,\ldots,\omega_n,1) + \frac{1}{2}M_{n+1}(\omega_1,\ldots,\omega_n,-1),$$

valid for almost all $\omega \in \Omega$.[1] In fact, because Ω is finite and P assigns positive measure to each ω_j, the stated equality must hold for all ω. Moreover, since $\mathscr{F}_0 = \{\varnothing,\Omega\}$, the preceding discussion continues to hold for $n = 0$ if we define $M_0 = \mathrm{E}M_1$.

Since $M_n(\omega) = \frac{1}{2}M_n(\omega) + \frac{1}{2}M_n(\omega)$, the following holds for all $0 \le n \le N-1$ and all $\omega \in \Omega$:

(8.81) $$M_{n+1}(\omega_1,\ldots,\omega_n,1) - M_n(\omega) = M_n(\omega) - M_{n+1}(\omega_1,\ldots,\omega_n,-1).$$

---

[1]While this calculation is intuitively clear, you should prove its validity by first checking it for M_{n+1} of the form $M_{n+1}(\omega_1,\ldots,\omega_{n+1}) = \prod_{j=1}^{n+1} h_j(\omega_j)$, and then appealing to a monotone class argument.

Let $d_j := M_{j+1} - M_j$ so that $M_{n+1}(\omega) - M_0 = \sum_{j=0}^n d_j(\omega)$, and apply the preceding as follows:

$$M_{n+1}(\omega) - M_0 = \sum_{j=0}^n \left[d_j(\omega)\mathbf{1}_{\{1\}}(\omega_{j+1}) + d_j(\omega)\mathbf{1}_{\{-1\}}(\omega_{j+1})\right]$$

$$(8.82) \qquad = \sum_{j=0}^n (M_{j+1}(\omega_1, \ldots, \omega_j, 1) - M_j(\omega))\left[\mathbf{1}_{\{1\}}(\omega_{j+1}) - \mathbf{1}_{\{-1\}}(\omega_{j+1})\right]$$

$$= \sum_{j=0}^n (M_{j+1}(\omega_1, \ldots, \omega_j, 1) - M_j(\omega))\left[S_{j+1}(\omega) - S_j(\omega)\right].$$

This proves (8.75) with $H_j(\omega) := M_j(\omega_1, \ldots, \omega_{j-1}, 1) - M_{j-1}(\omega)$ and $M_0 := \mathrm{E}M_1$. [Note that H is previsible.] □

We are ready to prove the second half of the Black–Scholes formula.

Proof of the Black–Scholes Formula (Sufficiency). The process $Y_n = \mathrm{E}[(S_N - C)^+ \mid \mathscr{F}_n]$ ($0 \le n \le N$) is a non-negative Doob martingale. Also it has the property that $Y_N = (S_N - C)^+$ almost surely, and hence for all ω (why?). Thanks to the martingale representation theorem, we can find a previsible process A such that

$$(8.83) \qquad Y_n = \mathrm{E}Y_1 + \sum_{j=1}^{n-1} A_j(S_j - S_{j-1}).$$

It follows that A is a hedging strategy with $y_0 = \mathrm{E}Y_1$. By the martingale property, $\mathrm{E}Y_1 = \mathrm{E}Y_2 = \cdots = \mathrm{E}Y_N$. This implies that $y_0 = \mathrm{E}[(S_N - C)^+]$, which proves the theorem. □

6.6. Rademacher's Theorem. A function $f : (0, 1] \to \mathbf{R}$ is *Lipschitz continuous* if there exists a constant $A > 0$ such that for all $x, y \in (0, 1]$,

$$(8.84) \qquad |f(x) - f(y)| \le A|x - y|.$$

The optimal choice of A is called the *Lipschitz constant* of f.

If f' exists and is continuous, then one can perform a one-term Taylor expansion to note that f is Lipschitz continuous. The following theorem of Rademacher (1919) asserts a remarkable converse.

Theorem 8.43. *If $f : (0, 1] \to \mathbf{R}$ is Lipschitz-continuous then it is differentiable almost everywhere.*

Proof. Let $((0, 1], \mathscr{B}((0, 1]), \mathrm{P})$ denote the Steinhaus probability space, so that P is Lebesgue measure, and "a.e." is the same thing as "a.s." Also let $\{\mathscr{F}_n^0\}_{n=1}^\infty$ and $\{\mathscr{F}_n\}_{n=1}^\infty$ respectively denote the dyadic intervals and filtration (p. 142).

We associate to all dyadic intervals $Q \in \mathscr{F}_n^0$ two numbers, $\ell(Q)$ and $r(Q)$: $\ell(Q)$ is the left end-point of Q and $r(Q)$, the right one. For instance, if $Q = (k2^{-n}, (k+1)2^{-n}]$, then $\ell(Q) = k2^{-n}$ and $r(Q) = (k+1)2^{-n}$.

Define

$$(8.85) \qquad X_n(\omega) := \sum_{Q \in \mathscr{F}_n^0} \frac{f(r(Q)) - f(\ell(Q))}{r(Q) - \ell(Q)} \mathbf{1}_Q(\omega) \qquad \forall \omega \in (0,1].$$

This is a difference quotient because the sum consists of exactly one term, and the Lipschitz continuity of f ensures that $\sup_n |X_n|$ is a bounded random variable. From here on, the proof splits into two steps.

Step 1. X is a martingale with respect to $\mathscr{F}$. To prove this, write

$$(8.86) \qquad X_n(\omega) = \sum_{J \in \mathscr{F}_{n-1}^0} \sum_{\substack{Q \in \mathscr{F}_n^0: \\ Q \subset J}} \frac{f(r(Q)) - f(\ell(Q))}{2^{-n}} \mathbf{1}_Q(\omega).$$

By Corollary 8.8, if $\omega \in J \in \mathscr{F}_{n-1}^0$ and $Q \in \mathscr{F}_n$ is a subset of J, then $\mathrm{P}(Q \mid \mathscr{F}_{n-1})(\omega)$ is the classical probability $\mathrm{P}(Q \mid J)$, which is $\frac{1}{2}$. If $\omega \notin J$ then $\mathrm{P}(Q \mid \mathscr{F}_{n-1})(\omega) = 0$. Thus,

$$(8.87) \qquad \begin{aligned} \mathrm{E}[X_n \mid \mathscr{F}_{n-1}] &= \frac{1}{2} \sum_{J \in \mathscr{F}_{n-1}^0} \sum_{\substack{Q \in \mathscr{F}_n^0: \\ Q \subset J}} \frac{f(r(Q)) - f(\ell(Q))}{2^{-n}} \mathbf{1}_J \\ &= \sum_{J \in \mathscr{F}_{n-1}^0} \frac{f(r(J)) - f(\ell(J))}{2^{-n+1}} \mathbf{1}_J \\ &= X_{n-1}. \end{aligned}$$

According to the martingale convergence theorem, all bounded martingales converge a.s. and in $L^1(\mathrm{P})$ (p. 134). Therefore, we can find X_∞ such that $X_n \to X_\infty$ a.s. [P] and in $L^1(\mathrm{P})$.

Step 2. The Conclusion. Suppose $I, J \in \mathscr{F}_n^0$ for the same $n \geq 1$, and I lies to the left of J; that is, every $u \in I$ is less than every $v \in J$. Then we denote this by $I < J$.

For all $Q \in \mathscr{F}_n^0$, $\int_Q X_n(\omega)\,d\omega = f(r(Q)) - f(\ell(Q))$. Therefore, for all $J \in \mathscr{F}_n^0$,

$$(8.88) \qquad f(r(J)) - f(0) = \sum_{\substack{I \in \mathscr{F}_n^0: \\ I < J}} \int_I X_n(\omega)\,d\omega = \int_0^{r(J)} X_n(\omega)\,d\omega.$$

Given any $x \in (0,1]$ and $n \geq 1$, we can find a unique $J \in \mathscr{F}_0^n$ such that $\omega \in \mathscr{F}_0^n$. If A denotes the Lipschitz constant of f, then

(8.89) $$|f(x) - f(r(J))| \leq A|x - r(J)| \leq \frac{A}{2^n}.$$

Therefore, $|f(x) - f(0) - \int_0^{r(J)} X_n(\omega)\,d\omega| \leq A2^{-n}$. Also,

(8.90) $$\left| \int_0^{r(J)} X_n(\omega)\,d\omega - \int_0^x X_n(\omega)\,d\omega \right| \leq \int_x^{x+2^{-n}} |X_n(\omega)|\,d\omega.$$

By the dominated convergence theorem, the right-hand side goes to zero as $n \to \infty$. Therefore, by the monotone convergence theorem,

(8.91) $$f(x) - f(0) = \int_0^x X_\infty(\omega)\,d\omega \qquad \forall x \in (0,1].$$

Rademacher's theorem follows from Lebesgue's differentiation theorem (Theorem 8.39), and $f' = X_\infty$ almost everywhere. $\square$

6.7. Random Patterns. Suppose $X_1, X_2, \ldots$ are i.i.d. random variables with $P\{X_1 = 1\} = p$ and $P\{X_1 = 0\} = q$, where $q := 1 - p$ and $0 < p < 1$. It is possible to use the Kolmogorov zero-one law and deduce that the infinite sequence $X_1, X_2, \ldots$ will a.s. contain a zero, say. In fact, with probability one, any predescribed finite pattern of zeroes and ones will appear infinitely often in the sequence $\{X_1, X_2, \ldots\}$. Let N denote the first k such that the sequence $\{X_1, \ldots, X_k\}$ contains a predescribed, non-random pattern. Then, we wish to know EN.

The simplest patterns are "0" and "1." So let N denote the smallest k such that $\{X_1, \ldots, X_k\}$ contains a "0." It is not hard to convince yourself that $EN = 1/q$ because $P\{N = j\} = p^{j-1}q$ for $j = 1, 2, \ldots$. But this calculation uses too much of the structure of the pattern "0." Next is a more robust argument, due to Li (1980): Consider the process

(8.92) $$Y_n := \frac{1}{q}\mathbf{1}_{\{X_1=0\}} + \frac{1}{q}\mathbf{1}_{\{X_2=0\}} + \cdots + \frac{1}{q}\mathbf{1}_{\{X_n=0\}}.$$

Define $\mathscr{F}_n$ to be the σ-algebra defined by $\{X_i\}_{i=1}^n$ for every $n \geq 1$. Then, for all $n \geq 1$,

(8.93) $$E[Y_{n+1} \mid \mathscr{F}_n] = Y_n + \frac{1}{q}P(X_{n+1} = 0 \mid \mathscr{F}_n) = Y_n + 1.$$

Therefore, $\{Y_n - n\}_{n=1}^\infty$ is a mean-zero martingale (check!). By the optional stopping theorem, $E[N \wedge n] = EY_{N \wedge n}$ for all $n \geq 1$. But $N < \infty$ a.s., and both $\{N \wedge n\}_{n=1}^\infty$ and $\{Y_{N \wedge n}\}_{n=1}^\infty$ are increasing. Therefore, we can apply the monotone convergence theorem to deduce that $EY_N = EN$. Because $Y_N = (1/q)$ almost surely, $EN = 1/q$, as we know already.

The advantage of the second proof is that it can be applied to other patterns. Suppose, for instance, the pattern is a sequence of ℓ ones, where $\ell \geq 1$ is an integer. Consider

$$
\begin{aligned}
Z_{n,\ell} := {} & \frac{1}{p^\ell}\mathbf{1}_{\{X_1=1,\ldots,X_\ell=1\}} + \frac{1}{p^\ell}\mathbf{1}_{\{X_2=1,\ldots,X_{\ell+1}=1\}} \\
& + \cdots + \frac{1}{p^\ell}\mathbf{1}_{\{X_{n-\ell+1}=1,\ldots,X_n=1\}} + \frac{1}{p^{\ell-1}}\mathbf{1}_{\{X_{n-\ell+2}=1,\ldots,X_n=1\}} \\
& + \frac{1}{p^{\ell-2}}\mathbf{1}_{\{X_{n-\ell+3}=1,\ldots,X_n=1\}} + \cdots + \frac{1}{p}\mathbf{1}_{\{X_n=1\}}.
\end{aligned}
\tag{8.94}
$$

Then, you should check that $\{Z_{n,\ell} - n\}_{n=1}^\infty$ is a martingale. As before, we have $\mathrm{E}Z_{N,\ell} = \mathrm{E}N$, and now we note that $Z_{N,\ell} = (1/p^\ell) + (1/p^{\ell-1}) + \cdots + (1/p)$ a.s. Therefore,

$$
\mathrm{E}N = \sum_{k=1}^{\ell} \frac{1}{p^k} = \frac{1}{q}\left(\frac{1}{p^\ell} - 1\right).
\tag{8.95}
$$

Therefore, set $\ell = 2$ to find that

$$
\mathrm{E}N = \frac{1+p}{p^2} \quad \text{for the pattern "11."}
\tag{8.96}
$$

Our next result is another example of this kind.

Lemma 8.44. *If the pattern is "01" then* $\mathrm{E}N = 1/(pq)$.

Define

$$
W_n := \frac{1}{pq}\mathbf{1}_{\{X_1=0,X_2=1\}} + \cdots + \frac{1}{pq}\mathbf{1}_{\{X_{n-1}=0,X_n=1\}} + \frac{1}{q}\mathbf{1}_{\{X_n=0\}}.
\tag{8.97}
$$

One can prove that $\{W_n - n\}_{n=1}^\infty$ is a martingale. Lemma 8.44 is proved by using the preceding martingale methods.

Suppose we wished to know which of the two patterns, "01" and "11," is more likely to come first. To answer this, we first note that $\{W_n - Z_{n,2}\}_{n=1}^\infty$ is a martingale, since it is the difference of two martingales.

Define T to be the smallest integer $k \geq 1$ such that the sequence $\{X_1, \ldots, X_k\}$ contains either "01" or "11." Then, we argue as before and find that $\mathrm{E}[W_T - Z_{T,2}] = 0$. But $W_T - Z_{T,2} = q^{-1}$ on $\{$"01" comes up first$\}$, and $W_T - Z_{T,2} = -(1/p) - (1/p^2) = -(p+1)/p^2$ on $\{$"11" comes up first$\}$. Therefore,

$$
\begin{aligned}
0 &= \mathrm{E}[W_T - Z_{T,2}] \\
&= \frac{1}{q}\mathrm{P}\{\text{"01" comes up first}\} - \frac{p+1}{p^2}\mathrm{P}\{\text{"11" comes up first}\}.
\end{aligned}
\tag{8.98}
$$

Solve to find that

$$
\mathrm{P}\{\text{"01" comes up before "11"}\} = 1 - p^2.
\tag{8.99}
$$

6.8. Random Quadratic Forms. Let $\{X_i\}_{i=1}^{\infty}$ be a sequence of i.i.d. random variables. For a given double array $\{a_{i,j}\}_{i,j=1}^{\infty}$ of real numbers, we wish to consider the "quadratic form" process,

(8.100) $$Q_n := \sum\sum_{1 \leq i,j \leq n} a_{i,j} X_i X_j \qquad \forall n \geq 1.$$

Define $a_{i,j}^* := (a_{i,j} + a_{j,i})/2$. A little thought shows that we do not alter the value of Q_n if we replace $a_{i,j}$ by $a_{i,j}^*$. Therefore, we can assume that $a_{i,j} = a_{j,i}$, and suffer no loss in generality.

The quadratic form process $\{Q_n\}_{n=1}^{\infty}$ arises in many disciplines. For instance, in mathematical statistics, $\{Q_n\}_{n=1}^{\infty}$ belongs to an important family of processes called "U-statistics."

Theorem 8.45 (Varberg, 1966). *Suppose* $\mathrm{E}X_1 = 0$, $\mathrm{E}[X_1^2] = 1$, *and* $\mathrm{E}[X_1^4] < \infty$. *If* $\sum_{i=1}^{\infty}\sum_{j=1}^{\infty} a_{i,j}^2 < \infty$, *then* $\lim_{n\to\infty}(Q_n - \sum_{1 \leq i \leq n} a_{i,i})$ *exists and is finite a.s.*

Proof. Let $A_n := \sum_{1 \leq i \leq n} a_{i,i}$. Then, a direct computation reveals that $\{Q_n - A_n\}_{n=1}^{\infty}$ is a mean-zero martingale. We plan to prove that the collection $\{Q_n - A_n\}_{n=1}^{\infty}$ is bounded in $L^2(\mathrm{P})$. Because boundedness in $L^2(\mathrm{P})$ implies boundedness in $L^1(\mathrm{P})$, the martingale convergence theorem does the rest.

Thanks to the symmetry of the $a_{i,j}$'s we can write

(8.101) $$Q_n - A_n = 2U_n + V_n,$$

where

(8.102) $$U_n := \sum\sum_{1 \leq i < j \leq n} a_{i,j} X_i X_j \quad \text{and} \quad V_n := \sum_{1 \leq i \leq n} a_{i,i}\left[X_i^2 - 1\right].$$

Because $(x+y)^2 \leq 2(x^2 + y^2)$ for all $x, y \in \mathbf{R}$, it follows that $(Q_n - A_n)^2 \leq 8U_n^2 + 2V_n^2$. Therefore,

(8.103) $$\mathrm{E}\left[(Q_n - A_n)^2\right] \leq 8\mathrm{E}[U_n^2] + 2\mathrm{E}[V_n^2].$$

By independence, $\mathrm{E}[V_n^2] = \sum_{1 \leq i \leq n} a_{i,i}^2 \mathrm{Var}(X_1^2)$, and this is bounded in n. Similarly, $\sup_n \mathrm{E}[U_n^2] = \sum\sum_{1 \leq i < j < \infty} a_{i,j}^2 < \infty$. $\square$

Problems

8.1. We say that $X_n \in L^1(\mathrm{P})$ converges to $X \in L^1(\mathrm{P})$ *weakly in* $L^1(\mathrm{P})$ if for all bounded random variables Z, $\lim_{n\to\infty} \mathrm{E}[X_n Z] = \mathrm{E}[XZ]$. Show that $X_n \to X$ weakly in $L^1(\mathrm{P})$ if for any sub-σ-algebra $\mathscr{G} \subseteq \mathscr{F}$, $\mathrm{E}[X_n \mid \mathscr{G}]$ converges to $\mathrm{E}[X \mid \mathscr{G}]$ weakly in $L^1(\mathrm{P})$. Conversely, prove that if $X_n \to X$ in $L^1(\mathrm{P})$, then for any sub-σ-algebra $\mathscr{G} \subseteq \mathscr{F}$, $\mathrm{E}[X_n \mid \mathscr{G}] \to \mathrm{E}[X \mid \mathscr{G}]$ in $L^1(\mathrm{P})$.

8.2. Construct three random variables U, V, W such that $\mathrm{E}[\mathrm{E}(U \mid V) \mid W] \neq \mathrm{E}[\mathrm{E}(U \mid W) \mid V]$ with positive probability.

8.3. Suppose X and Y are independent real (say) random variables, and $f : \mathbf{R}^2 \to \mathbf{R}$ is bounded and measurable. If $g(x) := \mathrm{E}[f(x, Y)]$ for all $x \in \mathbf{R}$, then prove that $g(X) = \mathrm{E}[f(X, Y) \mid X]$ a.s.

8.4. Suppose $X, Y \in L^1(P)$ satisfy $E[X \mid Y] = Y$ and $E[Y \mid X] = X$ a.s. Prove then that $X = Y$ a.s. (HINT: Prove first that $E[X - Y; X < q < Y] = 0$ for all $q \in \mathbf{Q}$; see also Doob (1953, p. 314).)

8.5. Let $((0, 1], \mathscr{B}((0, 1]), P)$ denote the Steinhaus probability space, and consider $X(\omega) = \omega$ for $0 < \omega \leq 1$. Compute and compare $E[X \mid \mathscr{S}_1]$ and $E[X \mid \mathscr{S}_2]$ where $\mathscr{S}_1$ is the σ-algebra generated by $\sigma((0, 1/2])$, and $\mathscr{S}_2$ is the sub-σ-algebra of $\mathscr{B}((0, 1])$ that is generated by $\mathscr{B}((0, 1/2])$. This is due to J. A. Turner.

8.6 (Conditional Variance). Suppose $\mathscr{G} \subseteq \mathscr{F}$ is a σ-algebra. If $X \in L^2(P)$, then define $\mathrm{Var}(X \mid \mathscr{G}) = E\{(X - E[X \mid \mathscr{G}])^2 \mid \mathscr{G}\}$. Prove that $\mathrm{Var}(X \mid \mathscr{G}) \leq E\{(X - \xi)^2 \mid \mathscr{G}\}$ a.s. for every $X \in L^2(\Omega, \mathscr{F}, P)$ and $\xi \in L^2(\Omega, \mathscr{G}, P)$. Derive (8.4) as a consequence.

8.7 (Conditional Chebyshev). Prove that if $\mathscr{G}$ is a sub-σ-algebra of $\mathscr{F}$ and X is a non-negative random variable on $(\Omega, \mathscr{F}, P)$, then $\lambda P(X \geq \lambda \mid \mathscr{G}) \leq E[X \mid \mathscr{G}]$ a.s. for all $\lambda > 0$.

8.8 (Corollary 8.8, p. 125, Continued). Let (X, Y) be an absolutely continuous random variable with piecewise-continuous density $f(x, y)$. Prove that for every non-negative measurable $h : \mathbf{R} \to \mathbf{R}$,

$$(8.104) \qquad E[h(X) \mid Y] = \frac{\int_{-\infty}^{\infty} h(x) f(x, Y) \, dx}{\int_{-\infty}^{\infty} f(x, Y) \, dx} \qquad \text{a.s.}$$

Even though $P\{Y = y\} = 0$ for all $y \in \mathbf{R}$, use the preceding to justify the classical definition, $P(X \leq a \mid Y = y) = \int_{-\infty}^{a} f(x, y) \, dx \, / \int_{-\infty}^{\infty} f(x, y) \, dx$ for all $a \in \mathbf{R}$.

8.9 (Censoring). Let $\mathscr{F}$ denote a filtration, and consider $\mathscr{F}$-stopping times $S \leq T$ a.s. For any fixed event $A \in \mathscr{F}_S$ define $\tau = S \mathbf{1}_A + T \mathbf{1}_{A^c}$. Prove that τ is an $\mathscr{F}$-stopping time.

8.10. Verify Proposition 8.9.

8.11. Carefully prove Lemmas 8.25 and 8.26. In addition, construct an example that shows that the difference of two stopping times, even if non-negative, need not be a stopping time.

8.12. The optional stopping theorem (p. 130) assumes that T is almost surely bounded. Construct an example to show that this assumption cannot be dropped altogether.

8.13. Prove Lemma 8.29.

8.14 (Gambler's Ruin). If $\{X_i\}_{i=1}^{\infty}$ are i.i.d. random variables with $P\{X_1 = 1\} = 1 - P\{X_1 = -1\} = p \neq \frac{1}{2}$, then verify that, in the proof of Theorem 8.34, $\{\zeta^{S_n}\}_{n=1}^{\infty}$ is indeed a bounded mean-one martingale. Also compute ET in the case $p = 1/2$.

8.15. Let $\{X_n\}_{n=1}^{\infty}$ be independent random variables such that for all $k \geq 1$, $h_k(t) = E e^{t X_k}$ exists and is finite for all $t \in (-t_0, t_0)$ for a fixed $t_0 > 0$. Prove that whenever $|t| < t_0$, $M_n(t) = e^{t S_n} / \prod_{k=1}^{n} h_k(t)$ defines a mean-one martingale. [As usual, S_n denotes $\sum_{i=1}^{n} X_i$.]

8.16 (Likelihood Ratios). Suppose f and g are two strictly positive probability density functions on $\mathbf{R}$. Prove that if $\{X_i\}_{i=1}^{\infty}$ are i.i.d. random variables with probability density f, then $\prod_{j=1}^{n} [g(X_j)/f(X_j)]$ defines a mean-one martingale. When does it converge, and in what sense(s)?

8.17 (Pólya's Urns). An urn initially contains R red and B black balls. Except for their colors, the balls are identical. A ball is chosen at random. If it comes up red (resp. black), then it is replaced with two red (resp. black) balls. Let X_n denote the number of red balls in the urn after n draws. Prove that the fraction $f_n = X_n/(n + R + B)$ of red balls has an almost-sure limit.

8.18. Prove that Definition 8.12(iii) is equivalent to $E[X_{n+k} \mid \mathscr{F}_n] \geq X_n$ a.s. for all $k, n \geq 1$.

8.19. Prove that Doob's decomposition (p. 128) is a.s.-unique, as long as we insist that $\{Z_n\}_{n=1}^{\infty}$ is previsible and $Z_1 = 0$.

8.20. Let $\{X_n\}_{n=1}^{\infty}$ be a martingale. Prove that X is bounded in $L^1(P)$ iff $\sup_n E[X_n^+] < \infty$.

8.21. Prove that if X is a martingale, then

$$(8.105) \qquad P\left\{ \max_{1 \leq j \leq n} |X_j| \geq \lambda \right\} \leq \frac{E(|X_n|^p)}{\lambda^p} \qquad \forall p \geq 1, \, n \geq 1, \, \lambda > 0.$$

Also, prove that Doob's inequalities imply Kolmogorov's maximal inequality (p. 74).

8.22 (Doob's L^p Inequality). Suppose ξ and ζ are a.s. non-negative random variables such that

(8.106) $$P\{\xi > a\} \leq \frac{1}{a} E[\zeta; \xi \geq a] \qquad \forall a > 0.$$

Prove that for all $p > 1$,

(8.107) $$\|\xi\|_p \leq \left(\frac{p}{p-1}\right) \|\zeta\|_p.$$

Use this show the *strong L^p-inequality of Doob*: If X is a non-negative submartingale and $X_n \in L^p(P)$ for all $n \geq 1$ and some $p > 1$, then

(8.108) $$E\left[\max_{1 \leq j \leq n} X_j^p\right] \leq \left(\frac{p}{p-1}\right)^p E[X_n^p].$$

Use this to prove Corollary 8.41.

8.23 (Pitman's L^2 Inequality; Problem 8.22, Continued). Suppose $\{X_i\}_{i=1}^n$ and $\{M_k\}_{k=1}^n$ are processes that satisfy: (i) $M_k = X_k$ whenever $M_k \neq M_{k-1}$; and (ii) $E[\sum_{k=2}^n M_{k-2}(X_k - X_{k-1})]$ is non-negative. Prove that $E[M_n^2] \leq 4E[X_n^2]$. Use this to conclude (8.108) in the case that $p = 2$ (Pitman, 1981).

8.24 (Problem 8.22, Continued; Square Functions). Let $\{\mathscr{F}_i\}_{i=1}^\infty$ denote a filtration, and suppose $\{X_i\}_{i=1}^\infty$ is a martingale such that $X_i \in L^2(P)$ for all $i \geq 1$.

  (1) Prove that $X_n^2 - A_n$ defines a mean-zero martingale where $A_n = \sum_{i=1}^n E[d_i^2 \mid \mathscr{F}_{i-1}]$, $d_i = X_i - X_{i-1}$, $X_0 = 0$, and $\mathscr{F}_0$ is the trivial σ-algebra. The process A is called the *square function* of X.
  (2) Prove that $E[\sup_{i \leq n} X_i^2] \leq 4E[A_n]$ for all $n \geq 1$.
  (3) Conclude that $\lim_n X_n(\omega)$ exists for almost all $\omega \in \{A_\infty < \infty\}$.
  (4) Explore the case where the d_i's are independent.

8.25 (Theorem 8.30, Continued). Suppose $S_n = X_1 + \cdots + X_n$ defines a random walk with $EX_1 = \mu$ and $VarX_1 = \sigma^2 < \infty$. Let $\mathscr{F}_n = \sigma(\{X_i\}_{i=1}^n)$ ($n \geq 1$), and consider an $\mathscr{F}$-stopping time T that has a finite mean. Prove *Wald's second identity*, $VarS_T = VarX_1 ET$. (HINT: Problem 8.22.)

8.26. Consider two random variables X and Y, both of which are defined on a common probability space $(\Omega, \mathscr{F}, P)$. Define

(8.109) $$X_n = \sum_{j=-\infty}^\infty \left(\frac{j}{2^n}\right) \mathbf{1}_{\{X \in [j2^{-n}, (j+1)2^{-n})\}} \qquad \forall n \geq 1.$$

Prove that for any $Y \in L^1(P)$, $\lim_{n \to \infty} E[Y \mid X_n] = E[Y \mid X]$ a.s. and in $L^1(P)$.

8.27. Suppose that $Y \in L^1(P)$ is real-valued, and that X is a random variable that takes values in $\mathbf{R}^n$. Prove that there exists a Borel measurable function f such that $E[Y \mid X] = f(X)$ almost surely.

8.28. Suppose that $\{X_i\}_{i=1}^\infty$ are independent mean-zero random variables in $L^2(P)$ that are bounded; that is, that there exists a constant B such that almost surely, $|X_n| \leq B$ for all n. Prove that if $S_n = X_1 + \cdots + X_n$, then for all $\lambda > 0$ and $n \geq 1$,

(8.110) $$P\left\{\max_{1 \leq j \leq n} |S_j| \leq \lambda\right\} \leq \frac{(B + \lambda)^2}{VarS_n}.$$

(Khintchine and Kolmogorov, 1925).

8.29 (Martingale Convergence in L^p). Refine the martingale convergence theorem by showing that $\lim_{n \to \infty} X_n$ exists in $L^p(P)$ whenever X is bounded in L^p for some $p > 1$. In addition, prove that if $X_\infty = \lim_{n \to \infty} X_n$, then $X_n = E[X_\infty \mid \mathscr{F}_n]$ a.s. for all $n \geq 1$. (HINT: Use Problem 8.22.)

8.30 (Double-or-Nothing). Let $\{\gamma_i\}_{i=1}^\infty$ denote a sequence of i.i.d. random variables with $P\{\gamma_1 = 0\} = P\{\gamma_1 = 1\} = \frac{1}{2}$. Consider the stochastic process X, where $X_1 := 1$, and $X_n := 2X_{n-1}\gamma_n$ for all $n \geq 2$. Prove that X is an L^1-bounded martingale that does not converge in $L^1(P)$. Consequently, Problem 8.29 can fail for $p = 1$. Compute the almost-sure limit of X_n.

8.31. Let $\{X_n\}_{n=1}^\infty$ be independent standard normal random variables and define $S_n = \sum_{i=1}^n X_i$ ($n \geq 1$). Prove that $M_n = (n+1)^{-1/2}\exp\{S_n^2/(2n+2)\}$ defines a mean-one martingale (Woodroofe, 1975, Problem 12.10, p. 344).

8.32 (Problem 8.31, Continued). Define M_n as in Problem 8.31. Use only the martingale convergence theorem (p. 134) and the CLT (p. 100) to prove that $\lim_{n\to\infty} M_n = 0$ a.s. Derive the following precursory formulation of the LIL (Steinhaus, 1922):

$$(8.111) \qquad S_n = o\left((n\ln n)^{1/2}\right) \qquad \text{a.s. as } n \to \infty.$$

8.33. Suppose X is a submartingale with bounded increments; i.e., there exists a non-random finite constant B such that almost surely, $|X_n - X_{n-1}| \leq B$ for all $n \geq 2$. Then prove that $\lim_n X_n$ exists a.s. on the set $\{\sup_m |X_m| < \infty\}$.

8.34. Suppose $\{\mathscr{F}_n\}_{n=1}^\infty$ is a filtration of σ-algebras, and $Y \in L^1(P)$ is fixed. Define $M_n = E[Y \mid \mathscr{F}_n]$ to be the corresponding Doob martingale. Prove that for all finite stopping times T, $M_T = E[Y \mid \mathscr{F}_T]$ a.s. (Dubins and Freedman, 1966).

8.35. Prove that X is a martingale if and only if $EM_T = EM_1$ for all bounded stopping times T. Characterize super- and submartingales similarly.

8.36. Let $\{X_n\}_{n=0}^\infty$ be a non-negative supermartingale that attains the value zero at some a.s.-finite time. Prove that $\lim_{k\to\infty} X_k = 0$ a.s.

8.37. Follow the proof of Theorem 8.40 and prove that $c_1 A(f) \leq Mf \leq c_2 A(f)$ where c_1 and c_2 are positive and finite constants that do not depend on f, and $A(f) := \sup_n M_n + \sup_n N_n$.

8.38. The following is a variant of Problem 8.17. First choose and fix $\lambda \in (0, 1)$. Then consider random variables $X_n \in [0, 1]$, adapted to $\mathscr{F}_n$, such that a.s. for all $n \geq 1$,

$$(8.112) \qquad \begin{aligned} P\{X_{n+1} = \lambda + (1-\lambda)X_n \mid \mathscr{F}_n\} &= X_n \\ P\{X_{n+1} = (1-\lambda)X_n \mid \mathscr{F}_n\} &= 1 - X_n. \end{aligned}$$

Prove that $X_\infty := \lim_{n\to\infty} X_n$ exists a.s. and in $L^p(P)$ for all $p \geq 1$, and that X_∞ is zero or one almost surely. Compute $P\{X_\infty = 1\}$.

8.39. Suppose that $\{X_i\}_{i=1}^\infty$ are i.i.d. with $P\{X_1 = 1\} = P\{X_1 = -1\} = 1/2$. As before, let $S_n := X_1 + \cdots + X_n$.

(1) Prove that $E\exp(tS_n) \leq \exp(nt^2/2)$ for all $n \geq 1$ and $t \in \mathbf{R}$.

(2) Prove that (8.41) continues to hold in the present setting.

(3) Prove the following half of the LIL for (± 1) random variables:
$$\limsup_{n\to\infty} \frac{S_n}{(2n\ln\ln n)^{1/2}} \leq 1 \qquad \text{a.s.}$$

(4) Suppose $\{Y_i\}_{i=1}^\infty$ are i.i.d. with $P\{Y_1 = 0\} = P\{Y_1 = 1\} = 1/2$ and $T_n := Y_1 + \cdots + Y_n$. Prove that
$$\limsup_{n\to\infty} \frac{T_n - \frac{n}{2}}{(2n\ln\ln n)^{1/2}} \leq \frac{1}{2} \qquad \text{a.s.}$$
Check that this is one-half of the LIL (p. 138) for T_n.

8.40 (Uniform Integrability). Consider a martingale $\{X_n\}_{n=1}^\infty$ for which T is an a.s.-finite stopping time. Suppose, in addition, that $\{X_{T\wedge n}\}_{n=1}^\infty$ is uniformly integrable; see Problem 4.28 on page 51. Then prove that $EX_T = EX_1$.

8.41 (Uniform Integrability). Let $\{X_n\}_{n=1}^\infty$ be a uniformly integrable martingale with respect to some filtration $\mathscr{F}$; see Problem 4.28 on page 51. Prove that $X_\infty = \lim_{n\to\infty} X_n$ exists and is finite a.s. Prove also that outside a null set $X_n = E[X_\infty \mid \mathscr{F}_n]$ for all $n \geq 1$.

8.42 (Theorem 8.39, Continued). Prove that if $f:[0,1)^k \to \mathbf{R}$ is integrable ($k \geq 2$), then for almost all $x \in [0,1)^k$

(8.113) $$\lim_{\delta \downarrow 0} \frac{1}{\delta^k} \int_{x_k}^{x_k+\delta} \cdots \int_{x_1}^{x_1+\delta} f(u)\,du_1 \cdots du_k = f(x).$$

8.43. Let X_1 be uniformly distributed on $(0,1)$. Conditionally on X_1, define X_2 to be uniformly distributed on $(0, X_1)$; i.e., $\mathrm{P}\{X_2 \in A \mid X_1\} = m(A \cap [0, X_1])/X_1$, where m denotes the Lebesgue measure. Iteratively define

(8.114) $$\mathrm{P}\{X_n \in A \mid X_1, \ldots, X_{n-1}\} = \frac{m(A \cap [0, X_{n-1}])}{X_{n-1}}.$$

Explore the structure of $\{X_n\}_{n=1}^\infty$, and the behavior of X_n for large n.

8.44 (Patterns). Verify Lemma 8.44. Also, find the probability that we see ℓ consecutive ones before k consecutive zeros.

8.45 (U-Statistics). Prove that $\mathrm{E}[U_n V_n] = 0$ in the proof of Theorem 8.45. From this conclude that

(8.115) $$\mathrm{E}\left[(Q_n - A_n)^2\right] = 4 \sum\sum_{1 \leq i < j \leq n} a_{i,j}^2 + \mathrm{Var}(X_1^2) \sum_{1 \leq i \leq n} a_{i,i}^2.$$

8.46 (Runs in Random Permutations; Hard). Let Π_n denote a random permutation of $\{1, \ldots, n\}$, all permutations being equally likely. A block of ascending elements of Π_n is a *run* if it is not a sub-block of a longer block of ascending elements. For example, if $\Pi_3 = \{7, 6, 8, 3, 1, 2, 5, 4\}$, then it has five runs: $\{7\}$; $\{6, 8\}$; $\{3\}$; $\{1, 2, 5\}$; and $\{4\}$. Prove that if R_n denotes the number of runs of Π_n, then $nR_n - \binom{n+1}{2}$ defines a mean-zero martingale, and $\mathrm{Var}\,R_n = O(n)$ as $n \to \infty$. Conclude that

(8.116) $$\lim_{n \to \infty} \frac{R_n}{n} = \frac{1}{2} \quad \text{a.s.}$$

8.47 (Reversed Doob Martingales; Hard). Let $\mathscr{F}_1 \supseteq \mathscr{F}_2 \supseteq \cdots$ be a decreasing family of sub-σ-algebras of $\mathscr{F}$. The family $\{\mathscr{F}_n\}_{n=1}^\infty$ is called a *reversed* (or "backward") filtration. Prove that if $Y \in L^1(\mathrm{P})$ then $\lim_{n \to \infty} \mathrm{E}[Y \mid \mathscr{F}_n]$ exists a.s. and in $L^1(\mathrm{P})$.

8.48 (Exchangeability; Problem 8.47, Continued; Hard). Random variables $X_1, X_2, \ldots$ are said to be *exchangeable* if the distribution of $(X_1, \ldots, X_n)$ is the same as that of $(X_{\pi(1)}, \ldots, X_{\pi(n)})$ for every permutation π of $\{1, \ldots, n\}$ and all $n \geq 1$. Define $S_n = X_1 + \cdots + X_n$ for all $n \geq 1$ and let $\mathscr{E}_n$ denote the σ-algebra generated by $\{S_k\}_{k=n}^\infty$. If $\mathrm{E}|X_1| < \infty$ then:

(1) Compute $\mathrm{E}[X_i \mid \mathscr{E}_n]$ for all $1 \leq i \leq n$.
(2) Prove that $Z := \lim_{n \to \infty}(S_n/n)$ exists a.s. and in $L^1(\mathrm{P})$.

8.49 (Lévy's Equivalence Theorem; Hard). Let $\{X_n\}_{n=1}^\infty$ be independent. Prove that $S_n = \sum_{i=1}^n X_i$ converges almost surely if and only if S_n converges in probability (Lévy, 1937, Théorème 44, p. 139). (HINT: Begin by proving that $\exp(itS_n)/\mathrm{E}\exp(itS_n)$ defines a mean-one "complex" martingale for any $t \in \mathbf{R}$.)

8.50 (Azuma–Hoeffding Inequality; Hard). Let $\{X_i\}_{i=0}^n$ be a mean-zero martingale. Suppose there exist non-random constants $\{c_i\}_{i=1}^n$ such that $|X_i - X_{i-1}| \leq c_i$ a.s. ($i = 1, \ldots, n$). Prove that

(8.117) $$\mathrm{P}\left\{\max_{0 \leq i \leq n} |X_i| \geq z\right\} \leq 2\exp\left(-\frac{z^2}{2\sum_{i=1}^n c_i^2}\right) \quad \forall z > 0.$$

(HINT: Consult Problems 4.30 (p. 51) and 6.33 (p. 87).)

8.51 (Problem 8.40, Continued; Hard). Find an example of a mean-zero martingale $\{X_n\}_{n=1}^\infty$ and an a.s.-finite stopping time T such that $\mathrm{E}X_T \neq 0$.

8.52 (Problem 8.22, Continued; Hard). Suppose ξ and ζ are a.s. non-negative random variables such that for all $a > 0$,

(8.118) $$P\{\xi > a\} \leq \frac{1}{a} E[\zeta;\, \xi \geq a].$$

Prove that $\xi \in L^1(P)$ as long as $E[\zeta \ln_+ \zeta] < \infty$. Here, $\ln_+ y = \ln(y \wedge e)$. Use this to prove the *strong L^1-inequality* of Doob: *If X is a non-negative submartingale, then* $\sup_n E(|X_n| \ln_+ |X_n|) < \infty$ *implies* $E \sup_n |X_n| < \infty$. (HINT: Prove first that if $0 \leq x \leq y$, then $x \ln y \leq x \ln_+ x + (y/e)$.)

8.53 (Hard). Prove that if $f : \mathbf{R} \to \mathbf{R}$ is convex, then f' exists almost everywhere. [In fact, f'' exists a.e., but this is a little more difficult.]

8.54 (de Finetti's Theorem; Problem 8.48, Continued; Hard). Suppose $\{X_i\}_{i=1}^\infty$ is an exchangeable sequence of zeros and ones, $S_n := X_1 + \cdots + X_n$, and $\mathscr{E}_n = \sigma(S_n, S_{n+1}, \ldots)$ for all $n \geq 1$.

(1) Prove that $P(X_1 = \cdots = X_k = 1 \mid \mathscr{E}_n) = \binom{n-k}{n-S_n} / \binom{n}{S_n}$ a.s.

(2) Use Stirling's formula to conclude the theorem of de Finetti (1937):
$$P(X_1 = \cdots = X_k = 1,\, X_{k+1} = \cdots = X_n = 0 \mid Z) = Z^k(1-Z)^{n-k} \quad \text{a.s.}$$

8.55 (Problem 8.52, Continued; Harder). Problem 8.52 cannot be improved. Let $\{X_n\}_{n=1}^\infty$ denote i.i.d. mean-zero random variables, and define $S_n = \sum_{i=1}^n X_i$ for all $n \geq 1$. Prove that $E \sup_n |S_n/n|$ and $E\{|X_1| \ln_+ |X_1|\}$ converge and diverge together (Burkholder, 1962).

8.56 (Problem 8.39, Continued; Harder). Prove the other half of the LIL (p. 138) for (± 1) random variables: $\limsup_{n\to\infty} S_n/a_n \geq 1$ a.s., where $a_n := (2n \ln \ln n)^{1/2}$. You may use the following argument (de Acosta, 1983, Lemma 2.4):

(1) Prove that it suffices to show that for all $c > 0$,
$$\liminf_{n\to\infty} \frac{1}{\ln \ln n} \ln P\left\{S_n \geq c^{1/2} a_n\right\} \geq -c.$$

(2) To establish (1) choose $p_n \to \infty$ such that n divides p_n, and then prove that
$$P\left\{S_n \geq c^{1/2} a_n\right\} \geq \left(P\left\{S_{p_n} \geq c^{1/2} a_n p_n/n\right\}\right)^{n/p_n}.$$

(3) Use the central limit theorem and the preceding with $p_n \sim \alpha n/(\ln \ln n)$ to prove (1). Conclude the proof of the LIL for (± 1) random variables.

(HINT: For part (2) first write $S_n = S_{p_n} + (S_{2p_n} - S_{p_n}) + \cdots + (S_n - S_{(n-1)p_n/n})$. Next observe that if each of these n/p_n terms is greater than $p_n \lambda/n$, then $S_n \geq \lambda$. Finally choose λ judiciously. For part (3) optimize over the choice of α.)

8.57 (Problem 8.24, Continued; Harder). Suppose $\{X_n\}_{n=1}^\infty$ is a martingale with respect to some filtration $\{\mathscr{F}_n\}_{n=1}^\infty$. Suppose also that $d_n = X_n - X_{n-1}$ satisfies $|d_n| \leq \alpha$ for all $n \geq 1$, where $X_0 = 0$, $\mathscr{F}_0 = \{\varnothing, \Omega\}$, and α is a non-random positive constant.

(1) Prove that for all $x \in \mathbf{R}$, $e^x \leq 1 + x + \frac{1}{2} x^2 e^{|x|}$. Use this to prove that for all $t \in \mathbf{R}$ and all $i = 1, 2, \ldots$,
$$E\left[e^{td_i} \mid \mathscr{F}_{i-1}\right] \leq 1 + \frac{t^2 e^{\alpha|t|} E[d_i^2 \mid \mathscr{F}_{i-1}]}{2} \leq \exp\left(\frac{t^2 e^{\alpha|t|} E[d_i^2 \mid \mathscr{F}_{i-1}]}{2}\right) \quad \text{a.s.}$$

(2) Let $\{A_n\}_{n=1}^\infty$ denote the square function of X. Then conclude that given a non-random $t \in \mathbf{R}$ the following defines a non-negative supermartingale:
$$M_n = \exp\left(tX_n - \frac{t^2 e^{\alpha|t|} A_n}{2}\right) \quad \forall n \geq 1.$$

Moreover, verify that $EM_n \leq 1$ for all $n \geq 1$.

(3) Prove that if $X_n \geq 0$, then $\lim_{n\to\infty} X_n/A_n$ exists and is finite a.s. Prove also that
$$\lim_{n\to\infty} \frac{X_n(\omega)}{A_n(\omega)} = 0 \quad \text{for almost all } \omega \in \{A_\infty = \infty\}.$$

Notes

(1) Martingales were first introduced and studied by Ville (1939). The current powerful theory was formed by Doob (1940, 1949) shortly thereafter.

(2) Our proof of the martingale convergence theorem (p. 134) is due to Isaac (1965). Aside from this and the original proof of Doob (1940) there are other nice proofs of Doob's martingale convergence theorem. For example, see Chatterji (1968), Helms and Loeb (1982), Lamb (1973), and C. Ionescu Tulcea and A. Ionescu Tulcea (1963).

(3) An enormous literature is devoted to the study of the law of the iterated logarithm and its variants. An excellent starting point is the theorem of Strassen (1967). It implies that if $X_1, X_2, \ldots$ are i.i.d., $EX_1 = 0$, and $\text{Var} X_1 = 1$, then on some suitable probability space there exist i.i.d. $N(0,1)$ random variables $\{G_i\}_{i=1}^{\infty}$ such that

$$\lim_{n \to \infty} \frac{\left|\sum_{i=1}^n X_i - \sum_{i=1}^n G_i\right|}{(n \log \log n)^{1/2}} = 0 \quad \text{a.s.}$$

In particular, this shows that the general LIL follows from the one proved here. Moreover, if the X_i's have higher moments than two, then the rate of approximation can be improved upon. This is the starting point of a theory of "strong approximations." Csörgő and Révész (1981) is an excellent treatment. Two scholarly reviews of the LIL are Feller (1945), for the classical theory, and Bingham (1986), for the more modern advances.

(4) Equation (8.45) has the following improvement, due to Laplace (1805, pp. 490–493):

$$P\{X_1 \geq \lambda\} = \frac{e^{-\lambda^2/2}}{\lambda\sqrt{2\pi}} \cdot \left[1 + \cfrac{\lambda^2}{1 + 2 \cdot \cfrac{\lambda^2}{1 + 3 \cdot \cfrac{\lambda^2}{1 + 4 \cdots}}}\right]^{-1} \quad \forall \lambda > 0.$$

(5) Theorem 8.39 is also known as the *Lebesgue density theorem*. It states that the antiderivative of every $f \in L^1(dx)$ is f a.e. On the other hand, it is the case that "most" continuous functions are nowhere differentiable (Banach, 1931; Mazurkiewicz, 1931; Paley, Wiener, and Zygmund, 1933; Kahane, 1997, 2000, 2001).

(6) The material on option-pricing (§6.5) is based in part on the discussions of Baxter and Rennie (1996, Chapter 2) and Williams (1991, Section 15.2). There you will learn, among other things, that there are in fact hedging strategies that never sell short. This demonstration requires only a little more effort than the proof described here, and is worth looking at.

(7) The notion of Lipschitz continuity is due to the work of Lipschitz (1876) on differential equations.

(8) One can streamline the method of §6.7; see Li (1980) and Gerber and Li (1981).

(9) Problem 8.46 is, in essence, borrowed from the exciting book of Mahmoud (2000, pp. 48–51) on sorting. It can be shown that

$$\frac{R_n - (n/2)}{\sqrt{n}} \Rightarrow N(0, 1/12).$$

(*ibid.*, Proposition 1.10, p. 51).

(10) Problem 8.48 is due to de Finetti (1937), but the proof outlined here is borrowed from Doob (1949). Aldous (1985) presents a masterly modern-day account of exchangeability and related topics.

(11) When the increments of X are independent, Problem 8.50 is due to Hoeffding (1963). The general case is due to Azuma (1967), and is proved by the same argument.

(12) Problem 8.54 states that all exchangeable sequences of zeros and ones are "conditionally i.i.d." The proof outlined here is motivated by Exercise 6.3 of Durrett (1996, p. 271). For a detailed historical account see Cifarelli and Regazzini (1996).

Remarkably enough, de Finetti's theorem has consequences in diverse subjects such as the philopophy of statistics (de Finetti, 1937; Kyburg and Smokler, 1980), statistical mechanics (Georgii, 1988), and geometry of Hilbert spaces (Bretagnolle and Dacunha-Castelle, 1969).

(13) Problem 8.57 generalizes the "law of large numbers" of Dubins and Freedman (1965). The central ideas used here come from a paper of de Acosta (1983).

Chapter 9

Brownian Motion

The theory of random functions always makes the impression of a much greater degree of artificiality than corresponds to the facts.

−Raymond E. A. C. Paley and Norbert Wiener

On March 29, 1900, a doctoral student of J. H. Poincaré by the name of Louis Jean Baptiste Alphonse Bachelier presented his thesis to the Faculty of Sciences of the Academy of Paris. Louis Bachelier's work was chiefly concerned with finding "a formula which expresses the likelihood of a market fluctuation" (Bachelier, 1964, p. 17). Bachelier's solution to this problem required the introduction of a number of novel ideas, one of which was today's "Brownian motion." See also the English translation in the volume edited by Cootner (Bachelier, 1964).

In 1828, the botanist Robert Brown noted empirically that the grains of pollen in water undergo erratic motion. Brown himself admitted to not having a scientific explanation for this phenomenon. And it was years later, in 1905, that an explanation was found by Albert Einstein. The key idea in Einstein's solution was the introduction of a stochastic process that Einstein called "the Brownian motion." Unaware of the earlier work of Bachelier in economics, Einstein had rediscovered that Brownian motion is related concretely to the diffusion of particles. As a main application of his theory, Einstein found a very good estimate for Avogadro's constant.

Einstein's theory was tacitly based on the assumption that the Brownian motion process exists. Nearly two decades later, Wiener (1923a) proved the validity of Einstein's assumption. In the present context, the contributions of von Smoluchowski (1918) and Perrin (1913) are also particularly noteworthy.

From a mathematical point of view, Bachelier's work went further than Einstein's. However, we introduce the latter's work because it is easier to describe. Thus, we begin with a modern statement of *Einstein's postulates*: Brownian motion $\{W(t)\}_{t\geq 0}$ is a random function of t (= "time") such that:

(P-a) $W(0) = 0$, and for any given time $t > 0$, the distribution of $W(t)$ is normal with mean zero and variance t.

(P-b) For any $0 < s < t$, $W(t) - W(s)$ is independent of $\{W(u)\}_{0 \leq u \leq s}$. Think of s as the current time. Then, this condition is saying that "given the value of W at the present time, the future is independent of the past." This is called the *Markov property*.

(P-c) The random variable $W(t) - W(s)$ has the same distribution as $W(t - s)$. That is, Brownian motion has stationary increments.

(P-d) The random path $t \mapsto W(t)$ is continuous with probability one.

Remark 9.1. One can also have a Brownian motion B that starts at an arbitrary point $x \in \mathbf{R}$ by defining

(9.1) $$B(t) := x + W(t),$$

where W is a Brownian motion started at the origin. One can check directly that B has all the properties of W, except that $B(t)$ has mean x for all $t \geq 0$, and $B(0) = x$. Unless stated to the contrary, our Brownian motions always start at the origin.

So why are (P-a) through (P-d) postulates and not facts? The sticky point is the a.s.-continuity (P-d). In fact, Lévy (1937, Théorème 54.2, p. 181) has proven that if in (P-a) we replace the normal by any other distribution, then either there is no process that satisfies (P-a)–(P-c), or else (P-d) fails to hold.

In summary, while the predictions of theoretical physics were correct, a more solid understanding of Brownian motion required a rather in-depth undertaking such as that of N. Wiener. Since Wiener's work Brownian motion has been studied by multitudes of mathematicians. This and the next chapter aim to whet your appetite to learn more about this elegant theory.

1. Gaussian Processes

Let us temporarily leave aside the question of the existence of Brownian motion, and first study normal distributions, Gaussian random variables, and Gaussian processes. Before proceeding further, you may wish to recall §5.2 (p. 11), as well as Examples 3.18 (p. 28) and 7.12 (p. 97), where normal random variables and their characteristic functions have been introduced.

1. Gaussian Processes

1.1. Normal Random Variables.

Definition 9.2. An **R**-valued random variable Y is said to be *centered* if $Y \in L^1(P)$ and $EY = 0$. An $\mathbf{R}^n$-valued random variable $Y = (Y_1, \ldots, Y_n)'$ is said to be centered if each Y_i is. If, in addition, $E\{|Y_i|^2\} < \infty$ for all $i = 1, \ldots, n$, then the *covariance matrix* $Q = (Q_{i,j})$ of Y is the matrix whose (i, j)th entry is the covariance of Y_i and Y_j; i.e., $Q_{i,j} = E[Y_i Y_j]$.

Suppose that $X = (X_1, \ldots, X_n)'$ is a centered n-dimensional random variable in $L^2(P)$. Let $\alpha \in \mathbf{R}^n$ denote a constant (column) vector, and note that $\alpha'X = \alpha \cdot X = \sum_{i=1}^n \alpha_i X_i$ is a centered **R**-valued random variable in $L^2(P)$ with variance

$$(9.2) \qquad \mathrm{Var}(\alpha'X) = \sum_{i=1}^n \sum_{j=1}^n \alpha_i E[X_i X_j] \alpha_j = \alpha' Q \alpha,$$

where $Q = (Q_{i,j})$ is the covariance matrix of X. Since the variance of any random variable is non-negative, we have the following.

Lemma 9.3. *If Q denotes the covariance matrix of a centered $L^2(P)$-valued random variable $X = (X_1, \ldots, X_n)'$ in $\mathbf{R}^n$, then Q is a symmetric non-negative definite matrix. Moreover, the diagonal terms of Q are given by $Q_{j,j} = \mathrm{Var} X_j$.*

Definition 9.4. An $\mathbf{R}^n$-valued random variable $X = (X_1, \ldots, X_n)'$ is *centered normal* (or centered Gaussian) if for all $\alpha \in \mathbf{R}^n$,

$$(9.3) \qquad Ee^{i\alpha \cdot X} = e^{-\frac{1}{2}\alpha' Q \alpha},$$

where Q is a symmetric non-negative definite real matrix. The matrix A is called the *covariance matrix* of X.

We have seen in Lemma 9.3 that covariance matrices are symmetric and non-negative definite. The following implies that the converse is true also.

Theorem 9.5. *Let Q be a symmetric non-negative definite $(n \times n)$ matrix of real numbers. Then there exists a centered normal random variable $X = (X_1, \ldots, X_n)$ whose covariance matrix is Q. If Q is non-singular, then the distribution of X is absolutely continuous with respect to the n-dimensional Lebesgue measure and has the density of Example 3.18 (p. 28) with Q replacing Σ there. Finally, an $\mathbf{R}^n$-valued random variable $X = (X_1, \ldots, X_n)$ is centered Gaussian if and only if $\alpha'X$ is a mean-zero normal random variable for all $\alpha \in \mathbf{R}^n$.*

Proof (Sketch). Let $\{\lambda_i\}_{i=1}^n$ denote the n eigenvalues of Q. The λ_j's are real and non-negative. Let $\{v_i\}_{i=1}^n$ denote the respective orthonormal eigenvectors, and view them as column vectors. Then the $(n \times n)$ matrix

$P = (v_1, \ldots, v_n)$ is orthogonal. Moreover, we can write $Q = P'\Lambda P$, where Λ is the diagonal matrix of the eigenvalues $\lambda_1, \ldots, \lambda_n$.

Next, let $\{Z_i\}_{i=1}^n$ denote n independent standard normal random variables. It is not difficult to see that $Z = (Z_1, \ldots, Z_n)'$ is a centered $\mathbf{R}^n$-valued random variable whose covariance is the identity matrix. Define $X := P'\Lambda^{1/2}Z$, where $\Lambda^{1/2}$ denotes the diagonal matrix whose jth diagonal entry is $\lambda_j^{1/2}$. Since $X = (X_1, \ldots, X_n)'$ is a linear combination of centered random variables, it too is a centered $\mathbf{R}^n$-valued random variable.

Define

$$(9.4) \qquad A_k = \sum_{l=1}^n \alpha_l \left(P'\Lambda^{1/2}\right)_{l,k} \qquad \forall k = 1, \ldots, n.$$

Then $\alpha \cdot X = \sum_{k=1}^n Z_k A_k$ for all $\alpha \in \mathbf{R}^n$. By independence,

$$(9.5) \qquad \mathrm{E} e^{i\alpha \cdot X} = \prod_{k=1}^n \mathrm{E} e^{iZ_k A_k} = e^{-\frac{1}{2}\sum_{k=1}^n A_k^2}.$$

One can check readily that $\sum_{k=1}^n A_k^2 = \alpha'Q\alpha$. Therefore, we have constructed a centered Gaussian process X that has covariance matrix Q. To check that Q is indeed the matrix of the covariances of X, make another round of computations to see that $\mathrm{E}[X_i X_j] = Q_{i,j}$.

Next we suppose that Q is non-singular. Let μ denote the distribution of X. We have shown that $\widehat{\mu}(t) = \exp(-\frac{1}{2}t'Qt)$. Because $\widehat{\mu}$ is absolutely integrable on $\mathbf{R}^n$, the inversion theorem (see Problem 7.12, p. 112) implies that the probability density $f = d\mu/dx$ exists and is given by the formula

$$(9.6) \qquad f(x) = \frac{1}{(2\pi)^n} \int_{\mathbf{R}^n} e^{-it\cdot x - \frac{1}{2}t'Qt} \, dt \qquad \forall x \in \mathbf{R}^n.$$

Write $Q := P'\Lambda^{1/2}\Lambda^{1/2}P$, and change variables $[s = \Lambda^{\frac{1}{2}}Pt]$. This transforms the preceding n-dimensional integral into a product of n one-dimensional integrals. Each of the latter integrals can be computed painlessly by completing the square. Therefrom follows the form of f.

To complete this proof, we derive the assertion about linear combinations. First suppose $\alpha'X$ is a centered Gaussian variable in $\mathbf{R}$. We have seen already that its variance is $\alpha'Q\alpha$, where Q denotes the covariance matrix of X. In particular, thanks to Example 7.12 (p. 97), $\mathrm{E}e^{i\alpha \cdot X} = \exp(-\frac{1}{2}\alpha'Q\alpha)$. Thus, if $\alpha'X$ is a mean-zero normal variable in $\mathbf{R}$ for all $\alpha \in \mathbf{R}^n$, then X is centered Gaussian. The converse is proved similarly. $\square$

Remark 9.6. According to Theorem 9.5, the covariance matrix of a centered normal random variable determines its distribution. However, it can happen that X_1 and X_2 are normally distributed even though (X_1, X_2) is not; see Problem 9.4 below. This demonstrates that in general the normality of $(X_1, \ldots, X_n)$ is a stronger property than the normality of the individual X_j's.

The following important corollary asserts that for normal random vectors independence and uncorrelatedness are one and the same.

Corollary 9.7. *Let $(X_1, \ldots, X_n, Y_1, \ldots, Y_m)$ be a centered normal random variable such that $\mathrm{Cov}(X_i, Y_j) = 0$ for all $i = 1, \ldots, n$ and $j = 1, \ldots, m$. Then $(X_1, \ldots, X_n)$ and $(Y_1, \ldots, Y_m)$ are independent.*

1.2. Brownian Motion as a Gaussian Process.

Definition 9.8. We say that a real-valued process X is *centered Gaussian* if $(X(t_1), \ldots, X(t_k))$ is a centered normal random variable in $\mathbf{R}^k$ for all $0 \le t_1, t_2, t_3, \ldots, t_k$. The function $Q(s,t) := \mathrm{E}[X(s)X(t)]$ is called the *covariance function* of the process X.

For the time being, we assume that Brownian motion exists, and derive some of its elementary properties. We establish the existence later on.

Theorem 9.9. *If $W := \{W(t)\}_{t \ge 0}$ denotes a Brownian motion, then it is a centered Gaussian process with covariance function $Q(s,t) := s \wedge t$. Conversely, any a.s.-continuous centered Gaussian process that has covariance function Q and starts at 0 is a Brownian motion. Furthermore:*

(1) (Quadratic Variation) *For each $t > 0$, as $n \to \infty$,*

$$V_n(t) := \sum_{j=0}^{n-1} \left[W\left(\left(\frac{j+1}{n}\right)t\right) - W\left(\left(\frac{j}{n}\right)t\right) \right]^2 \xrightarrow{\mathrm{P}} t.$$

(2) (The Markov Property) *For any $T > 0$, the process*

$$\{W(t+T) - W(T)\}_{t \ge 0}$$

is a Brownian motion that is independent of $\sigma(\{W(r)\}_{0 \le r \le T})$.

Remark 9.10. (1) Any function with bounded variation has zero quadratic variation. Therefore, Part 1 implies that Brownian motion has unbounded variation a.s.

(2) Because $W(t+T) = [W(t+T) - W(T)] + W(T)$, the Markov property tells us that given the values of W before time T, the "post-T" process $t \mapsto W(t+T)$ is a "conditionally independent" Brownian motion that starts at $W(T)$. The dependence of the post-T process on the past is "local" since it depends only on the last value $W(T)$.

Theorem 9.9 has the following useful consequence.

Corollary 9.11. Let $\mathscr{F}_s$ be the σ-algebra generated by $\{W(u)\}_{0 \le u \le s}$. Then, Brownian motion is a continuous-time $\mathscr{F}$-martingale. That is, $\mathrm{E}[W(t) \mid \mathscr{F}_s] = W(s)$ a.s. for all $t \ge s \ge 0$.

Proof. We know that $W(t) - W(s)$ has mean zero and is independent of $\mathscr{F}_s$. Therefore, $\mathrm{E}[W(t) - W(s) \mid \mathscr{F}_s] = 0$ a.s. The result follows from the obvious fact that $W(s)$ is $\mathscr{F}_s$-measurable. □

Proof of Theorem 9.9 (Sketch). First, let us find the covariance function of W, assuming that W is indeed a centered Gaussian process: If $t \ge s \ge 0$, then because $W(t) - W(s)$ is a mean-zero random variable that is independent of $W(s)$,

$$(9.7) \quad \begin{aligned} Q(s,t) = \mathrm{E}\left[W(s)W(t)\right] &= \mathrm{E}\left[(W(t) - W(s) + W(s))W(s)\right] \\ &= \mathrm{E}\left[|W(s)|^2\right] = s. \end{aligned}$$

In other words, $Q(s,t) = s \wedge t$ for all $s, t \ge 0$. Next we will prove that W is a centered Gaussian process.

By the independence of the increments of W, for all $0 = t_0 \le t_1 \le t_2 \le \cdots \le t_n$, and $\alpha_1, \ldots, \alpha_n \in \mathbf{R}$,

$$(9.8) \quad \mathrm{E}e^{i \sum_{k=1}^n \alpha_k (W(t_k) - W(t_{k-1}))} = \prod_{k=1}^n \mathrm{E}e^{i\alpha_k(W(t_k) - W(t_{k-1}))}.$$

For each $k = 1, \ldots, n$, $W(t_k) - W(t_{k-1})$ is a mean-zero normal random variable with variance $t_k - t_{k-1}$; its characteristic function is computed in Example 7.12, p. 97. This leads to

$$(9.9) \quad \mathrm{E}e^{i \sum_{k=1}^n \alpha_k (W(t_k) - W(t_{k-1}))} = e^{-\frac{1}{2}\alpha' M \alpha},$$

where the matrix M is described by $M_{i,j} = 0$ if $i \ne j$, and $M_{j,j} = t_j - t_{j-1}$. In other words, the vector $(W(t_j) - W(t_{j-1}); 1 \le j \le n)$ is a centered normal random variable in $\mathbf{R}^n$. Now for any $\beta = (\beta_1, \ldots, \beta_n) \in \mathbf{R}^n$,

$$(9.10) \quad \sum_{k=1}^n \beta_k W(t_k) = \sum_{k=1}^n \alpha_k [W(t_k) - W(t_{k-1})],$$

where $\beta_k := \alpha_k + \cdots + \alpha_n$. Therefore, $\sum_{k=1}^n \beta_k W(t_k)$ is a centered normal random variable in $\mathbf{R}$. That is, W is a centered Gaussian process.

Next we prove that if $G = \{G(t)\}_{t \ge 0}$ is an a.s.-continuous centered Gaussian process with covariance function Q, and if $G(0) = 0$, then G is a Brownian motion. Because the remaining conditions of Brownian motion are easily verified for G, it suffices to show that whenever $t > s$, $G(t) - G(s)$ is independent of $\{G(u)\}_{0 \le u \le s}$. We fix $0 \le u_1 \le \cdots \le u_k \le s$ and prove that $G(t) - G(s)$ is independent of $(G(u_1), \ldots, G(u_n))$. However, the distribution of the $(n+1)$-dimensional random vector $(G(t) - G(s), G(u_1), \ldots, G(u_n))$

2. Brownian Motion on $[0,1)$

is the same as that of $(W(t) - W(s), W(u_1), \ldots, W(u_n))$. This is because everything reduces to the same calculations involving the function Q. This proves that G is a Brownian motion.

Problem 9.6 contains ample hints for proving that $V_n(t) \to t$ in probability as $n \to \infty$.

Note that $t \mapsto W(t+T) - W(T)$ is a continuous centered Gaussian process. We verify that: (a) This is Brownian motion; and (b) it is independent of $\sigma(\{W(r)\}_{0 \le r \le T})$. For the former, suppose $0 \le s \le t$, and notice that

(9.11)
$$\begin{aligned}\mathrm{E}\left[\left(W(s+T) - W(T)\right)\left(W(t+T) - W(T)\right)\right] \\ = \mathrm{E}[W(s+T)W(t+T)] - \mathrm{E}[W(T)W(t+T)] \\ - \mathrm{E}[W(T)W(s+T)] + \mathrm{E}\left[(W(T))^2\right] \\ = (s+T) - T - T + (T) = s = s \wedge t.\end{aligned}$$

In particular, $t \mapsto W(t+T) - W(T)$ is a Brownian motion. The assertion about independence is proved similarly: If $s \le T$, then

(9.12)
$$\begin{aligned}\mathrm{E}\left[W(s)\left(W(T+t) - W(T)\right)\right] \\ = \mathrm{E}\left[W(s)W(T+t)\right] - \mathrm{E}\left[W(s)W(T)\right] = s - s = 0.\end{aligned}$$

Corollary 9.7 proves the independence of $t \mapsto W(t+T) - W(T)$ from $\sigma(\{W(r)\}_{0 \le r \le T})$. □

2. Wiener's Construction: Brownian Motion on $[0,1)$

It remains to prove that Brownian motion exists. We begin by reducing the scope of our ultimate goal: If we can show the existence of Brownian motion indexed by $[0,1)$, then we have the following more general existence result.

Lemma 9.12. *Suppose $\{B_i\}_{i=0}^{\infty}$ are independent Brownian motions indexed by $[0,1)$. Then, the following recursive definition describes a Brownian motion W indexed by $[0,\infty)$:*

(9.13)
$$W(t) := \begin{cases} B_0(t) & \text{if } t \in [0,1), \\ B_0(1) + B_1(t-1) & \text{if } t \in [1,2), \\ \vdots \\ \sum_{k=0}^{j-1} B_k(1) + B_j(t-j) & \text{if } t \in [j, j+1), \\ \vdots \end{cases}$$

Proof. The defined process W is a continuous centered Gaussian process because it is a finite sum of continuous centered Gaussian processes. It remains to compute the covariance of W: If $s \le t$, then either we can find

$j \geq 0$ such that $j \leq s \leq t < j+1$, or $j \leq s \leq j+1 \leq \ell \leq t < \ell+1$ for some $\ell > j$. In the first case, $\mathrm{E}[W(s)W(t)]$ is equal to

(9.14)
$$\mathrm{E}\left[\left(\sum_{k=0}^{j-1} B_k(1) + B_j(t-j)\right)\left(\sum_{k=0}^{j-1} B_k(1) + B_j(s-j)\right)\right]$$
$$= \mathrm{E}\left[\sum_{k=0}^{j-1} (B_k(1))^2\right] + \mathrm{E}\left[(B_j(s-j)B_j(t-j))\right]$$
$$= j + (s-j) = s = s \wedge t.$$

In the second case, one obtains the same final answer. In any event, W has the correct covariance function, and is therefore a Brownian motion. □

The simplest construction of Brownian motion indexed by $[0,1)$—in fact, $[0,1]$—is the following minor variant of the original construction of Norbert Wiener. Throughout, let $\{X_i\}_{i=0}^\infty$ denote a sequence of i.i.d. normal random variables with mean zero and variance one; the existence of this sequence is guaranteed by Theorem 6.17 on page 70. Then, formally speaking, the Brownian motion $\{W(t)\}_{0 \leq t \leq 1}$ is the limit of the following sequence:

(9.15) $\quad W_n(t) = tX_0 + \dfrac{\sqrt{2}}{\pi} \sum_{j=1}^n \dfrac{\sin(j\pi t)}{j} X_j \qquad \forall 0 \leq t \leq 1,\ n = 1, 2, \ldots.$

In order to prove the existence properly, we first need to complete the probability space. Informally, this means that we declare all subsets of null sets measurable and null; this can always be done at no cost. See Theorem 3.20 on page 29. Now we can prove Wiener's theorem (1923a).

Wiener's Theorem. *If the underlying probability space is complete, then $W(t) = \lim_{n \to \infty} W_{2^n}(t)$ exists a.s., and the convergence is uniform for all $t \in [0,1]$. The process W is a Brownian motion indexed by $[0,1]$.*

Proof. We split the proof into three steps.

Step 1. Uniform Convergence. For $n = 1, 2, \ldots$ and $t \geq 0$ define

(9.16) $\quad S_n(t) := \sum_{k=1}^n \dfrac{\sin(k\pi t)}{k} X_k.$

Thus, $W_n(t) = tX_0 + (\sqrt{2}/\pi)S_n(t)$. Stated more carefully, we have two processes $S_n(t,\omega)$ and $W_n(t,\omega)$, and as always we do not show the dependence on ω.

We will prove that S_{2^n} forms a Cauchy sequence a.s. and in $L^2(\mathrm{P})$, uniformly in $t \in [0,1]$. Define

(9.17) $\quad \mathscr{E}_n(t) := S_{2^{n+1}}(t) - S_{2^n}(t).$

2. Brownian Motion on $[0,1)$

Note that

(9.18) $$|\mathscr{E}_n(t)|^2 = \left(\sum_{j=2^n+1}^{2^{n+1}} \frac{\sin(j\pi t)}{j} X_j\right)^2 \leq \left|\sum_{j=2^n+1}^{2^{n+1}} \frac{e^{ij\pi t}}{j} X_j\right|^2.$$

Therefore,

(9.19) $$\begin{aligned}|\mathscr{E}_n(t)|^2 &\leq \sum_{j=2^n+1}^{2^{n+1}} \sum_{k=2^n+1}^{2^{n+1}} \frac{e^{i(j-k)\pi t}}{jk} X_j X_k \\ &= \sum_{k=2^n+1}^{2^{n+1}} \frac{X_j^2}{j^2} + 2\sum_{l=1}^{2^n-1} \sum_{k=2^n+1}^{2^{n+1}-l} \frac{e^{il\pi t}}{k(l+k)} X_k X_{l+k} \\ &\leq \sum_{k=2^n+1}^{2^{n+1}} \frac{X_j^2}{j^2} + 2\sum_{l=1}^{2^n-1} \left|\sum_{k=2^n+1}^{2^{n+1}-l} \frac{X_k X_{l+k}}{k(k+l)}\right|.\end{aligned}$$

The right-hand side is independent of $t \geq 0$, so we can take expectations and apply Minkowski's inequality to obtain

(9.20) $$\begin{aligned}\mathrm{E}\left[\sup_{t\geq 0} |\mathscr{E}_n(t)|^2\right] &\leq \sum_{k=2^n+1}^{2^{n+1}} \frac{1}{j^2} + 2\sum_{l=1}^{2^n-1} \left\|\sum_{k=2^n+1}^{2^{n+1}-l} \frac{X_k X_{l+k}}{k(k+l)}\right\|_2 \\ &= \sum_{k=2^n+1}^{2^{n+1}} \frac{1}{j^2} + 2\sum_{l=1}^{2^n-1} \sqrt{\sum_{k=2^n+1}^{2^{n+1}-l} \left\|\frac{X_k X_{l+k}}{k(k+l)}\right\|_2^2}.\end{aligned}$$

(Why?) The final squared-$L^2(\mathrm{P})$-norm is equal to $k^{-2}(k+l)^{-2}$. On the other hand, by monotonicity, $\sum_{k=2^n+1}^{2^{n+1}} k^{-2} \leq 2^{-n}$ and $\sum_{l=1}^{2^n-1}(l+2^n)^{-1} \leq 1$. Therefore,

(9.21) $$\mathrm{E}\left[\sup_{t\geq 0} |\mathscr{E}_n(t)|^2\right] \leq 2^{-n} + 2 \cdot 2^{-n/2}.$$

It follows that $\sum_n \sup_{t\geq 0} |S_{2^{n+1}}(t) - S_{2^n}(t)| < \infty$ a.s. Thus, as $n \to \infty$, $S_{2^n}(t)$ converges uniformly in $t \geq 0$ to the limiting random process

(9.22) $$S_\infty(t) := \sum_{j=1}^\infty \frac{\sin(j\pi t)}{j} X_j.$$

In particular, $W(t) = \lim_{n\to\infty} W_{2^n}(t)$ exists uniformly in $t \geq 0$ almost surely.

Step 2. Continuity and Distributional Properties. The random map $t \mapsto W_{2^n}(t)$ defined in (9.15) is obviously continuous. Because W is an a.s.-uniform limit of continuous functions, it is a.s. continuous; see Step 3 below for a technical note on this issue. Moreover, since W_{2^n} is a mean-zero

Gaussian process, so is W (Problem 9.2). Since $W(0) = 0$, it remains to prove that

(9.23) $$\mathrm{E}\left[|W(t) - W(s)|^2\right] = t - s \qquad \forall 0 \le s \le t.$$

But the independence of the X's, together with Lemma 6.8 (p. 67), yields

(9.24)
$$\mathrm{E}\left[|W(t) - W(s)|^2\right] = (t-s)^2 + \frac{2}{\pi^2}\mathrm{E}\left[(S_\infty(t) - S_\infty(s))^2\right]$$
$$= (t-s)^2 + \frac{2}{\pi^2}\sum_{j=1}^{\infty}\left(\frac{\sin(j\pi t) - \sin(j\pi s)}{j}\right)^2.$$

Define $f(x) = \mathbf{1}_{[\pi s, \pi t]}(x) + \mathbf{1}_{[-\pi t, -\pi s]}(x)$ $(x \in [-\pi, \pi])$ and $\phi_n(x) = e^{inx}/\sqrt{2\pi}$ $(x \in [-\pi, \pi]; n = 0, \pm 1, \pm 2, \ldots)$. Then,

(9.25)
$$\mathrm{E}\left[|W(t) - W(s)|^2\right] = (t-s)^2 + \frac{1}{\pi}\sum_{j=1}^{\infty}\left|\int_{-\pi}^{\pi} f(x)\phi_j(x)\,dx\right|^2$$
$$= \frac{1}{2\pi}\sum_{j=-\infty}^{\infty}\left|\int_{-\pi}^{\pi} f(x)\phi_j(x)\,dx\right|^2.$$

By the Riesz–Fischer theorem (Theorem A.5, p. 205), the right-hand side is equal to $(2\pi)^{-1}\int_{-\pi}^{\pi} |f(x)|^2\,dx = (t-s)$. This yields (9.23).

Step 3. Technical Wrap-up. Now we tie up a subtle loose end: The uniform limit $W(t) = \lim_{n \to \infty} W_{2^n}(t)$ is known to exist, but only with probability one. This is insufficient because we need to define $W(t, \omega)$ for all ω. Thus, we *define*

(9.26) $$W(t) := \limsup_{n \to \infty} W_{2^n}(t).$$

The process W is well defined and continuous a.s. The remainder of the calculations of Step 2 goes through, since by redefining a random variable on a set of measure zero we do not change its distribution (unscramble this!). Finally, the completion is needed to ensure that the event C that W is continuous is measurable; in Step 1, we showed that C^c is a subset of a null set. Because the underlying probability is complete, C^c is null. $\square$

3. Nowhere-Differentiability

We are in a position to prove the following striking theorem of Paley, Wiener, and Zygmund (1933). Throughout, W denotes a Brownian motion.

Theorem 9.13. *Suppose the underlying probability space is complete. Then, Brownian motion is nowhere differentiable almost surely.*

3. Nowhere-Differentiability

Proof. For any $\lambda > 0$ and $n \geq 1$, consider the event

$$(9.27) \qquad E_\lambda^n = \left\{ \exists s \in [0,1] : \sup_{t \in [s-2^{-n}, s+2^{-n}]} |W(s) - W(t)| \leq \lambda 2^{-n} \right\}.$$

(Why is this measurable?) We intend to show that $\sum_n \mathrm{P}(E_\lambda^n) < \infty$ for all $\lambda > 0$.

Indeed, suppose there exists $s \in [0,1]$ such that for all t within 2^{-n} of s, $|W(s) - W(t)| \leq \lambda 2^{-n}$. Then there must exist a possibly random $j = 0, \ldots, 2^n - 1$ such that $s \in D(j;n)$, where

$$(9.28) \qquad D(j;n) := \left[j2^{-n}, (j+1)2^{-n} \right].$$

Thus, $|W(s) - W(t)| \leq \lambda 2^{-n}$ for all $t \in D(j;n)$. By the triangle inequality, we can deduce that $|W(u) - W(v)| \leq 2\lambda 2^{-n} = \lambda 2^{-n+1}$ for all $u, v \in D(j;n)$. Subdivide $D(j;n)$ into four smaller dyadic intervals, and note that the successive differences in the values of W (at the endpoints of the subdivided intervals) are at most $\lambda 2^{-n+1}$. This leads us to the following:

$$(9.29) \qquad E_\lambda^n \subseteq \bigcup_{j=0}^{2^n-1} \bigcap_{\ell=0}^{3} \left\{ |\Delta_{j,\ell}^n| \leq \lambda 2^{-n+1} \right\},$$

where

$$(9.30) \qquad \Delta_{j,\ell}^n = W\left(j2^{-n} + (\ell+1)2^{-(n+2)} \right) - W\left(j2^{-n} + \ell 2^{-(n+2)} \right).$$

Thanks to the independent-increments property of Brownian motion (P-b),

$$(9.31) \qquad \mathrm{P}(E_\lambda^n) \leq \sum_{j=0}^{2^n-1} \prod_{\ell=0}^{3} \mathrm{P}\left\{ |\Delta_{j,\ell}^n| \leq \lambda 2^{-n+1} \right\},$$

On the other hand, by the stationary-increments property of W, $\Delta_{j,\ell}^n$ has a normal distribution with mean zero and variance $2^{-(n+2)}$ ((P-a) and (P-c)). Thus, for all $\beta > 0$,

$$(9.32) \qquad \mathrm{P}\left\{ |\Delta_{j,\ell}^n| \leq \beta \right\} = \int_{-\beta 2^{(n+2)/2}}^{\beta 2^{(n+2)/2}} \frac{e^{-x^2/2}}{\sqrt{2\pi}} \, dx \leq \beta 2^{(n+2)/2}.$$

Apply this with $\beta = \lambda 2^{-n+1}$ to deduce that $\mathrm{P}(E_\lambda^n) \leq 256 \lambda^4 2^{-n}$. In particular, $\sum_n \mathrm{P}(E_\lambda^n) < \infty$, as was asserted earlier. By the Borel–Cantelli lemma, for any $\lambda > 0$, the following holds with probability one: For all but a finite number of n's,

$$(9.33) \qquad \inf_{0 \leq s \leq 1} \sup_{|t-s| \leq 2^{-n}} \frac{|W(s) - W(t)|}{|s-t|} \geq \inf_{0 \leq s \leq 1} \sup_{|t-s| \leq 2^{-n}} \frac{|W(s) - W(t)|}{2^{-n}} \geq \lambda.$$

Thus, if $W'(s)$ existed for some $s \in [0,1]$, then $|W'(s)| \geq \lambda$ a.s. Because $\lambda > 0$ is arbitrary, this proves that $|W'(s)| = \infty$ a.s. This contradicts the differentiability of W at some $s \in [0,1]$. Thanks to scaling (Theorem 9.9),

W is a.s. nowhere differentiable in $[0,c]$ for any $c > 0$. Therefore W is a.s. nowhere differentiable.

Technical Aside in the Proof. In fact, we have proven that there exists a null set N such that $\mathsf{D} \subseteq N$, where D is the collection of all ω's for which $t \mapsto W(t,\omega)$ is differentiable somewhere. The collection D need not be measurable. But this is immaterial to us since we can complete the underlying probability space at no cost (Theorem 3.20, p. 29). In the completed probability space D *is* a null set, and our task is complete. $\square$

4. The Brownian Filtration and Stopping Times

Recall the Markov property of Brownian motion (Theorem 9.9): Given any fixed $T > 0$, the "post-T" process $t \mapsto W(T+t) - W(T)$ is a Brownian motion that is independent of $\sigma(\{W(u)\}_{0 \leq u \leq T})$. Intuitively, this states that given the value of $W(T)$, the process after time T is independent of the process before time T.

The *strong Markov property* states that the Markov property holds for a large class of random times T that are called stopping times. We have encountered such times when studying martingales in discrete time, and their continuous-time definition is formally the same.

Throughout, W is a Brownian motion with some prescribed starting point, say $W(0) = x$.

Definition 9.14. A filtration $\mathscr{A} = \{\mathscr{A}_t\}_{t \geq 0}$ is a collection of sub-σ-algebras of $\mathscr{F}$ such that $\mathscr{A}_s \subseteq \mathscr{A}_t$ if $s \leq t$. If $\mathscr{A}$ is a filtration, then a measurable function $T : \Omega \to [0,\infty]$ is a *stopping time* (or $\mathscr{A}$-stopping time) if $\{T \leq t\}$ is $\mathscr{A}_t$-measurable for all $t \geq 0$. Given a stopping time T, we can define $\mathscr{A}_T$ by

$$(9.34) \qquad \mathscr{A}_T = \left\{ A \in \mathscr{F} : A \cap \{T \leq t\} \in \mathscr{A}_t \ \forall t \geq 0 \right\}.$$

T is called a *simple stopping time* if there exist $0 \leq \tau_0, \tau_1, \ldots < \infty$ such that $T(\omega) \in \{\tau_0, \tau_1, \ldots\}$ for all $\omega \in \Omega$.

Define

$$(9.35) \qquad \mathscr{F}_t^0 = \sigma\left(\{W(u)\}_{0 \leq u \leq t}\right).$$

In light of our development of martingale theory this definition is quite natural. Here are some of the properties of $\mathscr{F}_T^0$ when T is a stopping time.

Lemma 9.15. *If T is a finite $\mathscr{F}^0$-stopping time, then $\mathscr{F}_T^0$ is a σ-algebra, and T is $\mathscr{F}_T^0$-measurable. Furthermore, if $S \leq T$ is another stopping time, then $\mathscr{F}_S^0 \subseteq \mathscr{F}_T^0$.*

4. The Brownian Filtration and Stopping Times

Proof. For each $t \geq 0$,

(9.36) $\qquad A^c \cap \{T \leq t\} = \{T \leq t\} \setminus (A \cap \{T \leq t\}) \in \mathscr{F}_t^0.$

Consequently, $\mathscr{F}_T^0$ is closed under complementation. Because $\mathscr{F}_t^0$ is a monotone class, it is a σ-algebra.

For all $a \in (0, \infty)$ and $t \geq 0$,

(9.37) $\qquad T^{-1}([0, a]) \cap \{T \leq t\} = \{T \leq a \wedge t\} \in \mathscr{F}_{a \wedge t}^0 \subseteq \mathscr{F}_t^0.$

This proves that T is $\mathscr{F}_T^0$-measurable.

Finally, we suppose $A \in \mathscr{F}_S^0$, and note that for any $t \geq 0$, $A \cap \{T \leq t\} = A \cap \{S \leq t\} \cap \{T \leq t\}$. Since $A \cap \{S \leq t\}$ and $\{T \leq t\}$ are both in $\mathscr{F}_t^0$, this proves that $A \cap \{T \leq t\} \in \mathscr{F}_t^0$, whence we have $A \in \mathscr{F}_T^0$. This proves the remaining assertion that $\mathscr{F}_S^0 \subseteq \mathscr{F}_T^0$. $\square$

For technical reasons, we need to modify the filtration $\mathscr{F}^0$. For every $t \geq 0$ let $\bar{\mathscr{F}}_t$ denote the completion of $\mathscr{F}_t^0$. It can be checked that $\{\bar{\mathscr{F}}_t\}_{t \geq 0}$ is a filtration of σ-algebras. Note that any $\mathscr{F}^0$- or $\bar{\mathscr{F}}$-stopping time is also an $\bar{\mathscr{F}}$-stopping time. We alter the filtration $\bar{\mathscr{F}}$ as well to obtain the *Brownian filtration*.

Definition 9.16. A filtration $\{\mathscr{A}_t\}_{t \geq 0}$ is *right-continuous* if for all $t \geq 0$, $\mathscr{A}_t = \cap_{\epsilon > 0} \mathscr{A}_{t+\epsilon}$. The Brownian filtration $\{\mathscr{F}_t\}_{t \geq 0}$ is defined as the smallest right-continuous filtration that contains $\{\bar{\mathscr{F}}_t\}_{t \geq 0}$. That is, $\mathscr{F}_t := \cap_{s > t} \bar{\mathscr{F}}_s$ for all $t \geq 0$.

Next we construct interesting stopping times.

Proposition 9.17. *If $A \subseteq \mathbf{R}$ is either open or closed, then the first hitting time $T_A := \inf\{t \geq 0 : W(t) \in A\}$, where $\inf \varnothing := \infty$, is a stopping time with respect to the Brownian filtration $\mathscr{F}$.*

Remark 9.18. If you know what F_σ- and G_δ-sets are, then convince yourself that when A is of either variety, then T_A is a stopping time. You may ask further, "What about T_A when A is measurable but is neither F_σ nor G_δ?" The answer is given by a quite deep theorem of Hunt (1957): T_A *is a stopping time for all Borel sets A.*

Proof of Proposition 9.17. We prove this proposition in two steps.

Step 1. T_A is a stopping time when A is open. Suppose A is open. We wish to prove that $\{T_A \leq t\} \in \mathscr{F}_t$ for all $t \geq 0$. It suffices to prove that $\{T_A < t\} \in \mathscr{F}_t$ for all $t \geq 0$, because the right-continuity of $\mathscr{F}$ ensures that $\{T_A \leq t\} = \cap_{\epsilon > 0}\{T_A < t + \epsilon\} \in \mathscr{F}_t$. But $\{T_A < t\}$ is the event that there exists a time s before t at which $W(s) \in A$. Let C denote the collection of all ω such that $t \mapsto W(t, \omega)$ is continuous. We know that $\mathrm{P}(\mathsf{C}) = 1$. And since A is open, $\{T_A < t\} \cap \mathsf{C}$ is the event that there exists a rational $s \leq t$ at which $W(s) \in A$. That is,

(9.38)
$$\{T_A < t\} \cap \mathsf{C} = \bigcup_{s \leq t : s \in \mathbf{Q}} \{W(s) \in A\} \cap \mathsf{C}$$
$$= \bigcup_{s \leq t : s \in \mathbf{Q}} \{W(s) \in A\} \setminus \left(\bigcup_{s \leq t : s \in \mathbf{Q}} \{W(s) \in A\} \cap \mathsf{C}^c \right).$$

Because $\bar{\mathscr{F}}_t$ is complete, so is $\mathscr{F}_t$. Therefore, all subsets of null sets are $\mathscr{F}_t$-measurable null sets. It follows from this that $\cup_{s \leq t : s \in \mathbf{Q}} \{W(s) \in A\} \cap \mathsf{C}^c \in \mathscr{F}_t$. On the other hand, it is clear that $\cup_{s \leq t : s \in \mathbf{Q}} \{W(s) \in A\} \in \mathscr{F}_t^0 \subseteq \mathscr{F}_t$, and this implies that $\{T_A < t\} \cap \mathsf{C} \in \mathscr{F}_t$. Since $\{T_A \leq t\} \cap \mathsf{C}^c \subseteq \mathsf{C}^c$ is null, it is $\mathscr{F}_t$-measurable, thanks to completeness. To summarize, we have $\{T_A < t\} = (\{T_A < t\} \cap \mathsf{C}) \cup (\{T_A < t\} \cap \mathsf{C}^c) \in \mathscr{F}_t$, as desired.

Step 2. T_A *is a stopping time when A is closed.* For each $n = 1, 2, \ldots$, define A_n to be the set of all $x \in \mathbf{R}$ such that the Euclidean distance between x and the set A is $< n^{-1}$. It is clear that A_n is open, and $\{T_A \leq t\} \cap \mathsf{C} = \cap_n \{T_{A_n} \leq t\} \cap \mathsf{C}$, where C was defined in Step 1. By Step 1, $\{T_A \leq t\} \cap \mathsf{C}$ is in $\mathscr{F}_t$. Because $\mathscr{F}_t$ is complete, it follows that $\{T_A \leq t\} \in \mathscr{F}_t$ also. □

For all random variables $T : \Omega \to [0, \infty)$ we define the random variable $W(T)$ as follows:

(9.39)
$$W(T)(\omega) := W(T(\omega), \omega).$$

Proposition 9.19. *If T is a finite stopping time, then $W(T)$ is measurable with respect to $\mathscr{F}_T$.*

The proof of this proposition relies on a simple though important approximation scheme.

Lemma 9.20. *Given any finite $\mathscr{F}$-stopping time T, one can construct a non-increasing sequence of simple stopping times $T_1 \geq T_2 \geq \cdots$ such that $\lim_n T_n(\omega) = T(\omega)$ for all $\omega \in \Omega$. In addition, $\mathscr{F}_T = \cap_n \mathscr{F}_{T_n}$.*

Proof. Here is a recipe for the T_n's:

(9.40)
$$T_n(\omega) = \sum_{k=0}^{\infty} \left(\frac{k+1}{2^n} \right) \mathbf{1}_{[k2^{-n}, (k+1)2^{-n})}(T(\omega)).$$

Since every interval of the form $[k2^{-n}, (k+1)2^{-n})$ is obtained by splitting into two an interval of the form $[j2^{-n+1}, (j+1)2^{-n+1})$, we have $T_n \geq T_{n+1} \geq \cdots \geq T$.

To check that T_n is a stopping time, note that $\{T_n \leq (k+1)2^{-n}\} = \{T \leq (k+1)2^{-n}\} \in \mathscr{F}_{(k+1)2^{-n}}$, since T is a stopping time. Now given any

5. The Strong Markov Property

$t \geq 0$, we can find k and n such that $t \in [k2^{-n}, (k+1)2^{-n})$. Therefore, $\{T_n \leq t\} = \{T_n \leq k2^{-n}\} = \{T \leq k2^{-n}\} \in \mathscr{F}_{k2^{-n}} \subseteq \mathscr{F}_t$. This proves that the T_n's are non-increasing simple stopping times. Moreover, T_n converges to T since $0 \leq T_n - T \leq 2^{-n}$. It remains to prove that $\mathscr{F}_T \supseteq \cap_n \mathscr{F}_{T_n}$; see Proposition 9.17 but replace $\mathscr{F}^0$ by $\mathscr{F}$ everywhere.

If $A \in \cap_{k=1}^{\infty} \mathscr{F}_{T_k}$, then $A \cap \{T_n \leq t\} \in \mathscr{F}_t$ for all $n \geq 1$ and $t \geq 0$. Therefore,

$$(9.41) \qquad A \cap \{T \leq t\} = \bigcap_{\epsilon > 0} \bigcup_{m=1}^{\infty} \bigcap_{n=m}^{\infty} (A \cap \{T_n \leq t + \epsilon\}) \in \bigcap_{\epsilon > 0} \mathscr{F}_{t+\epsilon}.$$

The lemma follows from the right-continuity of the filtration $\mathscr{F}$. □

Proof of Proposition 9.19. First suppose that T is a simple $\mathscr{F}$-stopping time which takes values in $\{\tau_0, \tau_1, \ldots\}$. In this case, given any Borel set A and any $t \geq 0$,

$$(9.42) \qquad \{W(T) \in A\} \cap \{T \leq t\} = \bigcup_{\substack{n \geq 0: \\ \tau_n \leq t}} \{W(\tau_n) \in A\} \cap \{T = \tau_n\} \in \mathscr{F}_t.$$

For a general finite stopping time T, we can find simple stopping times $T_n \downarrow T$ (Lemma 9.20) with $\mathscr{F}_T = \cap_n \mathscr{F}_{T_n}$. Let C denote the collection of ω's for which $t \mapsto W(t, \omega)$ is continuous and recall that $\mathsf{P}(\mathsf{C}) = 1$. Then, for any open set $A \subseteq \mathbf{R}$,

$$(9.43) \qquad \begin{aligned} &\{W(T) \in A\} \cap \mathsf{C} \cap \{T \leq t\} \\ &= \bigcap_{\epsilon \in \mathbf{Q}_+} \bigcup_{m=1}^{\infty} \bigcap_{n=m}^{\infty} \{W(T_n) \in A\} \cap \mathsf{C} \cap \{T_n \leq t + \epsilon\}. \end{aligned}$$

Since T_n is a finite simple stopping time, $\{W(T_n) \in A\} \cap \{T_n \leq t\} \in \mathscr{F}_t$. In particular, the completeness of $\mathscr{F}_t$ shows that the above, and hence, $\{W(T) \in A\} \cap \{T \leq t\}$ are also in $\mathscr{F}_t$. The collection of all $A \in \mathscr{B}(\mathbf{R})$ such that $\{W(T) \in A\} \cap \{T \leq t\} \in \mathscr{F}_t$ is a monotone class that contains all open sets. It follows from the monotone class theorem that for all $A \in \mathscr{B}(\mathbf{R})$ and $t \geq 0$,

$$(9.44) \qquad \{W(T) \in A\} \cap \{T \leq t\} \in \mathscr{F}_t.$$

This proves the proposition. □

5. The Strong Markov Property

We are finally in a position to state and prove the strong Markov property of Brownian motion.

Theorem 9.21. *If T is a finite $\mathscr{F}$-stopping time, where $\mathscr{F}$ denotes the Brownian filtration, then $\{W(T+t) - W(T)\}_{t \geq 0}$ is a Brownian motion that is independent of $\mathscr{F}_T$.*

Proof. We prove this first for simple stopping times, and then approximate, using Lemma 9.20, a general stopping time with simple ones.

Step 1. Simple Stopping Times. If T is a simple stopping time, then there exist $\tau_0 \leq \tau_1 \leq \cdots$ such that $T \in \{\tau_0, \tau_1, \ldots\}$ a.s. Now for any $A \in \mathscr{F}_T$, and for all $B_1, \ldots, B_m \in \mathscr{B}(\mathbf{R})$,

$$(9.45) \quad \begin{aligned} &\mathrm{P}\left(A \cap \left\{ {}^{\forall} i \leq m : W(T + t_i) - W(t_i) \in B_i \right\}\right) \\ &= \sum_{k=0}^{\infty} \mathrm{P}\left(A \cap \left\{ {}^{\forall} i \leq m : W(\tau_k + t_i) - W(\tau_k) \in B_i \,,\, T = \tau_k \right\}\right). \end{aligned}$$

But $A \cap \{T = \tau_k\} = A \cap \{T \leq \tau_k\} \cap \{T \leq \tau_{k-1}\}^c$ is in $\mathscr{F}_{\tau_k}$ since $A \in \mathscr{F}_T$. Therefore, by the Markov property (Theorem 9.9),

$$(9.46) \quad \begin{aligned} &\mathrm{P}\left(A \cap \left\{ {}^{\forall} i \leq m : W(T + t_i) - W(t_i) \in B_i \right\}\right) \\ &= \sum_{k=0}^{\infty} \mathrm{P}\left\{ {}^{\forall} i \leq m : W(\tau_k + t_i) - W(\tau_k) \in B_i \right\} \mathrm{P}\left(\{T = \tau_k\} \cap A\right) \\ &= \mathrm{P}\left\{ {}^{\forall} i \leq m : W(t_i) \in B_i \right\} \mathrm{P}(A). \end{aligned}$$

This proves the theorem in the case that T is a simple stopping time. Indeed, to deduce that $t \mapsto W(t + T) - W(T)$ is a Brownian motion, simply set $A = \mathbf{R}$. The asserted independence also follows since $A \in \mathscr{F}_T$ is arbitrary.

Step 2. The General Case. In the general case, we approximate T by simple stopping times as in Lemma 9.20. Namely, we find $T_n \downarrow T$—all simple stopping times—such that $\cap_n \mathscr{F}_{T_n} = \mathscr{F}_T$. Now for any $A \in \mathscr{F}_T$, and for all open $B_1, \ldots, B_m \subseteq \mathbf{R}$,

$$(9.47) \quad \begin{aligned} &\mathrm{P}\left(A \cap \left\{ {}^{\forall} i \leq m : W(T + t_i) - W(t_i) \in B_i \right\}\right) \\ &= \lim_{n \to \infty} \mathrm{P}\left(A \cap \left\{ {}^{\forall} i \leq m : W(T_n + t_i) - W(t_i) \in B_i \right\}\right) \\ &= \lim_{n \to \infty} \mathrm{P}\left\{ {}^{\forall} i \leq m : W(t_i) \in B_i \right\} \mathrm{P}(A). \end{aligned}$$

In the first equation we used the fact that the B's are open and W is continuous, while in the second equation we used the fact that $A \in \mathscr{F}_{T_n}$ for all n, together with the result of Step 1 applied to T_n. This proves the theorem. $\square$

6. The Reflection Principle

The following "reflection principle" is a prime example of how the strong Markov property (Theorem 9.21) can be applied to make nontrivial computations for the Brownian motion.

Theorem 9.22. *If t is non-random and positive, then $\sup_{0 \le s \le t} W(s)$ has the same distribution as $|W(t)|$. Equivalently, for all $a \ge 0$ and $t \ge 0$,*

$$(9.48) \qquad P\left\{\sup_{0 \le s \le t} W(s) \ge a\right\} = \sqrt{\frac{2}{\pi t}} \int_a^\infty \exp\left(-\frac{z^2}{2t}\right) dz.$$

Proof. Define $T_a = \inf\{s \ge 0 : W(s) \ge a\}$ where $\inf \varnothing = \infty$. Thanks to Proposition 9.17, T_a is an $\bar{\mathscr{F}}$- and hence an $\mathscr{F}$-stopping time.

Step 1. T_a is finite a.s. By scaling (Theorem 9.9), for any $t > 0$, the event $\{W(t) \ge \sqrt{t}\}$ has probability $(2\pi)^{-1/2} \int_1^\infty e^{-x^2/2} dx = c > 0$. Consequently,

$$(9.49) \qquad P\left\{\limsup_{t \to \infty} \frac{W(t)}{\sqrt{t}} \ge 1\right\} \ge c > 0.$$

(Why?). Among other things, this and the zero-one law (Problem 9.11) together imply that $\limsup_{t \to \infty} W(t) = \infty$ a.s. Since W is continuous a.s., it must then hit a at some finite time a.s. Therefore, with probability one, T_a is finite and $W(T_a) = a$.

Step 2. Reflection. Note that $\{\sup_{0 \le s \le t} W(s) \ge a\} = \{T_a \le t\} \in \mathscr{F}_t$. Moreover, $P\{T_a \le t\}$ is equal to

$$(9.50) \qquad \begin{aligned} & P\{T_a \le t, W(t) \ge a\} + P\{T_a \le t, W(t) < a\} \\ &= P\{W(t) \ge a\} + P\{T_a \le t, W(T_a + (t - T_a)) - W(T_a) < 0\} \\ &= P\{W(t) \ge a\} \\ & \quad + E\Big[P\{W(T_a + (t - T_a)) - W(T_a) < 0 \mid \mathscr{F}_{T_a}\}; T_a \le t\Big], \end{aligned}$$

since T_a is $\mathscr{F}_{T_a}$-measurable (Proposition 9.17). On the other hand, by the strong Markov property (Theorem 9.21), $P\{W(T_a + (t - T_a)) - W(T_a) < 0 \mid \mathscr{F}_{T_a}\}$ is the conditional probability that a Brownian motion independent of $\mathscr{F}_{T_a}$ is below zero at time $t - T_a$, given the value of T_a. Independence and symmetry (Theorem 9.9) together imply that the said probability is a.s. equal to $P\{W(T_a + (t - T_a)) - W(T_a) > 0 \mid \mathscr{F}_{T_a}\}$ (why a.s.?). [In other words, we have reflected the post-T_a process to get another Brownian motion, whence the term "reflection principle."] Therefore, we make this change and backtrack in the preceding display to deduce that $P\{T_a \le t\}$ is

equal to
$$P\{W(t) \geq a\}$$
$$+ E\left[P\{W(T_a + (t - T_a)) - W(T_a) > 0 \,|\, \mathscr{F}_{T_a}\}\,;\, T_a \leq t\right]$$
(9.51)
$$= P\{W(t) \geq a\} + P\{T_a \leq t, W(T_a + (t - T_a)) - W(T_a) > 0\}$$
$$= P\{W(t) \geq a\} + P\{T_a \leq t, W(t) > a\}$$
$$= 2P\{W(t) \geq a\},$$

because $P\{W(t) = a\} = 0$. The latter is manifestly equal to the integral in the statement of the theorem. In addition, thanks to symmetry (Theorem 9.9),

(9.52)
$$2P\{W(t) \geq a\} = P\{W(t) \geq a\} + P\{-W(t) \geq a\}$$
$$= P\{|W(t)| \geq a\}.$$

This completes our proof. □

The reflection principle has the following peculiar consequences: While we expect Brownian motion to reach a level a at some finite time, this time has infinite expectation. That is,

Corollary 9.23. *Let $T_a = \inf\{s \geq 0 : W(s) = a\}$ denote the first time Brownian motion reaches the level a. Then for all $a \neq 0$, $T_a < \infty$ a.s. but $ET_a = \infty$.*

Proof. (Sketch) We have seen already that T_a is a.s. finite; let us sketch a proof that T_a has infinite expectation. Without loss of generality, we can assume that $a > 0$ (why?). Then, thanks to Theorem 9.22,

(9.53)
$$P\{T_a \geq t\} = \int_{-a/\sqrt{t}}^{a/\sqrt{t}} \frac{e^{-y^2/2}}{\sqrt{2\pi}}\, dy.$$

See Problem 9.13 for more details. The preceding formula demonstrates that

(9.54)
$$P\{T_a \geq t\} \sim a \left(\frac{2}{\pi t}\right)^{1/2} \quad \text{as } t \to \infty.$$

Therefore, $\sum_{n=1}^{\infty} P\{T_a \geq n\} = \infty$, and Lemma 6.8 (p. 67) finishes the proof. □

Problems

Throughout, W denotes a Brownian motion.

9.1. Prove the following: If X and Y are respectively $\mathbf{R}^n$- and $\mathbf{R}^m$-valued random variables, then X and Y are independent if and only if

(9.55)
$$Ee^{iu \cdot X + iv \cdot Y} = Ee^{iu \cdot X} Ee^{iv \cdot Y} \quad \forall u \in \mathbf{R}^n, v \in \mathbf{R}^m.$$

Use this to prove Corollary 9.7.

9.2. Suppose for every $n = 1, 2, \ldots, G^n = (G_1^n, \ldots, G_k^n)$ is an $\mathbf{R}^k$-valued centered normal random variable. Suppose further that $Q_{i,j} = \lim_{n \to \infty} \mathrm{E}[G_i^n G_j^n]$ exists and is finite. Then prove that Q is a symmetric nonnegative definite matrix, and that G^n converges weakly to a centered normal random variable $G = (G_1, \ldots, G_k)$ whose covariance matrix is Q.

9.3 (Linear Regression). Suppose $G = (G_1, \ldots, G_n)$ is an $\mathbf{R}^n$-valued centered normal random variable, and let $\mathscr{G}$ denote the σ-algebra generated by $(G_1, \ldots, G_m)$, where $m < n$. Prove that, conditionally on $\mathscr{G}$, $(G_{m+1}, \ldots, G_n)$ is a centered normal random variable. Find the conditional mean, as well as the covariance matrix.

9.4. Construct an example of two random variables X_1 and X_2 such that each of them is standard normal, although (X_1, X_2) is not a Gaussian random variable. (HINT: If $X = \pm 1$ with probability $1/2$, and Z is an independent $N(0, 1)$, then $X|Z|$ is an $N(0, 1)$ also.)

9.5. Prove that if W denotes Brownian motion, then $\{\int_0^t W(s)\, ds\}_{t \geq 0}$ is a continuously differentiable Gaussian process. Use this as a guide to construct a k-times continuously differentiable Gaussian process.

9.6. Let $t > 0$ be fixed and define $V_n(t)$ as in Theorem 9.9. Prove:

(1) The first two moments of $V_n(t)$ are respectively t and $(2t^2/n) + t^2$. Use this to verify that $V_n(t)$ converges to t in probability.

(2) There exists a constant $A > 0$ such that for all $n \geq 1$,
$$\|V_n(t) - t\|_4 \leq \frac{A}{\sqrt{n}}.$$
Use this to prove that $V_n(t)$ converges to t almost surely.

9.7. Prove that $\{-W(t)\}_{t \geq 0}$, $\{tW(1/t)\}_{t \geq 0}$, and $\{c^{-1/2}W(ct)\}_{t \geq 0}$ are Brownian motions for any fixed, non-random $c > 0$.

9.8 (Heat equation). Let p_t denote the density function of $W(t)$. Compute $p_t(x)$, and verify that it solves the partial differential equation,

(9.56) $$\frac{\partial p_t}{\partial t}(x) = \frac{1}{2}\frac{\partial^2 p_t}{\partial x^2}(x) \qquad \forall t > 0, \ x \in \mathbf{R}.$$

Also, prove that p_t solves $p_{t+s}(x) = \int_{-\infty}^{\infty} p_t(y-x) p_s(y)\, dy$ (Bachelier, 1900, p. 29 and pp. 39–40).

9.9 (Khintchine's LIL). If W denotes a Brownian motion, then prove the following LILs of Khintchine (1933): With probability one,

(9.57) $$\limsup_{t \to \infty} \frac{W(t)}{(2t \ln \ln t)^{1/2}} = \limsup_{t \to 0} \frac{W(t)}{(2t \ln \ln(1/t))^{1/2}} = 1.$$

9.10 (Brownian Bridge). Given a Brownian motion $\{W(t)\}_{t \geq 0}$ define the *Brownian bridge* to be the process $B(t) = W(t) - tW(1)$. Prove that for all $0 \leq t_1 < t_2 < \cdots < t_m < 1$ and $u_1, \ldots, u_m \in \mathbf{R}$,

(9.58) $$\lim_{\epsilon \to 0} \mathrm{E}\left[e^{i \sum_{j=1}^m u_j W(t_j)} \,\Big|\, |W(1)| \leq \epsilon\right] = \mathrm{E} e^{i \sum_{j=1}^m u_j B(t_j)}.$$

This justifies the assertion that *Brownian bridge is Brownian motion conditioned to be zero at time one*. (HINT: B is independent of $W(1)$.)

9.11 (Blumenthal's Zero-One Law). If W denotes a Brownian motion, define the *tail σ-algebra* $\mathscr{T}$ as follows: First, for any $t \geq 0$, define $\mathscr{T}_t$ to be the P-completion of $\sigma\{W(u);\ u \geq t\}$. Then, define $\mathscr{T} = \cap_{t \geq 0} \mathscr{T}_t$.

(1) Prove that $\mathscr{T}$ is trivial; i.e., $\mathrm{P}(A) \in \{0, 1\}$ for all $A \in \mathscr{T}$.

(2) Let $\mathscr{F}_t^0 = \sigma(\{W(u)\}_{u \leq t})$, define $\bar{\mathscr{F}}_t$ to be the P-completion of $\mathscr{F}_t^0$, and let $\mathscr{F}_t$ be the right-continuous extension. That is, $\mathscr{F}_t = \cap \bar{\mathscr{F}}_s$, where the intersection is taken over all rational $s > t$. Prove *Blumenthal's zero-one law*: $\mathscr{F}_0$ is trivial; i.e., $\mathrm{P}(A) \in \{0, 1\}$ for all $A \in \mathscr{F}_0$.

9.12. Follow the next three steps in order to refine Wiener's Theorem (p. 166).
   (1) Check that $\{W_m(t) - W_{2^n}(t)\}_{m=2^n}^{\infty}$ is a martingale for all fixed $n \geq 1$ and $t \in [0,1]$.
   (2) Conclude that $m \mapsto \sup_{0 \leq t \leq 1} |W_m(t) - W_{2^n}(t)|^2$ is a submartingale as m varies over $2^n, \ldots, 2^{n+1}$.
   (3) Prove that a.s., $W(t) = \lim_n W_n(t)$ uniformly over all $t \in [0,1]$.

9.13. Given $a \neq 0$, define $T_a = \inf\{s \geq 0 : W(s) = a\}$.
   (1) Prove that the density function f_{T_a} of T_a is given by

$$(9.59) \qquad f_{T_a}(x) = \frac{|a|e^{-a^2/(2x)}}{x^{3/2}\sqrt{2\pi}} \qquad \forall a \in \mathbf{R},\ x > 0.$$

   For a bigger challenge compute the characteristic function of T_a. The distribution of f_{T_1} is called the *stable distribution with index* $1/2$.
   (2) Show that the stochastic process $\{T_a\}_{a \geq 0}$ has i.i.d. increments.

9.14. Choose and fix $a, b \geq 0$. Find the probability that Brownian motion does not take the value zero at any time in the interval $(a, a+b)$. Use this to find the distribution function of

$$(9.60) \qquad L := \sup\{t \leq 1 : W(t) = 0\}.$$

(HINT: L is not a stopping time; condition on $W(a)$.)

9.15. Let $\{W(t)\}_{t \in [0,1)}$ denote a Brownian motion on $[0,1)$. We wish to prove that the maximum of W is achieved at a unique place a.s.
   (1) Prove that $\mathrm{P}\{\sup_{t \in [a,b]} W(t) = x\} = 0$ for all $x \in \mathbf{R}$ and $0 < a < b$.
   (2) Prove that $\mathrm{P}\{\sup_{s \in [0,a]} W(s) = \sup_{t \in [a,b]} W(t)\} = 0$ for all $0 < a < b$.
   (3) Combine these to prove the existence of an almost surely unique $\rho \in (0,1)$ such that $W(\rho) = \sup_{t \in [0,1]} W(t)$.

9.16 (Simple Walk inside Brownian Motion). Define $T_0 := 0$ and

$$T_{k+1} := \inf\{s > T_k : |W(s) - W(T_k)| = 1\} \qquad \forall k = 1, 2, \ldots.$$

   (1) Prove that the T_j's are stopping times.
   (2) Prove that the vectors $(W(T_{k+1}) - W(T_k), T_{k+1} - T_k)$ $(k \geq 0)$ are i.i.d. Therefore, the process $\{W(T_k)\}_{k=0}^{\infty}$ is an embedding of a simple random walk inside Brownian motion (Knight, 1962). In fact, *every* mean-zero finite-variance random walk can be embedded inside Brownian motion via stopping times (Skorohod, 1961), but this is considerably more difficult to prove.

9.17 (The Forgery Theorem). Let W denote a Brownian motion on $[0,1)$, and consider a non-random continuous function $f : [0,1) \to \mathbf{R}$. Then prove that for all $\epsilon > 0$,

$$(9.61) \qquad \mathrm{P}\left\{\sup_{t \in [0,1)} |W(t) - f(t)| \leq \epsilon\right\} > 0.$$

Think of the graph of f as describing a "signature." Then, this shows that Brownian motion can forge this signature to within ϵ with positive probability (Lévy, 1951).

9.18 (Heat Semigroup). Let W denote Brownian motion. Suppose $f : \mathbf{R} \to \mathbf{R}$ is measurable, and $\mathrm{E}|f(x + W(t))| < \infty$ for all $t \geq 0$ and $x \in \mathbf{R}$. Define the *heat operator*, $(H_t f)(x) := \mathrm{E}f(x + W(t))$, where $t \geq 0$ and $x \in \mathbf{R}$. Prove that $\{H_t\}_{t > 0}$ has the following "semigroup" property:

$$H_t(H_s f) = H_{s+t} f \qquad \forall t, s \geq 0.$$

9.19 (White Noise; Hard). Let $f : \mathbf{R} \to \mathbf{R}$ be non-random, differentiable, and zero outside $[a,b]$. Then we can define $\int f\, dW$ to be $-\int W(s) f'(s)\, ds$. Prove then that $\|\int f\, dW\|_2^2 = \|f\|_{L^2(m)}^2$, where m denotes the Lebesgue measure on $\mathbf{R}$. Use this "L^2-isometry" to construct $\int f\, dW$ for all $f \in L^2(m)$.

(1) Prove that for all $f, g \in L^2(m)$,
$$\mathrm{E}\left[\int f\,dW \int g\,dW\right] = \int_{-\infty}^{\infty} f(x)g(x)\,dx.$$

(2) Let $G(f) = \int f\,dW$ ($f \in L^2(m)$), and prove that $G := \{G(f)\}_{f \in L^2(m)}$ is a centered Gaussian process. The process G is the so-called *white noise*, as well as the Wiener integral.

(3) Prove that if $\{\phi_i\}_{i=1}^{\infty}$ is an orthonormal basis for $L^2(m)$, then $\{G(\phi_i)\}_{i=1}^{\infty}$ are i.i.d. standard normal random variables (Wiener, 1923a,b).

9.20 (Hard). Suppose W is a Brownian motion and $f: \mathbf{R} \to \mathbf{R}$ is infinitely differentiable and $\mathrm{E}|f(W(t))| < \infty$ for all $t \geq 0$. Prove that $\{f(W(t))\}_{t \geq 0}$ is a martingale if and only if there exist $a, b \in \mathbf{R}$ such that $f(x) = a + bx$ for all $x \in \mathbf{R}$.

9.21 (Hard). Our proof of Theorem 9.13 can be refined to produce a stronger statement. Indeed, suppose $\alpha > \frac{1}{2}$ is fixed, and then prove that for all $s \in [0,1]$,

(9.62) $$\limsup_{t \to s} \frac{|W(s) - W(t)|}{|s-t|^\alpha} = \infty \quad \text{a.s.}$$

9.22 (Hard). Choose and fix some $\alpha > \frac{1}{2}$, and define W to be a Brownian motion. Prove that there exist finite random times $\sigma_1 > \sigma_2 > \ldots$, decreasing to zero, such that $W(\sigma_j) = \sigma_j^\alpha$ for all $j \geq 1$. Is there an $\alpha < \frac{1}{2}$ for which this property holds? (HINT: Problem 9.21.)

9.23 (Reflection Principle, Hard). Let $\{X_n\}_{n=1}^{\infty}$ be i.i.d. taking the values ± 1 with probability $\frac{1}{2}$ each. Define $S_n = \sum_{i=1}^{n} X_i$ ($n \geq 1$). Prove that $\max_{1 \leq i \leq n} S_i$ has the same distribution as $|S_n|$. This is due to André (1887).

9.24 (Hard). Prove that the zero-set $Z = \{t > 0 : W(t) = 0\}$ of Brownian motion W is a.s. uncountable, closed, and has no isolated points. Prove also that Z has zero Lebesgue measure.

9.25 (Problem 9.13, continued; Hard). Define $S(t) = \sup_{s \in [0,t]} W(s)$ and $X(t) = S(t) - W(t)$. Then, compute the density function of $X(t)$ for fixed $t > 0$. (HINT: Start by computing $\mathrm{P}\{S(t) > s, W(t) < w\}$.)

9.26 (Harder). Suppose $\{W_i\}_{i=0}^{\infty}$ is a collection of independent Brownian motions. Define the "two-parameter processes" $Z_1, Z_2, \ldots$, where

(9.63) $$Z_n(s,t) = sW_0(t) + \frac{\sqrt{2}}{\pi} \sum_{j=1}^{n} \frac{\sin(j\pi s)}{j} W_j(t) \quad \forall 0 \leq s, t \leq 1.$$

Prove that almost surely, $Z_{2^n}(s,t)$ converges to a limiting two-parameter process $\{Z(s,t)\}_{s,t \in [0,1]}$, uniformly for all $s, t \in [0,1]$. Prove that Z is an a.s.-continuous centered "Gaussian process" with covariance function $\mathrm{E}[Z(s,t)Z(u,v)] = (s \wedge u)(t \wedge v)$. Use this to prove that for any fixed $s > 0$, $t \mapsto s^{-1/2} Z(s,t)$ is a Brownian motion on $[0,1]$. The process $\{Z(s,t)\}_{s,t \in [0,1]}$ is the so-called *Brownian sheet* on $[0,1]^2$.

9.27 (Problem 9.16, continued; Harder). Prove that the embedding scheme of Problem 9.16 satisfies $\mathrm{E}T_1 = 1$ and $\mathrm{E}[T_1^2] < \infty$. Conclude that

$$|T_n - n| = O\left((n \ln \ln n)^{1/2}\right) < \infty \quad \text{as } n \to \infty \quad \text{a.s.}$$

Use this to prove that for all $\rho > \frac{1}{4}$,

(9.64) $$\lim_{n \to \infty} \frac{|W(T_n) - W(n)|}{n^\rho} = 0 \quad \text{a.s.}$$

Conclude that, on a suitable probability space, we can construct a simple walk $\{S_i\}_{i=1}^{\infty}$ and a Brownian motion $\{W(t)\}_{t \geq 0}$ such that

$$\max_{1 \leq k \leq n} |S_k - W(k)| = o(n^\rho) \quad \text{as } n \to \infty \quad \text{a.s.}$$

(HINT: $W^2(t) - t$ describes a mean-zero martingale.)

Notes

(1) Brownian motion is known also as the "Wiener process," or even sometimes as the "Bachelier–Wiener process."

(2) Although Bachelier's ideas are now regarded as revolutionary, they went largely unnoticed for nearly 60 years. Courtault, Kabanov, Bru, Crépel, Lebon, and Le Marchand (2000) contains an enthusiastic discussion of this issue. The said reference contains a number of other interesting facts about Bachelier.

(3) The book of Nelson (1967) contains a detailed account of the development of the physical theory of Brownian motion.

(4) The term "Avogadro's constant" is due to J. B. Perrin.

(5) The a.s.-continuity assumption of Theorem 9.9 is redundant; it is a consequence of the other assumptions there.

(6) The strong Markov property was introduced and utilized by Kinney (1953), Hunt (1956), Dynkin and Jushkevich (1956), and Blumenthal (1957). The phrase "strong Markov property" was coined by Dynkin and Jushkevich (1956).

(7) Parts of our modification to the Brownian filtration are unnecessary because Brownian motion is a.s. continuous. However, this requires a more advanced development of the Brownian motion.

(8) The reflection principle (p. 175) is due to Bachelier (1964, pp. 61–65). The central idea uses the method of André (1887) developed for the simple walk (Problem 9.23, p. 179).

Chapter 10

Terminus: Stochastic Integration

Reason's last step is the recognition that there are an infinite number of things which are beyond it.

–Blaise Pascal

1. The Indefinite Itô Integral

Given a "nice" stochastic process $H = \{H(s)\}_{s \geq 0}$, Itô (1944) constructed a natural integral $\int H\, dW = \int_0^\infty H(s)\, W(ds)$ despite the fact that W is nowhere differentiable a.s. (Theorem 9.13). In order to identify what "nice" means, it is best to go back and *redefine* what we mean by a stochastic process in continuous time.

Definition 10.1. A *measurable stochastic process* (also, *process* or *stochastic process*) $X = \{X(t)\}_{t \geq 0}$ is a product-measurable function $X : [0, \infty) \times \Omega \to \mathbf{R}$.

We often write $X(t)$ in place of $X(t, \omega)$; this is similar to what we did in discrete time.

This is a natural place to verify that Brownian motion is a stochastic process.

If H is nicely behaved, then it stands to reason that we should define $\int H\, dW$ as $\lim_{n \to \infty} \mathcal{I}_n(H)$, where

(10.1) $$\mathcal{I}_n(H) = \sum_{k=0}^{\infty} H\left(\frac{k}{2^n}\right)\left[W\left(\frac{k+1}{2^n}\right) - W\left(\frac{k}{2^n}\right)\right],$$

and the nature of the limit must be made precise. Clearly, $\mathcal{I}_n(H)$ is a well-defined random variable if, for instance, H has compact support. The following performs some of the requisite book-keeping about $\{\mathcal{I}_n(H)\}_{n=1}^\infty$.

Lemma 10.2. *If there exists $T > 0$ such that $H(s) = 0$ for all $s \geq T$ a.s., then $\mathcal{I}_n(H)$ is a.s. a finite sum and*

$$\mathcal{I}_{n+1}(H) - \mathcal{I}_n(H)$$
$$= -\sum_{j=0}^\infty \left[H\left(\frac{2j+1}{2^{n+1}}\right) - H\left(\frac{j}{2^n}\right) \right] \left[W\left(\frac{j+1}{2^n}\right) - W\left(\frac{2j+1}{2^{n+1}}\right) \right].$$

Proof. The sum is obviously finite. We derive the stated identity for $\mathcal{I}_{n+1}(H) - \mathcal{I}_n(H)$. Throughout we write $H_{k,n}$ in place of $H(k2^{-n})$, and $\Delta_{j,n}^{k,m} := W(j2^{-n}) - W(k2^{-m})$.

Consider the identity $\mathcal{I}_{n+1}(H) = \sum_{k=0}^\infty H_{k,n+1} \Delta_{k+1,n+1}^{k,n+1}$, and split the sum according to whether $k = 2j$ or $k = 2j+1$:

$$\mathcal{I}_{n+1}(H) = \sum_{j=0}^\infty H_{j,n} \Delta_{2j+1,n+1}^{j,n} + \sum_{j=0}^\infty H_{2j+1,n+1} \Delta_{j+1,n}^{2j+1,n+1}$$

(10.2)
$$= \sum_{j=0}^\infty H_{j,n} \left(\Delta_{2j+1,n+1}^{j,n} + \Delta_{j+1,n}^{2j+1,n+1} \right)$$

$$- \sum_{j=0}^\infty (H_{2j+1,n+1} - H_{j,n}) \Delta_{j+1,n}^{2j+1,n+1}.$$

Because $\Delta_{2j+1,n+1}^{j,n} + \Delta_{j+1,n}^{2j+1,n+1} = \Delta_{j+1,n}^{j,n}$, the first term is equal to $\mathcal{I}_n(H)$, whence follows the lemma. □

Definition 10.3. A process $H = \{H(s)\}_{t \geq 0}$ is *adapted* to the Brownian filtration $\mathscr{F}$ if $H(s)$ is $\mathscr{F}_s$-measurable for each $s \geq 0$. We say that H is a *compact-support process* if there exists a non-random $T \geq 0$ such that with probability one, $H(s) = 0$ for all $s \geq T$.

We also need the following technical definition.

Definition 10.4. Choose and fix $p \geq 1$. We say that H is *Dini-continuous in $L^p(\mathrm{P})$* if $H(s) \in L^p(\mathrm{P})$ for all $s \geq 0$ and

(10.3) $$\int_0^1 \frac{\psi_p(r)}{r} \, dr < \infty, \quad \text{where} \quad \psi_p(r) := \sup_{s,t: |s-t| \leq r} \|H(s) - H(t)\|_p.$$

The function ψ_p is called the *modulus of continuity* of H in $L^p(\mathrm{P})$.

If H is compact-support, continuous, and a.s.-bounded by a non-random quantity, then it is a.s. uniformly continuous in $L^p(\mathrm{P})$ for any $p \geq 1$; i.e.,

$\lim_{t \to 0} \psi_p(t) = 0$. Dini-continuity in $L^p(P)$ ensures that ψ_p converges to zero at some minimum rate. Here are a few examples:

Example 10.5. (a) Suppose H is a.s. differentiable with a derivative that satisfies $K := \sup_t \|H'(t)\|_p < \infty$. [Since $(s,\omega) \mapsto H(s,\omega)$ is product measurable, $\int |H'(r)|^p\, dr$ is a random variable, and hence $\|H'(t)\|_p$ are well defined.] By the fundamental theorem of calculus, if $t \geq s \geq 0$ then

$$\|H(s) - H(s)\|_p \leq \int_s^t \|H'(r)\|_p\, dr \leq K|t - s|.$$

Therefore, $\psi_p(r) \leq Kr$, and H is Dini-continuous in $L^p(P)$.

(b) Suppose $H(s) = f(W(s))$, where f is a non-random Lipschitz-continuous function; i.e., there exists L such that

$$|f(x) - f(y)| \leq L|y - x|.$$

It follows then that $\psi_p(r) = O(r^{1/2})$ as $r \to 0$, and this yields the Dini-continuity of H in $L^p(P)$ for any $p \geq 1$.

(c) Consider $H(s) = f(W(s), s)$, where $f(x,t)$ is a non-random function, twice continuously differentiable in each variable with bounded derivatives. Suppose, in addition, that there exists a non-random $T \geq 0$ such that $f(x,s) = 0$ for all $s \geq T$. Because $|H(s) - H(t)|$ is bounded above by

$$|f(W(s),s) - f(W(t),s)| + |f(W(t),s) - f(W(t),t)|,$$

by the fundamental theorem of calculus we can find a constant M such that for all $s, t \geq 0$,

$$|H(s) - H(t)| \leq M\left(|W(s) - W(t)| + |t - s|\right).$$

By Minkowski's inequality for all $p \geq 1$,

$$\|H(s) - H(s)\|_p \leq M(\|W(s) - W(t)\|_p + |t - s|)$$
$$= M\left(c_p|t - s|^{1/2} + |t - s|\right),$$

where $c_p := \|N(0,1)\|_p$. Therefore, $\psi_p(r) = O(r^{1/2})$ as $r \to 0$, whence follows the Dini-continuity of H in any $L^p(P)$ ($p \geq 1$).

Remark 10.6 (Cauchy Summability Test). Dini-continuity in $L^p(P)$ is equivalent to the summability of $\psi_p(2^{-n})$. Indeed, we can write

(10.4) $$\int_0^1 \frac{\psi_p(t)}{t}\, dt = \sum_{n=0}^{\infty} \int_{2^{-n-1}}^{2^{-n}} \frac{\psi_p(t)}{t}\, dt.$$

Because ψ_p is nondecreasing,

$$(10.5) \quad \frac{1}{2}\sum_{n=0}^{\infty}\psi_p\left(2^{-n-1}\right) \leq \int_0^1 \frac{\psi_p(t)}{t}\,dt \leq \sum_{n=0}^{\infty}\psi_p\left(2^{-n}\right).$$

We can now define $\int H\,dW$ for adapted compact-support processes that are Dini-continuous in $L^2(\mathrm{P})$. We will improve the construction later on.

Theorem 10.7. *Suppose H is an adapted compact-support stochastic process that is Dini-continuous in $L^2(\mathrm{P})$. Then $\lim_{n\to\infty}\boldsymbol{\mathcal{I}}_n(H)$ exists in $L^2(\mathrm{P})$. If we write $\int H\,dW$ for this limit, then*

$$(10.6) \quad \mathrm{E}\left[\int H\,dW\right] = 0 \quad \text{and} \quad \mathrm{E}\left[\left(\int H\,dW\right)^2\right] = \mathrm{E}\left[\int_0^\infty H^2(s)\,ds\right].$$

If $a,b \in \mathbf{R}$, and V is another adapted compact-support stochastic process that is Dini-continuous in $L^2(\mathrm{P})$, then with probability one,

$$(10.7) \quad \int (aH + bV)\,dW = a\int H\,dW + b\int V\,dW.$$

Definition 10.8. The second identity in (10.6) is called the *Itô isometry* (Itô, 1944).

Proof (Sketch). For $t \geq s$, $W(t) - W(s)$ is independent of $\mathscr{F}_s$ (Theorem 9.21, p. 174), and $H(u)$ is $\mathscr{F}_s$-measurable for $u \leq s$. Therefore, Lemma 10.2 implies the following. We use the notation introduced in the proof of Lemma 10.2 to simplify the type setting:

$$\|\boldsymbol{\mathcal{I}}_{n+1}(H) - \boldsymbol{\mathcal{I}}_n(H)\|_2^2$$

$$(10.8) \quad = \sum_{0\leq j\leq 2^n T-1} \left\|H\left(\frac{2j+1}{2^{n+1}}\right) - H\left(\frac{j}{2^n}\right)\right\|_2^2 \cdot \left\|\Delta_{j+1,n}^{2j+1,n+1}\right\|_2^2$$

$$= 2^{-n-1}\sum_{0\leq j\leq 2^n T-1}\left\|H\left(\frac{2j+1}{2^{n+1}}\right) - H\left(\frac{j}{2^n}\right)\right\|_2^2$$

$$\leq \sqrt{T}\,\psi_2\left(2^{-n-1}\right),$$

by Dini continuity. Consequently,

$$\|\boldsymbol{\mathcal{I}}_{N+M}(H) - \boldsymbol{\mathcal{I}}_N(H)\|_2 \leq \sum_{n=N+1}^{N+M-1}\|\boldsymbol{\mathcal{I}}_{n+1}(H) - \boldsymbol{\mathcal{I}}_n(H)\|_2$$

$$(10.9)$$

$$\leq \sqrt{T}\sum_{n=N+1}^{N+M-1}\psi_2\left(2^{-n-1}\right) \quad \forall N, M \geq 1.$$

It follows from Remark 10.6 that $\{\boldsymbol{\mathcal{I}}_n(H)\}_{n=1}^{\infty}$ is a Cauchy sequence in $L^2(\mathrm{P})$. This proves the asserted L^2-convergence.

1. The Indefinite Itô Integral

The basic properties of Brownian motion ensure that $\mathrm{E}\mathcal{I}_n(H) = 0$. By L^2-convergence we also have $\mathrm{E}[\int H\,dW] = 0$. Similarly, we can prove (10.6):

$$
\begin{aligned}
\mathrm{E}\left[\left(\int H\,dW\right)^2\right] &= \lim_{n\to\infty} \|\mathcal{I}_n(H)\|_2^2 \\
&= \lim_{n\to\infty} \sum_{k=0}^{\infty} \mathrm{E}\left[H^2\left(k2^{-n}\right)\right] 2^{-n} \\
&= \mathrm{E}\left[\int_0^{\infty} H^2(s)\,ds\right].
\end{aligned}
$$
(10.10)

The exchanges of limits and integrals are all justified by the compact-support assumption on H together with the continuity of the function $t \mapsto \|H(t)\|_2$.

Finally, (10.7) follows from the linearity of $H \mapsto \mathcal{I}_n(H)$ and the existence of $L^2(\mathrm{P})$-limits. □

We now drop many of the technical assumptions in Theorem 10.7.

Theorem 10.9 (Itô, 1944). *Suppose H is adapted and $\mathrm{E}[\int_0^{\infty} H^2(s)\,ds] < \infty$. Then one can define a stochastic integral $\int H\,dW$ that has mean zero and variance $\mathrm{E}[\int_0^{\infty} H^2(s)\,ds]$. Moreover, if V is another such integrand process, then for all $a, b \in \mathbf{R}$*

$$
\int (aH + bV)\,dW = a\int H\,dW + b\int V\,dW \quad \text{a.s.},
$$
$$
\mathrm{E}\left[\int H\,dW \cdot \int V\,dW\right] = \mathrm{E}\left[\int_0^{\infty} H(s)V(s)\,ds\right].
$$
(10.11)

The proof is function theoretic, and takes up the remainder of this section. If you wish to avoid such technical details, then you can do so without missing any of the central ideas: Simply skip to the next section.

Throughout, let m denote the Lebesgue measure on $(\mathbf{R}, \mathscr{B}(\mathbf{R}))$, and let $L^2(m \times \mathrm{P})$ denote the corresponding product L^2-space. We may note that

$$
\mathrm{E}\left[\int_0^{\infty} H^2(s)\,ds\right] = \|H\|_{L^2(m\times\mathrm{P})}^2,
$$
(10.12)

and $\mathrm{E}[\int H(s)V(s)\,ds]$ is the $L^2(m \times \mathrm{P})$ inner product between H and V.

The following is the key step in our construction of stochastic integrals.

Proposition 10.10. *Given any stochastic process $H \in L^2(m \times \mathrm{P})$ we can find processes $\{H_n\}_{n=1}^{\infty}$, all compact-support and Dini-continuous in $L^2(\mathrm{P})$, such that $\lim_{n\to\infty} H_n = H$ in $L^2(m \times \mathrm{P})$.*

Theorem 10.9 follows immediately from this.

Proof of Theorem 10.9. Proposition 10.10 asserts that there exist adapted compact-support processes $\{H_n\}_{n=1}^\infty$, that are Dini-continuous in $L^2(\mathrm{P})$ and converge to H in $L^2(m\times\mathrm{P})$. The Itô isometry (10.6) ensures that $\{\int H_n\,dW\}_{n=1}^\infty$ is a Cauchy sequence in $L^2(\mathrm{P})$, since $\{H_n\}_{n=1}^\infty$ is a Cauchy sequence in $L^2(m\times\mathrm{P})$. Consequently, $\int H\,dW = \lim_n \int H_n\,dW$ exists in $L^2(\mathrm{P})$. The properties of $\int H\,dW$ follow readily from those of $\int H_n\,dW$ and the $L^2(\mathrm{P})$-convergence that we have proved earlier. □

Let us conclude this section by proving the one remaining proposition.

Proof of Proposition 10.10. We proceed in three steps, each reducing the problem to a more restrictive class of processes H.

Step 1. Reduction to the Compact-Support Case. For each $n \geq 1$, let $H_n(t) := H(t)\mathbf{1}_{[0,n]}(t)$ and note that H_n is an adapted compact-support stochastic process. Moreover,

$$(10.13) \qquad \lim_{n\to\infty} \|H - H_n\|^2_{L^2(m\times\mathrm{P})} = \lim_{n\to\infty} \mathrm{E}\left[\int_n^\infty H^2(s)\,ds\right] = 0.$$

Therefore, we can, and will, assume without loss of generality that H is also compact-support.

Step 2. Reduction to the L^2-Bounded Case. Let us first extend the definition of H to all of $\mathbf{R}$ by assigning $H(t) := 0$ if $t < 0$. Next, define

$$(10.14) \qquad H_n(t) = n\int_{t-(1/n)}^t H(s)\,ds \qquad \forall t \geq 0,\ n \geq 1.$$

Check that H is an adapted process. Moreover, $H_n(t) = 0$ for all $t \geq T+1$, so that H_n is also compact-support. Next, we claim that H_n is bounded in $L^2(\mathrm{P})$. The Cauchy–Bunyakovsky–Schwarz inequality and the Fubini–Tonelli theorem together imply the following:

$$(10.15) \qquad \sup_{t\geq 0} \|H_n(t)\|_2^2 \leq n\int_0^\infty \|H(s)\|_2^2\,ds = n\|H\|^2_{L^2(m\times\mathrm{P})}.$$

It remains to prove that H_n converges in $L^2(m\times\mathrm{P})$ to H.

Since $\int_0^\infty H^2(s)\,ds < \infty$ a.s., the Lebesgue differentiation theorem (p. 140) implies that a.s., $H_n(t) \to H(t)$ for almost every $t \geq 0$. Therefore, $\lim_n H_n = H$ $(m\times\mathrm{P})$-almost surely by Fubini–Tonelli. According to the dominated convergence theorem, Step 2 follows if we prove that

$$(10.16) \qquad \sup_{n\geq 1}|H_n| \in L^2(m\times\mathrm{P}).$$

Note that $\sup_n |H_n| \leq \mathcal{M}H$, where the latter is the "maximal function,"

$$(10.17) \qquad (\mathcal{M}H)(t) = \sup_{n\geq 1}\left(n\int_{t-(1/n)}^t |H(s)|\,ds\right) \qquad \forall t \geq 0.$$

For each ω, $(\mathcal{M}H)(t+n^{-1})$ is the Hardy–Littlewood maximal function of H. Also, $\mathcal{M}H$ is an adapted process, whence

(10.18) $$\int_0^\infty |(\mathcal{M}H)(s)|^2 \, ds \leq 256 \int_0^\infty H^2(s) \, ds.$$

Confer with Corollary 8.41 on page 143. But $H(t) = 0$ for all $t \geq T$ and $\sup_t \|H(t)\|_2 < \infty$. Therefore, $\int_0^\infty H^2(s) \, ds < \infty$ for almost all ω. Moreover, we take first expectations (Fubini–Tonelli), and then square roots, to deduce that

(10.19) $$\|\mathcal{M}H\|_{L^2(m \times \mathrm{P})} \leq 16 \|H\|_{L^2(m \times \mathrm{P})} < \infty.$$

This reduces our problem to the one about the H's that are bounded in $L^2(\mathrm{P})$ and compact-support.

Step 3. The Conclusion. Finally, if H is bounded in $L^2(\mathrm{P})$ and compact-support, then we define H_n by (10.14) and note that H_n is differentiable, and $H_n'(t) = n\{H(t) - H(t - n^{-1})\}$. Therefore,

(10.20) $$\sup_t \|H_n'(t)\|_2 \leq 2n \sup_t \|H(t)\|_2 < \infty.$$

Part (b) of Example 10.5 implies the asserted Dini-continuity of H_n. On the other hand, the argument developed in Step 2 proves that $H_n \to H$ in $L^2(m \times \mathrm{P})$, whence follows the theorem. $\square$

2. Continuous Martingales in $L^2(\mathrm{P})$

The theories of continuous-time martingale and stochastic integration are intimately connected. Thus, before proceeding further, we take a side-step, and have a quick look at martingale theory in continuous time. To avoid unnecessary abstraction, $\mathscr{F}$ will denote the Brownian filtration throughout.

Definition 10.11. A process $M = \{M(t)\}_{t \geq 0}$ is a (continuous-time) *martingale* if:

(1) $M(t) \in L^1(\mathrm{P})$ for all $t \geq 0$.
(2) If $t \geq s \geq 0$ then $\mathrm{E}[M(t) \mid \mathscr{F}(s)] = M(s)$ a.s.

The process M is a *continuous L^2-martingale* if $t \mapsto M(t)$ is almost-surely continuous, and $M(t) \in L^2(\mathrm{P})$ for all $t \geq 0$.

Much of the theory of discrete-time martingales transfers to continuous $L^2(\mathrm{P})$-martingales. Here is a first sampler.

The Optional Stopping Theorem. *If M is a continuous L^2-martingale and $S \leq T$ are bounded $\mathscr{F}$-stopping times, then*

(10.21) $$\mathrm{E}[M(T) \mid \mathscr{F}_S] = M(S) \qquad a.s.$$

Proof. Throughout, choose and fix some non-random $K > 0$ such that $T \le K$ almost surely.

If S and T are simple stopping times, then the optional stopping theorem is a consequence of its discrete-time namesake (p. 130). In general, let $S_n \downarrow S$ and $T_n \downarrow T$ be the simple stopping times of Lemma 9.20 (p. 172). Note that the condition $S \le T$ imposes $S_m \le T_n + 2^{-m}$ for all $n, m \ge 1$. Because $T_n + 2^{-m}$ is a simple stopping time, it follows that

$$(10.22) \qquad \mathrm{E}\left[M(T_n + 2^{-m}) \,\middle|\, \mathscr{F}_{S_m}\right] = M(S_m) \quad \text{a.s.}$$

Moreover, this very argument implies that $\{M(T_i + 2^{-m})\}_{i=1}^{\infty}$ is a discrete-time martingale. Since $T_n \le T + 2^{-n} \le K + 2^{-n}$, Problem 8.22 on page 153 implies that

$$(10.23) \qquad \mathrm{E}\left[\sup_{n \ge 1} |M(T_n + 2^{-m})|^2\right] \le 4\mathrm{E}\left[|M(K+1)|^2\right] < \infty.$$

By the almost-sure continuity of M, $\lim_{n \to \infty} M(T_n + 2^{-m}) = M(T + 2^{-m})$ a.s. Convergence holds also in $L^2(\mathrm{P})$, thanks to the dominated convergence theorem. Therefore,

$$(10.24) \qquad \begin{aligned} &\|\mathrm{E}[M(T_n + 2^{-m}) \,|\, \mathscr{F}_{S_m}] - \mathrm{E}[M(T + 2^{-m}) \,|\, \mathscr{F}_{S_m}]\|_2^2 \\ &= \|\mathrm{E}\left[M(T_n + 2^{-m}) - M(T + 2^{-m}) \,|\, \mathscr{F}_{S_m}\right]\|_2^2 \\ &\le \mathrm{E}\left(\mathrm{E}\left[(M(T_n + 2^{-m}) - M(T + 2^{-m}))^2 \,\middle|\, \mathscr{F}_{S_m}\right]\right) \\ &\to 0 \qquad \text{as } n \to \infty. \end{aligned}$$

We have appealed to the conditional Jensen inequality (p. 120) for the first inequality, and the towering property of conditional expectations (Theorem 8.5, page 123) for the second identity.

By (10.22), $M(S_m) = \mathrm{E}[M(T + 2^{-m}) \,|\, \mathscr{F}_{S_m}]$ a.s. Because $\mathscr{F}_S \subseteq \mathscr{F}_{S_m}$, this and the towering property of conditional expectations together imply that $\mathrm{E}[M(S_m) \,|\, \mathscr{F}_S] = \mathrm{E}[M(T + 2^{-m}) \,|\, \mathscr{F}_S]$ a.s. Let $m \to \infty$ and appeal to the argument driving (10.24) to deduce that

$$(10.25) \qquad \begin{aligned} \mathrm{E}[M(T) \,|\, \mathscr{F}_S] &= \lim_{m \to \infty} \mathrm{E}\left[M(T + 2^{-m}) \,\middle|\, \mathscr{F}_{S_m}\right] \\ &= \lim_{m \to \infty} \mathrm{E}[M(S_m) \,|\, \mathscr{F}_S] \\ &= \mathrm{E}[M(S) \,|\, \mathscr{F}_S] \\ &= M(S), \end{aligned}$$

where the limits all hold in $L^2(\mathrm{P})$. This implies that $\mathrm{E}[M(T) \,|\, \mathscr{F}_S] = M(S)$ almost surely, as desired. $\square$

The following is a related result whose proof is relegated to the exercises.

Doob's Maximal Inequalities. *Let M denote a continuous L^2-martingale. Then for all $\lambda, t > 0$,*

$$(10.26) \qquad \lambda \mathrm{P}\left\{\sup_{0 \le s \le t} |M(s)| \ge \lambda\right\} \le \mathrm{E}\left[|M(t)|;\ \sup_{0 \le s \le t} |M(s)| \ge \lambda\right].$$

In particular, for all $p > 1$,

$$(10.27) \qquad \mathrm{E}\left[\sup_{0 \le s \le t} |M(s)|^p\right] \le \left(\frac{p}{p-1}\right)^p \mathrm{E}\left(|M(t)|^p\right).$$

3. The Definite Itô Integral

It is a natural time to mention the definite Itô integral. The latter is defined simply as $\int_0^t H\,dW = \int H\mathbf{1}_{[0,t)}\,dW$ for all adapted processes H such that $\mathrm{E}[\int_0^t H^2(s)\,ds] < \infty$ for all $t \ge 0$. This defines a collection of random variables $\int_0^t H\,dW$, one for each $t \ge 0$. The following is of paramount importance to us, since it says something about the properties of the random function $t \mapsto \int_0^t H\,dW$.

Theorem 10.12. *If H is an adapted process such that $\mathrm{E}[\int_0^t H^2(s)\,ds] < \infty$ then we can construct the process $\{\int_0^t H\,dW\}_{t \ge 0}$ such that it is a continuous L^2-martingale.*

Proof. According to Theorem 10.9, $\int_0^t H\,dW$ exists. Thus, we can proceed by verifying the assertions of the theorem. We do so in three steps.

Step 1. Reduction to H that is Dini-continuous in $L^2(\mathrm{P})$. Suppose we have proved the theorem for all processes H that are adapted and Dini-continuous in $L^2(\mathrm{P})$. In this first step we prove that this implies the remaining assertions of the theorem.

Let H be an adapted process such that $\mathrm{E}[\int_0^t H^2(s)\,ds] < \infty$ for all $t \ge 0$. We can find adapted processes H_n that are Dini-continuous in $L^2(\mathrm{P})$ and

$$(10.28) \qquad \lim_{n \to \infty} \mathrm{E}\left[\int_0^t (H_n(s) - H(s))^2\,ds\right] = 0.$$

Indeed, we can apply Proposition 10.10 to $H\mathbf{1}_{[0,t]}$, and use the recipe of the said proposition for H_n. Then apply the proposition to $H_n\mathbf{1}_{[0,t]}$. This shows that in fact H_n can even be chosen independently of t as well.

By (10.28) and the Itô isometry (10.6),

$$(10.29) \qquad \lim_{n \to \infty} \mathrm{E}\left[\left(\int_0^t H\,dW - \int_0^t H_n\,dW\right)^2\right] = 0.$$

Because $\int_0^t H_n\,dW - \int_0^t H_{n+k}\,dW = \int_0^t (H_n - H_{n+k})\,dW$ defines a continuous L^2-martingale, by Doob's maximal inequality (p. 189), for all non-random but fixed $T > 0$,

(10.30) $$\lim_{n \to \infty} \mathrm{E}\left[\sup_{0 \le t \le T}\left(\int_0^t H_n\, dW - \int_0^t H_{n+k}\, dW\right)^2\right] = 0.$$

In particular, for each $T > 0$ there exists a process $X = \{X(t)\}_{t \ge 0}$ and a subsequence $n' \to \infty$ such that

(10.31) $$\lim_{n' \to \infty} \sup_{0 \le t \le T}\left|\int_0^t H_{n'}\, dW - X(t)\right| = 0 \quad \text{a.s.}$$

Moreover, the same uniform convergence holds in $L^2(\mathrm{P})$, and along the original subsequence $n \to \infty$. Consequently, (10.29) implies that X is a particular construction of $t \mapsto \int_0^t H\, dW$ that is a.s.-continuous and adapted. In other words, X is adapted and a.s.-continuous, and also satisfies

(10.32) $$\mathrm{P}\left\{X(t) = \int_0^t H\, dW\right\} = 1 \quad \forall t \ge 0.$$

Finally, $X(t) \in L^2(\mathrm{P})$ for all $t \ge 0$, so it remains to prove that X is a martingale. But remember that we are assuming that $\{\int_0^t H_n\, dW\}_{t \ge 0}$ is a martingale.

By the conditional Jensen inequality (p. 120), and by $L^2(\mathrm{P})$-convergence,

(10.33) $$\left\|\mathrm{E}[X(t+s)\,|\,\mathscr{F}_s] - \mathrm{E}\left[\int_0^{t+s} H_n\, dW\,\bigg|\,\mathscr{F}_s\right]\right\|_2$$
$$= \left\|\mathrm{E}\left[X(t+s) - \int_0^{t+s} H_n\, dW\,\bigg|\,\mathscr{F}_s\right]\right\|_2$$
$$\le \left\|X(t+s) - \int_0^{t+s} H_n\, dW\right\|_2$$
$$\to 0 \quad \text{as } n \to \infty.$$

Consequently, $\lim_{n \to \infty} \int_0^t H_n\, dW = \mathrm{E}[X(t+s)\,|\,\mathscr{F}_s]$ in $L^2(\mathrm{P})$. But we have seen already that $\int_0^t H_n\, dW \to X(t)$ in $L^2(\mathrm{P})$. Therefore,

(10.34) $$\mathrm{E}[X(t+s)\,|\,\mathscr{F}_s] = X(s) \quad \text{a.s.}$$

That is, X is a martingale, as was claimed.

Step 2. A Continuous Martingale in the Dini-Continuous Case. Now we suppose that H is in addition Dini-continuous in $L^2(\mathrm{P})$, and prove the theorem in this special case. Together with Step 1 this completes the proof.

3. The Definite Integral

The argument is based on a trick. Define

(10.35)
$$\mathcal{J}_n(H)(t) = \sum_{0 \le k < 2^n t - 1} H\left(\frac{k}{2^n}\right)\left[W\left(\frac{k+1}{2^n}\right) - W\left(\frac{k}{2^n}\right)\right]$$
$$+ H\left(\frac{\lfloor 2^n t - 1 \rfloor}{2^n}\right)\left[W(t) - W\left(\frac{\lfloor 2^n t - 1 \rfloor}{2^n}\right)\right].$$

This is a minor variant of $\mathcal{I}_n(H\mathbf{1}_{[0,t]})$. Indeed, you should check that

(10.36)
$$\mathcal{J}_n(H)(t) - \mathcal{I}_n\left(H\mathbf{1}_{[0,t]}\right)$$
$$= H\left(\frac{\lfloor 2^n t - 1 \rfloor}{2^n}\right)\left[W(t) - W\left(\frac{\lfloor 2^n t - 1 \rfloor + 1}{2^n}\right)\right],$$

whose $L^2(\mathrm{P})$-norm goes to zero as $n \to \infty$. But $\mathcal{J}_n(H)$ is also a stochastic process that is: (a) Adapted; and (b) continuous—in fact, piecewise linear—in t. It is also a martingale. Here is why: Suppose $t \ge s \ge 0$. Then there exist integers $0 \le k \le K \le 2^n s - 1$ such that $s \in D(k; n) := [k2^{-n}, (k+1)2^{-n})$ and $t \in D(K; n)$. Then,

(10.37)
$$\mathcal{J}_n(H)(t) - \mathcal{J}_n(H)(s)$$
$$= \sum_{k \le j < K} H\left(\frac{j}{2^n}\right)\left[W\left(\frac{j+1}{2^n}\right) - W\left(\frac{j}{2^n}\right)\right]$$
$$+ H\left(\frac{K}{2^n}\right)\left[W(t) - W\left(\frac{K}{2^n}\right)\right] - H\left(\frac{k}{2^n}\right)\left[W(s) - W\left(\frac{k}{2^n}\right)\right],$$

where $\sum_{k \le j < k}(\cdots) := 0$ (in the case that $k = K$). Since W has independent increments, $\mathrm{E}[(\cdots) \mid \mathscr{F}_s] = 0$ a.s., where $(\cdots)$ is any of the terms of the preceding display in the parentheses. The adaptedness of H and Corollary 9.11 on page 164 together show that

(10.38)
$$\mathrm{E}\left[\mathcal{J}_n(H)(t) - \mathcal{J}_n(H)(s) \mid \mathscr{F}_s\right] = 0.$$

This proves the martingale property.

Step 3. The Conclusion. To finish the proof, suppose H is an adapted process that is Dini-continuous in $L^2(\mathrm{P})$. A calculation similar to that of Lemma 10.2 reveals that for any non-random $T > 0$,

(10.39)
$$\lim_{n \to \infty} \sup_{0 \le t \le T} \mathrm{E}\left[(\mathcal{J}_{n+1}(H)(t) - \mathcal{J}_n(H)(t))^2\right] = 0.$$

Therefore, by Doob's maximal inequality (p. 189),

(10.40)
$$\lim_{n \to \infty} \mathrm{E}\left[\sup_{0 \le t \le T} (\mathcal{J}_{n+1}(H)(t) - \mathcal{J}_n(H)(t))^2\right] = 0.$$

This implies that a subsequence of $\mathcal{J}_n(H)$ converges a.s. and uniformly for all $t \in [0, T]$ to some process X. Since $\mathcal{J}_n(H)$ is a continuous process, X is necessarily continuous a.s. Furthermore, the argument applied in Step 1 shows that here too X is a martingale. We have seen already that for any fixed $t \geq 0$,

(10.41) $$\mathcal{J}_n(H)(t) - \mathcal{I}_n(H\mathbf{1}_{[0,t]}) \to 0 \quad \text{in } L^2(\mathrm{P}).$$

Since $\mathcal{I}_n(H\mathbf{1}_{[0,t]}) \to \int_0^t H\, dW$ in $L^2(\mathrm{P})$, this proves that

(10.42) $$\mathrm{P}\left\{X(t) = \int_0^t H\, dW\right\} = 1 \quad \forall t \geq 0,$$

and whence the result. □

4. Quadratic Variation

We now elaborate a little on quadratic variation (Theorem 9.9, p. 163). Quadratic variation is a central theme in continuous-time martingale theory, but its study requires more time than we have. Therefore, we will develop only the portions for which we have immediate use.

Throughout, we define the *second-order* analogue of $\mathcal{I}_n$ (10.1),

(10.43) $$\mathcal{Q}_n(H) := \sum_{k=0}^{\infty} H\left(\frac{k}{2^n}\right) \left[W\left(\frac{k+1}{2^n}\right) - W\left(\frac{k}{2^n}\right)\right]^2.$$

Theorem 10.13. *Suppose H is adapted, compactly supported, and uniformly continuous in $L^2(\mathrm{P})$; i.e., $\lim_{r \to 0} \psi_2(r) = 0$. Then,*

(10.44) $$\lim_{n \to \infty} \mathcal{Q}_n(H) = \int_0^{\infty} H(s)\, ds \quad \text{in } L^2(\mathrm{P}).$$

Proof. To simplify the notation, we write for all integers $k \geq 0$ and $n \geq 1$,

(10.45) $$H_{k,n} := H(k2^{-n}) \quad \text{and} \quad d_{k,n} := W((k+1)2^{-n}) - W(k2^{-n}).$$

Recall next that we can find a non-random $T > 0$ such that for all $s \geq T$, $H(s) = 0$ a.s. Throughout, we keep such a T fixed.

Step 1. Approximating the Lebesgue Integral. We begin by proving that

(10.46) $$\sum_{k=0}^{\infty} H_{k,n} 2^{-n} \to \int_0^{\infty} H(s)\, ds \quad \text{in } L^2(\mathrm{P}) \text{ as } n \to \infty.$$

Note that

(10.47) $$\left|\sum_{k=0}^{\infty} H_{k,n} 2^{-n} - \int_0^{\infty} H(s)\, ds\right|$$
$$\leq \sum_{0 \leq k \leq 2^n T - 1} \int_{k2^{-n}}^{(k+1)2^{-n}} |H_{k,n} - H(s)|\, ds.$$

Therefore, by Minkowski's inequality,

(10.48) $$\left\| \sum_{k=0}^{\infty} H_{k,n} 2^{-n} - \int_0^{\infty} H(s)\, ds \right\|_2 \le \sum_{0 \le k \le 2^n T - 1} 2^{-n} \psi_2(2^{-n})$$
$$\le T \psi_2\left(2^{-n}\right),$$

which converges to zero as $n \to \infty$; Step 1 follows.

Step 2. Conclusion. Choose and fix $t \ge 0$, and for all $n \ge 1$ define

(10.49) $$\boldsymbol{\mathcal{D}}_n := \boldsymbol{\mathcal{Q}}_n(t) - \sum_{k=0}^{\infty} H_{k,n} 2^{-n}.$$

Note that $H_{k,n}$ is independent of $d_{k,n}$, and the latter has mean zero and variance 2^{-n}. Because $\boldsymbol{\mathcal{D}}_n = \sum_{k=0}^{\infty} H_{k,n} [d_{k,n}^2 - 2^{-n}]$, we first square and then take expectations to obtain

(10.50) $$\|\boldsymbol{\mathcal{D}}_n\|_2^2 = \sum_{0 \le k \le 2^n T - 1} \mathrm{E}\left[H_{k,n}^2\right] \mathrm{E}\left[\left(d_{k,n}^2 - 2^{-n}\right)^2\right]$$
$$+ 2 \sum_{0 \le j < k \le 2^n T - 1} \mathrm{E}\left[H_{k,n} H_{j,n} \left(d_{k,n}^2 - 2^{-n}\right)\left(d_{j,n}^2 - 2^{-n}\right)\right].$$

If $j < k$, then $d_{k,n}$ is independent of $H_{k,n} H_{j,n}(d_{j,n}^2 - 2^{-n})$, and has mean zero. Therefore,

(10.51) $$\|\boldsymbol{\mathcal{D}}_n\|_2^2 = \sum_{0 \le k \le 2^n T - 1} \mathrm{E}\left[H_{k,n}^2\right] \mathrm{E}\left[\left(d_{k,n}^2 - 2^{-n}\right)^2\right].$$

Next we observe that $[d_{k,n}^2 - 2^{-n}]$ has the same distribution as $2^{-n}(Z^2 - 1)$, where Z is standard normal. Because $\mathrm{E}[(Z^2-1)^2] = \mathrm{E}[Z^4] - 1 = 2$, it follows that

(10.52) $$\|\boldsymbol{\mathcal{D}}_n\|_2^2 = \frac{2}{4^n} \sum_{0 \le k \le 2^n T - 1} \|H_{k,n}\|_2^2.$$

Dini-continuity ensures that $t \mapsto \|H(t)\|_2$ is continuous, and hence bounded on $[0, T]$ by some constant K_T. Thus,

(10.53) $$\|\boldsymbol{\mathcal{D}}_n\|_2^2 \le \frac{2(2^n T - 1) K_T}{4^n} \to 0 \quad \text{as } n \to \infty.$$

This and Step 1 together imply the result. $\square$

5. Itô's Formula and Two Applications

Thus far, we have constructed the Itô integral, and studied some of its properties. In order to study the Itô integral further, we need to develop an operational calculus that mimics the calculus of the Lebesgue and/or

Riemann integral. To understand this better, we recall the chain rule of elementary calculus. Namely,

(10.54) $$(f \circ g)'(x) = f'(g(x))\, g'(x),$$

valid for all continuously differentiable functions f and g. In its integrated form—this is integration by parts—the chain rule states that for all $t \geq s \geq 0$,

(10.55) $$f(g(t)) - f(g(s)) = \int_s^t f'(g(u))g'(u)\, du.$$

For example, let $f(x) = x^2$ to find that

(10.56) $$g^2(t) - g^2(0) = \int_0^t g\, dg,$$

where $dg(s) = g'(s)\, ds$. What if g were replaced by Brownian motion? As a consequence of our next result we have

(10.57) $$W^2(t) - W^2(0) = \int_0^t W\, dW + \frac{t}{2} \quad \text{a.s.}$$

Compared with (10.56), this has an extra factor $(t/2)$.

Itô's Formula 1. *If $f : \mathbf{R} \to \mathbf{R}$ has two continuous derivatives, then for all $t \geq s \geq 0$, the following holds a.s.:*

(10.58) $$f(W(t)) - f(W(s)) = \int_s^t f'(W(r))\, W(dr) + \frac{1}{2} \int_s^t f''(W(r))\, dr.$$

Itô's formula is different from the chain rule for ordinary integrals because the nowhere differentiability of W forces us to replace the right-hand side of (10.55) with a stochastic integral plus a second-derivative term.

Remark 10.14. Itô's formula continues to hold even if we assume only that f'' exists almost everywhere and $\int_s^t (f'(W(r)))^2\, dr < \infty$ a.s. Of course then we have to make sense of the stochastic integral, etc.

Proof in the Case that f''' is Bounded and Continuous. We assume, without loss of too much generality, that $s = 0$.

The proof of Itô's formula starts out in the same manner as that of (10.55). Namely, by telescoping the sum we first write

(10.59) $$f\left(W\left(2^{-n}\lfloor 2^n t - 1\rfloor\right)\right) - f(0) = \sum_{0 \leq k \leq 2^n t - 1} \left[f\left(W\left(\frac{k+1}{2^n}\right)\right) - f\left(W\left(\frac{k}{2^n}\right)\right) \right].$$

To this we apply Taylor's expansion with remainder, and write

$$f\left(W\left(2^{-n}\lfloor 2^n t - 1\rfloor\right)\right) - f(0)$$

(10.60)
$$= \sum_{0 \le k \le 2^n t - 1} f'\left(W\left(\frac{k}{2^n}\right)\right) d_{k,n}$$

$$+ \frac{1}{2} \sum_{0 \le k \le 2^n t - 1} f''\left(W\left(\frac{k}{2^n}\right)\right) d_{k,n}^2 + \sum_{0 \le k \le 2^n t - 1} R_{k,n} d_{k,n}^3,$$

where $d_{k,n} := W((k+1)2^{-n}) - W(k2^{-n})$, and $|R_{k,n}| \le \sup_x |f'''(x)| := M < \infty$, uniformly for all k, n.

According to the proof of Theorem 10.7, the first term of the right-hand side of (10.60) converges in $L^2(\mathrm{P})$ to $\int_0^t f'(W(s))\, W(ds)$; see also Example 10.5. The second term, on the other hand, converges in $L^2(\mathrm{P})$ to $\frac{1}{2} \int_0^t f''(W(s))\, ds$; consult with Theorem 10.13. In addition, continuity and the dominated convergence theorem together imply the following:

(10.61) $$\lim_{n \to \infty} f\left(W(2^{-n}\lfloor 2^n t - 1\rfloor)\right) \to f(W(t)) \qquad \text{a.s. and in } L^2(\mathrm{P}).$$

It, therefore, suffices to prove that $\mathscr{E}_n \to 0$ in $L^1(\mathrm{P})$ as $n \to \infty$, where

(10.62) $$\mathscr{E}_n := \sum_{0 \le k \le 2^n t - 1} R_{k,n} d_{k,n}^3.$$

But as $n \to \infty$,

(10.63) $$\mathrm{E}|\mathscr{E}_n| \le M \sum_{0 \le k \le 2^n t - 1} \|d_{k,n}\|_3^3 \sim Mt \|N(0,1)\|_3^3\, 2^{-n/2}.$$

The proof follows. □

Next is an interesting refinement; it is proved by similar arguments involving Taylor series expansions that were used to derive Itô's formula 1.

Itô's Formula 2. *Let W denote Brownian motion with $W(0) = x_0$. If $f(x,t)$ is twice continuously differentiable in x and continuously differentiable in t, and if $\mathrm{E}[\int_0^t |\partial_x f(W(s),s)|^2\, ds] < \infty$ for all $t \ge 0$, then a.s.,*

(10.64)
$$f(W(t),t) = f(x_0,0) + \int_0^t \frac{\partial}{\partial x} f(W(s),s)\, W(ds)$$
$$+ \int_0^t \left[\frac{1}{2} \frac{\partial^2}{\partial x^2} f(W(s),s) + \frac{\partial}{\partial t} f(W(s),s)\right] ds.$$

This remains valid if f takes on complex values.

Of course, $\int H\, dW := \int \mathrm{Re}(H)\, dW + i \int \mathrm{Im}(H)\, dW$ whenever possible.

I will not prove this refinement. Instead, let us close this book with two fascinating consequences of Itô's formula 2.

5.1. Lévy's Theorem: A First Look at Exit Distributions. Choose and fix some $a > 0$, and let W denote a Brownian motion started somewhere in $(-a, a)$. We wish to know where W leaves the interval $(-a, a)$. The following remarkable answer is due to Lévy (1951):

Theorem 10.15. *Choose and fix some $a > 0$, and define*

(10.65) $$T_a := \inf \{s > 0 : W(s) = a \text{ or } -a\},$$

where $\inf \varnothing := \infty$. If $W(0) := x_0 \in (-a, a)$, then for all real numbers $\lambda \neq 0$,

(10.66) $$\mathrm{E}\exp(i\lambda T_a) = \frac{\cos\left(x_0\sqrt{2i\lambda}\right)}{\cos\left(a\sqrt{2i\lambda}\right)}.$$

Proof. We apply Itô's formula 2 with

(10.67) $$f(x, t) := \psi(x)e^{i\lambda t},$$

where $\lambda \neq 0$ is fixed, and the function ψ satisfies the following boundary-value problem:

(10.68) $$\begin{cases} \psi''(x) = 2i\lambda\psi(x) & \forall x \in (-a, a), \\ \psi(a) = \psi(-a) = 1. \end{cases}$$

By actually taking derivatives, etc., we find that the solution is

(10.69) $$\psi(x) = \frac{\cos\left(x\sqrt{2i\lambda}\right)}{\cos\left(a\sqrt{2i\lambda}\right)}.$$

It is possible to check that $\mathrm{E}[\int_0^t |\partial f(W(s), s)/\partial x|^2 \, ds]$ is finite for all $t > 0$. Moreover, the eigenvalue problem for ψ implies that f solves the partial differential equation,

(10.70) $$\frac{1}{2}\frac{\partial^2 f}{\partial x^2}(x, t) + \frac{\partial f}{\partial t}(x, t) = 0.$$

As a result, Itô's formula 2 tells us that $f(W(t), t) - f(x_0, 0)$ is a mean-zero (complex) martingale. By the optional stopping theorem (p. 187),

(10.71) $$\mathrm{E}\left[f(W(T_a \wedge t), T_a \wedge t)\right] = f(x_0, 0) = \psi(x_0).$$

Thanks to the dominated convergence theorem and the a.s.-continuity of W, we can let $t \to \infty$ to deduce that

(10.72) $$\mathrm{E}\left[f(W(T_a), T_a)\right] = \psi(x_0).$$

Because $W(T_a) = \pm a$ and $\psi(\pm a) = 1$,

(10.73) $$f(W(T_a), T_a) = e^{i\lambda T_a}.$$

This proves the theorem. □

5. Itô's Formula

5.2. Chung's Formula: A Second Look at Exit Distributions. Let us have a second look at Theorem 10.15 in the simplest setting where $x_0 := 0$ and $a := 1$. Define $T := T_1$ to find that the formula (10.66) simplifies to the following elegant form:

$$\text{(10.74)} \qquad \mathrm{E}\exp(i\lambda T) = \frac{1}{\cos\sqrt{2i\lambda}} \qquad \forall \lambda \in \mathbf{R} \setminus \{0\}.$$

In principle, the uniqueness theorem for characteristic functions tells us that the preceding formula determines the distribution of T. However, it is not always so easy to extract the right piece of information from (10.74). For instance, if it were not for (10.74), then we could not prove too easily that $1/\cos\sqrt{2i\lambda}$ is a characteristic function of a probability measure. Or for that matter, can you see from (10.74) that T has finite moments of all orders? (It does!)

The following theorem of Chung (1947) contains a different representation of the distribution of T that answers the previous question about the existence of the moments of T.

Theorem 10.16. *For all $t > 0$,*

$$\text{(10.75)} \qquad \mathrm{P}\{T > t\} = \frac{4}{\pi}\sum_{n=0}^{\infty}\frac{(-1)^n}{2n+1}\exp\left(-\frac{(2n+1)^2\pi^2 t}{8}\right).$$

Consequently, $\mathrm{P}\{T > t\} \sim (4/\pi)\exp(-\pi^2 t/8)$ as $t \to \infty$.

The preceding implies that $\mathrm{P}\{T > t\} \leq 2\exp(-\pi^2 t/8)$ for large values of t. In lieu of Lemma 6.8 (p. 67),

$$\text{(10.76)} \qquad \mathrm{E}\left[T^p\right] = p\int_0^\infty t^{p-1}\mathrm{P}\{T > t\}\,dt < \infty \qquad \forall p > 0.$$

Therefore, T has moments of all orders, as was asserted earlier. In fact, you might wish to carry out the computation a little further and produce the following neat formula for the pth moment of T:

$$\text{(10.77)} \qquad \mathrm{E}\left[T^p\right] = \frac{2^{3p+2}\Gamma(p+1)\beta(2p+1)}{\pi^{1+2p}} \qquad \forall p > 0,$$

where $\Gamma(p) = \int_0^\infty s^{p-1}e^{-s}\,ds$ denotes Euler's gamma function and β is the *Dirichlet beta function*, viz.,

$$\text{(10.78)} \qquad \beta(t) = \sum_{n=0}^{\infty}\frac{(-1)^n}{(2n+1)^t} \qquad \forall t \geq 1.$$

Theorem 10.16 implies also the following formula (Chung, 1947).

Corollary 10.17. *For all $x > 0$,*

$$(10.79) \quad P\left\{\sup_{0 \le s \le 1} |W(s)| \le x\right\} = \frac{4}{\pi} \sum_{n=0}^{\infty} \frac{(-1)^n}{2n+1} \exp\left(-\frac{(2n+1)^2 \pi^2}{8x^2}\right).$$

In particular, $P\{\sup_{0 \le s \le 1} |W(s)| \le x\} \sim (4/\pi) \exp(-\pi^2/(8x^2))$ as $x \to 0$.

Proof of Theorem 10.16 (Sketch). The proof follows three steps.

Step 1. A Cosine Formula. Choose and fix an integer $n \ge 1$, and define

$$(10.80) \quad f(x, t) = \cos\left(\frac{n\pi x}{2}\right) \exp\left(\frac{n^2 \pi^2 t}{8}\right) \quad \forall x \in \mathbf{R}, \, t \ge 0.$$

The function f solves the partial differential equation,

$$(10.81) \quad \frac{1}{2}\frac{\partial^2 f}{\partial x^2} + \frac{\partial f}{\partial t} = 0 \text{ subject to } f(\pm 1, t) = 0 \quad \forall t \ge 0.$$

This is a kind of *heat equation on $[-1, 1]$ with Dirichlet boundary conditions*. Barring technical conditions, Itô's formula 2 tells us that $f(W(t), t) - 1$ defines a mean-zero martingale. [This uses the fact that $f(0, 0) = 1$.] By the optional stopping theorem (p. 187), $E[f(W(T \wedge t), T \wedge t)] = 1$. Equivalently,

$$(10.82) \quad E[f(W(T), T); \, T \le t] + E[f(W(t), t); \, T > t] = 1.$$

Because $W(T) = \pm 1$ a.s., the first term in (10.82) vanishes a.s. Whence we obtain the following cosine formula:

$$(10.83) \quad E\left[\cos\left(\frac{n\pi W(t)}{2}\right); \, T > t\right] = \exp\left(-\frac{n^2 \pi^2 t}{8}\right).$$

Step 2. A Fourier Series. Let $L^2(-2, 2)$ denote the collection of all measurable functions $g : [-2, 2] \to \mathbf{R}$ such that $\int_{-2}^{2} g^2(x) \, dx < \infty$. Theorem A.5 (p. 205), after a little fidgeting with the variables, shows that $\frac{1}{2}$, $2^{-1/2} \sin(n\pi x/2)$, $2^{-1/2} \cos(m\pi x/2)$ $(n, m = 1, 2, \ldots)$ form an orthonormal basis for $L^2(-2, 2)$. In particular, any $\phi \in L^2(-2, 2)$ has the representation,

$$(10.84) \quad \phi(x) = \frac{A_0}{2} + \frac{1}{\sqrt{2}} \sum_{n=1}^{\infty} \left[A_n \cos\left(\frac{n\pi x}{2}\right) + B_n \sin\left(\frac{n\pi x}{2}\right)\right],$$

where:

- The infinite sums converge in $L^2(-2, 2)$;
- $A_0 = 2^{-1} \int_{-2}^{2} \phi(x) \, dx$;
- $A_n = 2^{-1/2} \int_{-2}^{2} \phi(x) \cos(n\pi x/2) \, dx$ for $n \ge 1$; and
- $B_n = 2^{-1/2} \int_{-2}^{2} \phi(x) \sin(n\pi x/2) \, dx$ for $n \ge 1$.

Step 3. Putting it Together. We can apply the result of Step 2 to the function $\phi(x) := \mathbf{1}_{(-1,1)}(x)$ to obtain

$$\mathbf{1}_{(-1,1)}(x) - \frac{1}{2} = \frac{2}{\pi} \sum_{n=0}^{\infty} \frac{(-1)^n}{2n+1} \cos\left(\frac{(2n+1)\pi x}{2}\right). \tag{10.85}$$

We "plug" in $x := W(t,\omega)$, multiply by $\mathbf{1}_{\{T(\omega)>t\}}$, and then apply expectations to find that

$$\begin{aligned}\mathrm{P}\{W(t) \in (-1,1)\,,\, T > t\} - \frac{1}{2}\mathrm{P}\{T > t\} \\ = \frac{2}{\pi} \sum_{n=0}^{\infty} \frac{(-1)^n}{2n+1} \mathrm{E}\left[\cos\left(\frac{(2n+1)\pi W(t)}{2}\right)\,;\, T > t\right].\end{aligned} \tag{10.86}$$

Since the left-hand side is equal to $\frac{1}{2}\mathrm{P}\{T > t\}$, the cosine formula of Step 1 completes our proof. □

The preceding proof is a sketch only because: (i) we casually treat the L^2-identity in (10.85) as a pointwise identity; and (ii) we exchange expectations with an infinite sum without actually justifying the exchange. With a little effort, these gaps can be filled.

Problems

10.1. In this exercise we construct a Dini-continuous process in $L^p(\mathrm{P})$ that is not a.s. continuous.

(1) Prove that if $0 < s < t$, then

$$\mathrm{P}\{W(s)W(t) < 0\} \leq \frac{1}{2}\sqrt{\frac{t-s}{t}}.$$

(2) Use this to prove that $H(s) := \mathbf{1}_{(0,\infty)}(W(e^{s \wedge 1}))$ is Dini-continuous in $L^2(\mathrm{P})$, but H is not a.s. continuous.

10.2. In this exercise you are asked to construct a rather general abstract integral that is due to Young (1970). See also McShane (1969).

A function $f : [0,1] \to \mathbf{R}$ is said to be *Hölder continuous* of order $\alpha > 0$ if there exists a finite constant K such that

$$|f(s) - f(t)| \leq K|t-s|^\alpha \quad \forall s, t \in [0,1]. \tag{10.87}$$

Let $\mathscr{C}^\alpha$ denote the collection of all such functions.

(1) Prove that $\mathscr{C}^\alpha$ contains only constants when $\alpha > 1$, whereas $\mathscr{C}^1$ includes but is not limited to all continuously differentiable functions.

(2) Prove that if $0 < \alpha \leq 1$, then $\mathscr{C}^\alpha$ is a complete normed linear space that is normed by

$$\|f\|_{\mathscr{C}^\alpha} := \sup_{\substack{s,t \in [0,1] \\ s \neq t}} \frac{|f(s) - f(t)|}{|s-t|^\alpha} + \sup_{t \in [0,1]} |f(t)|.$$

(3) Given two functions f and g, define for all $n \geq 1$,

$$\int_0^1 f\,\delta_n g = \sum_{k=0}^{2^n - 1} f\left(\frac{k}{2^n}\right)\left[g\left(\frac{k+1}{2^n}\right) - g\left(\frac{k}{2^n}\right)\right].$$

Suppose $f \in \mathscr{C}^\alpha$ and $g \in \mathscr{C}^\beta$ for some $0 \leq \alpha, \beta \leq 1$. Prove $\int_0^1 f \, \delta g := \lim_n \int_0^1 f \, \delta_n g$ exists whenever $\alpha + \beta > 1$. Note that when we let $g(x) = x$, we recover the Riemann integral of f; i.e., that $\int_0^1 f \, \delta g = \int_0^1 f(x) \, dx$.

(4) Prove that $\int_0^1 g \, \delta f$ is well defined, and
$$\int_0^1 f \, \delta g = f(1)g(1) - f(0)g(0) - \int_0^1 g \, \delta f.$$

The integral $\int f \, \delta g$ is called a *Young integral*. (HINT: Lemma 10.2.)

10.3. In this problem you are asked to derive Doob's maximal inequality (p. 189) and its variants. We say that M is a *submartingale* if it is defined as a martingale, except

(10.88) $\qquad \qquad \mathrm{E}[M(t) \mid \mathscr{F}_s] \geq M(s) \qquad$ a.s. whenever $t \geq s \geq 0$.

M is a *supermartingale* if $-M$ is a submartingale. A process M is said to be a *continuous L^2-submartingale* (respectively, continuous L^2-supermartingale) if it is a submartingale (respectively supermartingale), $\{M(t)\}_{t \geq 0}$ is a.s. continuous, and $M(t) \in L^2(\mathrm{P})$ for all $t \geq 0$. Prove:

(1) If Y is in $L^2(\mathrm{P})$, then $M(t) = \mathrm{E}[Y \mid \mathscr{F}_t]$ is a martingale. This is a *Doob martingale* in continuous time.

(2) If M is a martingale and ψ is convex, then $\psi(M)$ is a submartingale provided that $\psi(M(t)) \in L^1(\mathrm{P})$ for each $t \geq 0$.

(3) If M is a submartingale, ψ is a nondecreasing convex function, and $\psi(M(t)) \in L^1(\mathrm{P})$ for all $t \geq 0$, then $\psi(M)$ is a submartingale.

(4) The first Doob inequality on page 189 holds if $|M|$ is replaced by any a.s.-continuous submartingale. (HINT: Prove first that $\sup_{0 \leq s \leq t} M(s)$ is measurable.)

10.4. For all integers $n \geq 1$ define $\mu_n := \mathrm{E}\{(W(1))^n\}$, where W is Brownian motion. Use Itô's formula to prove that $\mu_{n+2} = (n+1)\mu_n$. Compute μ_n for all integers $n \geq 1$.

10.5. (Gambler's Ruin). If W denotes a Brownian motion, then for any $a \in \mathbf{R}$, define $T_a := \inf\{s \geq 0 : W(s) = a\}$ where $\inf \varnothing := \infty$. Recall that T_a is an $\mathscr{F}$-stopping time (Proposition 9.17, p. 171). If $a, b > 0$ then prove that

(10.89) $\qquad \qquad \mathrm{P}\{T_a < T_{-b}\} = \dfrac{b}{b+a},$

and compute $\mathrm{E}T_a$.

10.6. Prove Corollary 10.17.

10.7. Recall the Dirichlet beta function β from (10.78). Prove that $\beta(1) = \pi/4$ and $\beta(3) = \pi^3/32$. Few other evaluations of β are known explicitly. Even $\beta(2)$, the so-called "Catalan constant," does not have a simpler description.

10.8. Let W be a Brownian motion with $W(0) = 0$, and define T to be the first time W exits the interval $(-1, 1)$. Prove that $\mathrm{E}[T^\rho] < \infty$ whenever $-\infty < \rho < \infty$.

10.9. Let $T := \inf\{s > 0 : |W(s)| = 1\}$, where $\inf \varnothing := 0$. Prove that for all $\lambda > 0$,

(10.90) $\qquad \qquad \mathrm{E} \exp(-\lambda T) = \dfrac{1}{\cosh \sqrt{2\lambda}}.$

(Lévy, 1951).

10.10. Define

(10.91) $\qquad \qquad p(t;x) = \dfrac{1}{(2\pi t)^{1/2}} \exp\left(-\dfrac{x^2}{2t}\right) \qquad \forall x \in \mathbf{R}, \, t > 0.$

Prove that if W is Brownian motion, then $M(t) := p(t; W(t))$ defines a martingale.

10.11. (Hard). Let W be a Brownian motion with $W(0) = 0$ and define $T_{a,b}$ to be the first time W exits the interval $(-a, b)$, where $a, b > 0$ are fixed constants. Compute $\mathrm{E} \exp(i\lambda T_{a,b})$ for all $\lambda > 0$.

10.12 (Hard). Let W denote a Brownian motion, and $\beta > 0$ a fixed positive number. Define $W_\beta(t) = W(t) + \beta t$ to be the so-called Brownian motion *with drift* β. For any $a, b > 0$ define

(10.92) $$\tau_{a,-b} := \inf \left\{ s > 0 : W_\beta(s) = a \text{ or } -b \right\}.$$

Prove that

(10.93) $$P\left\{ W_\beta\left(\tau_{a,-b}\right) = -b \right\} = \frac{1 - e^{-2\beta a}}{e^{2\beta b} - e^{-2\beta a}}.$$

From this deduce the distribution of $-\inf_{t \geq 0} W_\beta(t)$. (HINT: For all $\alpha \in \mathbf{R}$ find a non-random function $h_\alpha : \mathbf{R}_+ \to \mathbf{R}$ such that $t \mapsto \exp(\alpha W(t) - h_\alpha(t))$ defines a mean-one martingale.)

Notes

(1) Instead of presenting a general theory of stochastic integration, we have discussed a special case that is: (i) broad enough to be applicable for our needs; and (ii) concrete enough so as to make the main ideas clear. Dellacherie and Meyer (1982) have written a definitive account of the general theory of processes. Their treatment includes a detailed description of the general theory of stochastic integration with respect to (semi-) martingales.

(2) Itô's theory of stochastic integrals uses the "left-hand rule," as can be seen clearly in (10.1). This "left-hand rule" is the hallmark of Itô's theory of stochastic integration. In general, it cannot be replaced by other rules—such as the midpoint- or the right-hand rule—without changing the resulting stochastic integral.

(3) For Theorem 10.15, and much more, see Knight (1981, Chapter 4).

(4) Theorem 10.16 can be extended to several dimensions Ciesielski and Taylor (1962).

One can use Corollary 10.17 in conjunction with the Poisson summation formula (Feller, 1966, p. 630) to deduce that

(10.94) $$P\left\{ \sup_{0 \leq s \leq 1} |W(s)| \leq x \right\} = \sqrt{\frac{2}{\pi}} \sum_{n=-\infty}^{\infty} \int_{(4n-1)x}^{(4n+1)x} \exp(-u^2/2) \, du.$$

Whereas (10.79) is useful for small values of x, the preceding is accurate when x is large.

The fact that the right-hand sides of (10.79) and (10.94) agree is one of the celebrated theta-function identities of analytic number theory.

(5) Now that we have reached the end of the book, let me close by suggesting that this is a natural place to start learning about W. K. Feller's theory of one-dimensional diffusions (1955a; 1955b; 1956). Modern and more pedagogic accounts include Bass (1998), Knight (1981, Chapter 4), and Revuz and Yor (1999, Chapters 3 and 7).

Appendix

The moving power of mathematical invention is not reasoning but imagination.

–Augustus de Morgan

1. Hilbert Spaces

Throughout, let **H** be a set, and recall that it is a (real) Hilbert space if it is linear and if there exists an inner product $\langle \cdot, \cdot \rangle$ on $\mathbf{H} \times \mathbf{H}$ such that $f \mapsto \langle f, f \rangle = \|f\|^2$ norms **H** into a complete space. We recall that inner product means that $\langle \alpha f + \beta g, h \rangle = \langle h, \alpha f + \beta g \rangle = \alpha \langle f, h \rangle + \beta \langle g, h \rangle$ for all $f, g, h \in \mathbf{H}$ and all $\alpha, \beta \in \mathbf{R}$.

Hilbert spaces come naturally equipped with a notion of angles: If $\langle f, g \rangle = 0$ then f and g are *orthogonal*.

Definition A.1. Given any $\mathbf{S} \subset \mathbf{H}$, we let $\mathbf{S}^\perp$ denote the collection of all elements of **H** that are orthogonal to all the elements of **S**. That is,

(A.1) $$\mathbf{S}^\perp = \{f \in \mathbf{H} : \langle f, g \rangle = 0 \; ^\forall g \in \mathbf{S}\}.$$

It is easy to see that $\mathbf{S}^\perp$ is itself a subspace of **H**, and that $\mathbf{S} \cap \mathbf{S}^\perp = \{0\}$. We now show that in fact **S** and $\mathbf{S}^\perp$ have a sort of complementary property.

Theorem A.2 (Orthogonal Decomposition)**.** *If **S** is a closed subspace of a Hilbert space **H**, then* $\mathbf{H} = \mathbf{S} + \mathbf{S}^\perp := \{f + g : f \in \mathbf{S}, g \in \mathbf{S}^\perp\}$.

In order to prove this, we need a lemma.

Lemma A.3. *If **X** is a closed and convex subset of a complete Hilbert space **H**, then there exists a unique $f \in \mathbf{X}$ such that $\|f\| = \inf_{g \in \mathbf{X}} \|g\|$.*

Proof. By definition we can find $f_n \in \mathbf{X}$ such that $\lim_{n\to\infty} \|f_n\|^2 = \inf_{h \in \mathbf{X}} \|h\|^2$. Recall the "parallelogram law":

(A.2) $$\|h+g\|^2 + \|h-g\|^2 = 2\left(\|h\|^2 + \|g\|^2\right) \qquad \forall h, g \in \mathbf{H}.$$

We apply this with $h := f_n$ and $g := f_m$ to find that for all $n, m \geq 1$,

(A.3)
$$\frac{\|f_n - f_m\|^2}{4} = \frac{1}{2}\left(\|f_n\|^2 + \|f_m\|^2 - 2\left\|\frac{f_n + f_m}{2}\right\|^2\right)$$
$$\leq \frac{1}{2}\left(\|f_n\|^2 + \|f_m\|^2 - 2\inf_{h \in \mathbf{X}}\|h\|^2\right).$$

The final inequality follows because $(f_n + f_m)/2 \in \mathbf{X}$, thanks to convexity. Let $n, m \to \infty$ to deduce that $\{f_n\}_{n=1}^\infty$ is a Cauchy sequence in $\mathbf{X}$. Because $\mathbf{X}$ is closed, it follows that $f_n \to f$ for some $f \in \mathbf{X}$, and hence $\|f\| = \inf_{h \in \mathbf{X}} \|h\|$. This verifies the existence of f.

For the uniqueness portion suppose there were two norm-minimizing functions $f, g \in \mathbf{X}$. By the parallelogram law,

(A.4) $$\frac{\|f-g\|^2}{4} = \inf_{h \in \mathbf{X}} \|h\|^2 - \left\|\frac{f+g}{2}\right\|^2 \leq 0.$$

(Why?). Thus, $f = g$. □

Proof of Theorem A.2. For all given $f \in \mathbf{H}$, the set $f + \mathbf{S}$ is closed and convex, where $f + \mathbf{S} := \{f + s : s \in \mathbf{S}\}$. In particular, $f + \mathbf{S}$ has a unique element $\mathscr{P}^\perp(f)$ of minimal norm (Lemma A.3); also define $\mathscr{P}(f) = f - \mathscr{P}^\perp(f)$.

Because $\mathscr{P}^\perp(f) \in f + \mathbf{S}$, it follows that $\mathscr{P}(f) \in \mathbf{S}$ for all $f \in \mathbf{H}$. Since $\mathscr{P}(f) + \mathscr{P}^\perp(f) = f$, it suffices to demonstrate that $\mathscr{P}^\perp(f) \in \mathbf{S}^\perp$ for all $f \in \mathbf{H}$. But by the definition of $\mathscr{P}^\perp$, $\|\mathscr{P}^\perp(f)\| \leq \|f - g\|$ for all $g \in \mathbf{S}$. Instead of g write $G = \alpha g + \mathscr{P}(f)$, where $\|g\| = 1$ and $\alpha \in \mathbf{R}$. Because $G \in \mathbf{S}$, we can deduce that for all $g \in \mathbf{S}$ with $\|g\| = 1$ and all $\alpha \in \mathbf{R}$,

(A.5)
$$\|\mathscr{P}^\perp(f)\|^2 \leq \|\mathscr{P}^\perp(f) - \alpha g\|^2$$
$$= \|\mathscr{P}^\perp(f)\|^2 - 2\alpha \langle \mathscr{P}^\perp(f), g \rangle + \alpha^2.$$

Let $\alpha = \langle \mathscr{P}^\perp(f), g \rangle$ to deduce that $\langle \mathscr{P}^\perp(f), g \rangle = 0$ for all $g \in \mathbf{S}$. This is the desired result. □

Theorem A.4. *To every bounded linear functional $\mathscr{L}$ on a Hilbert space $\mathbf{H}$ there corresponds a unique $\pi \in \mathbf{H}$ such that $\mathscr{L}(f) = \langle f, \pi \rangle$ for all $f \in \mathbf{H}$.*

Proof. If $\mathscr{L}(f) = 0$ for all $f \in \mathbf{H}$, then we define $\pi \equiv 0$, and we are done. If not, then $\mathbf{S} = \{f \in \mathbf{H} : \mathscr{L}(f) = 0\}$ is a closed subspace of $\mathbf{H}$ that does not span all of $\mathbf{H}$; i.e, there exists $g \in \mathbf{S}^\perp$ with $\|g\| = 1$ and $\mathscr{L}(g) > 0$;

this follows from the decomposition theorem for $\mathbf{H}$ (Theorem A.2). We will establish that $\pi := g\mathscr{L}(g)$ is the function that we seek, all the time remembering that $\mathscr{L}(g) \in \mathbf{R}$.

For all $f \in \mathbf{H}$ consider the function $h = \mathscr{L}(g)f - \mathscr{L}(f)g$, and note that $h \in \mathbf{S}$ since $\mathscr{L}(h) = 0$. Because $\pi \in \mathbf{S}^{\perp}$, this means that $\langle \pi, h \rangle = 0$. On the other hand, $\langle \pi, h \rangle = \mathscr{L}(g)\langle \pi, f \rangle - \mathscr{L}(g)\mathscr{L}(f)$. Since $\mathscr{L}(g) > 0$, we have $\mathscr{L}(f) = \langle \pi, f \rangle$ for all $f \in \mathbf{H}$. It remains to prove uniqueness, but this too is easy for if there were two of these functions, say π_1 and π_2, then for all $f \in \mathbf{H}$, $\langle f, \pi_1 - \pi_2 \rangle = 0$. In particular, let $f = \pi_1 - \pi_2$ to see that $\pi_1 = \pi_2$. □

2. Fourier Series

Throughout this section, we let $\mathbf{T} = [-\pi, \pi]$ denote the torus of length 2π, and consider some elementary facts about the trigonometric Fourier series on $\mathbf{T}$ that are based on the following functions:

$$(A.6) \qquad \phi_n(x) = \frac{e^{inx}}{\sqrt{2\pi}} \qquad \forall x \in \mathbf{T},\ n = 0, \pm 1, \pm 2, \ldots.$$

Let $L^2(\mathbf{T})$ denote the Hilbert space of all measurable functions $f : \mathbf{T} \to \mathbf{C}$ such that

$$(A.7) \qquad \|f\|_{\mathbf{T}}^2 := \int_{\mathbf{T}} |f(x)|^2\, dx < \infty.$$

As usual, $L^2(\mathbf{T})$ is equipped with the (semi-)norm $\|f\|_{\mathbf{T}}$ and inner product

$$(A.8) \qquad \langle f, g \rangle := \int_{\mathbf{T}} f(x)\overline{g(x)}\, dx.$$

Our goal is to prove the following theorem.

Theorem A.5. *The collection $\{\phi_n\}_{n \in \mathbf{Z}}$ is a complete orthonormal system in $L^2(\mathbf{T})$. Consequently, every $f \in L^2(\mathbf{T})$ can be written as*

$$(A.9) \qquad f = \sum_{n=-\infty}^{\infty} \langle f, \phi_n \rangle \phi_n,$$

where the convergence takes place in $L^2(\mathbf{T})$. Furthermore,

$$(A.10) \qquad \|f\|_{\mathbf{T}}^2 = \sum_{n=-\infty}^{\infty} |\langle f, \phi_n \rangle|^2.$$

The proof is not difficult, but requires some preliminary developments.

Definition A.6. A *trigonometric polynomial* is a finite linear combination of the ϕ_n's. An *approximation to the identity* is a sequence of integrable functions $\psi_0, \psi_1, \ldots : \mathbf{T} \to \mathbf{R}_+$ such that:

(i) $\int_{\mathbf{T}} \psi_n(x)\, dx = 1$ for all n.

(ii) There exists $\epsilon_0 > 0$ such that $\lim_{n \to \infty} \int_{-\epsilon}^{\epsilon} \psi_n(x)\, dx = 1$ for all $\epsilon \in [0, \epsilon_0]$.

Note that (a) all the ψ_n's are nonnegative; and (b) the preceding display shows that all of the area under ψ_n is concentrated near the origin when n is large. In other words, as $n \to \infty$, ψ_n looks more and more like a point mass.

For $n = 0, 1, 2, \ldots$ and $x \in \mathbf{T}$ consider

$$(\text{A.11}) \qquad \kappa_n(x) = \frac{(1+\cos x)^n}{\alpha_n}, \quad \text{where } \alpha_n = \int_{\mathbf{T}} (1+\cos(x))^n\, dx.$$

Lemma A.7. $\{\kappa_i\}_{i=0}^{\infty}$ *is an approximation to the identity.*

Proof. Choose and fix $\epsilon \in (0, \pi/2]$. Then,

$$(\text{A.12}) \qquad \int_{\epsilon}^{\pi} (1+\cos x)^n\, dx \leq \pi(1+\cos \epsilon)^n.$$

By symmetry, this estimates the integral away from the origin. To estimate the integral near the origin, we use a method of P.-S. Laplace and write

$$(\text{A.13}) \qquad \int_{0}^{\epsilon} (1+\cos x)^n\, dx = \int_{0}^{\epsilon} e^{ng(x)}\, dx,$$

where $g(x) := \ln(1+\cos x)$. Apply Taylor's theorem with remainder to deduce that for any $x \in [0, \epsilon]$ there exists $\zeta \in [0, x]$ such that $g(x) = \ln 2 - x^2/(1+\cos \zeta)$. But $\cos \zeta \geq 0$ because $0 < \zeta \leq \epsilon \leq \pi/2$. Thus, for all $n \geq 1$,

$$(\text{A.14}) \qquad \int_{0}^{\epsilon} (1+\cos x)^n\, dx \geq 2^n \int_{0}^{\epsilon} e^{-nx^2}\, dx \geq \frac{2^n}{\sqrt{n}} \int_{0}^{\epsilon} e^{-z^2}\, dz.$$

It follows from this and (A.12) that $\int_{\epsilon \leq |x| \leq \pi} (1+\cos x)^n\, dx = o(\alpha_n)$. This proves the lemma. $\square$

Proposition A.8. *If $f \in L^2(\mathbf{T})$ and $\epsilon > 0$, then there is a trigonometric polynomial T such that $\|T - f\|_{\mathbf{T}} \leq \epsilon$. [Trigonometric polynomials are dense in $L^2(\mathbf{T})$.]*

Proof. Since continuous functions (endowed with uniform topology) are dense in $L^2(\mathbf{T})$, it suffices to prove that trigonometric polynomials are dense in the space of all continuous function on $\mathbf{T}$ (why?). We first of all observe that the functions κ_n are trigonometric polynomials. Indeed, by the binomial theorem and the Euler formula $\cos x = \frac{1}{2}(e^{ix} + e^{-ix})$, $\kappa_n(x)$ is a linear

2. Fourier Series

combination of $\{\phi_j(x)\}_{j=-n}^n$. Next we note that the convolution $\kappa_n * f$ of κ_n and f is also a trigonometric polynomial, where

$$(\text{A.15}) \qquad (\kappa_n * f)(x) = \int_{\mathbf{T}} f(y) \kappa_n(x-y)\, dy.$$

Note that $(\kappa_n * f)(x) - f(x) = \int_{\mathbf{T}} \{f(y) - f(x)\} \kappa_n(y-x)\, dy$. We choose and fix $\epsilon \in (0, \pi)$ and split the last integral according to whether or not $|y - x| \leq \epsilon$. It follows that $|(\kappa_n * f)(x) - f(x)|$ is at most

$$(\text{A.16}) \qquad \sup_{\substack{y, u \in \mathbf{T}: \\ |y-u| \leq \epsilon}} |f(y) - f(u)| + 2 \sup_{w \in \mathbf{T}} |f(w)| \cdot \int_{\epsilon \leq |z| \leq \pi} \kappa_n(z)\, dz.$$

By Lemma A.7 the last term vanishes as $n \to \infty$. Thus, for all $\epsilon > 0$,

$$(\text{A.17}) \qquad \limsup_{n \to \infty} \sup_{x \in \mathbf{T}} |(\kappa_n * f)(x) - f(x)| \leq \sup_{\substack{y, u \in \mathbf{T}: \\ |y-u| \leq \epsilon}} |f(y) - f(u)|.$$

Let $\epsilon \to 0$ to see that the left-hand side is zero. Because

$$(\text{A.18}) \qquad \|(\kappa_n * f) - f\|_{\mathbf{T}}^2 \leq 2\pi \sup_{x \in \mathbf{T}} |(\kappa_n * f)(x) - f(x)|^2,$$

the proposition follows. $\square$

We are ready to prove Theorem A.5.

Proof of Theorem A.5. It is easy to see that $\{\phi_n\}_{n \in \mathbf{Z}}$ is an orthonormal sequence in $L^2(\mathbf{T})$; that is,

$$(\text{A.19}) \qquad \langle \phi_n, \phi_m \rangle = \begin{cases} 1 & \text{if } n = m, \\ 0 & \text{if } n \neq m. \end{cases}$$

To establish completeness suppose that $f \in L^2(\mathbf{T})$ is orthogonal to all ϕ_n's; i.e., $\langle f, \phi_n \rangle = 0$ for all $n \in \mathbf{Z}$. If ϵ and T are as in the preceding proposition, then: (i) $\langle f, T \rangle = 0$; and (ii) $\|f - T\|_{\mathbf{T}}^2 \leq \epsilon$. This last part, (ii), implies that

$$(\text{A.20}) \qquad \begin{aligned} \epsilon &\geq \|f - T\|_{\mathbf{T}}^2 \\ &= \|f\|_{\mathbf{T}}^2 + \|T\|_{\mathbf{T}}^2 - 2\langle f, T \rangle \\ &= \|f\|_{\mathbf{T}}^2 + \|T\|_{\mathbf{T}}^2. \end{aligned}$$

Since ϵ is arbitrary, $\|f\|_{\mathbf{T}} = 0$, from which we deduce that $f = 0$ almost everywhere. This proves completeness. The remainder is easy to prove, but requires the material from Chapter 4 [§6].

Let $\mathscr{P}_n$ and $\mathscr{P}_n^\perp$ respectively denote the projections onto $\mathbf{S}_n :=$ the linear span of $\{\phi_j\}_{j=-n}^n$ and $\mathbf{S}_n^\perp$. If $f \in L^2(\mathbf{T})$, then $\mathscr{P}_n f$ is the a.e.-unique function $g \in \mathbf{S}_n$ that minimizes $\|f - g\|_{\mathbf{T}}$. We can write $g = \sum_{j=-n}^n c_j \phi_j$

and expand the said L^2-norm to obtain the following optimization problem: Minimize over all $\{c_j\}_{j=-n}^n$,

$$(A.21) \qquad \left\| f - \sum_{j=-n}^n c_j \phi_j \right\|_\mathbf{T}^2 = \|f\|_\mathbf{T}^2 + \sum_{j=-n}^n c_j^2 - 2 \sum_{j=-n}^n c_j \langle f, \phi_j \rangle.$$

This is a calculus exercise and yields the optimal value of $c_j = \langle f, \phi_j \rangle$. It follows readily from this that:

(i) $\mathscr{P}_n f = \sum_{j=-n}^n \langle f, \phi_j \rangle \phi_j$;
(ii) $\mathscr{P}_n^\perp f = f - \mathscr{P}_n f$;
(iii) $\|\mathscr{P}_n f\|_\mathbf{T}^2 = \sum_{j=-n}^n |\langle f, \phi_j \rangle|^2$; and
(iv) $\|\mathscr{P}_n^\perp f\|_\mathbf{T}^2 = \|f\|_\mathbf{T}^2 - \sum_{j=-n}^n |\langle f, \phi_j \rangle|^2$.

The last inequality yields *Bessel's inequality*:

$$(A.22) \qquad \sum_{j=-\infty}^\infty |\langle f, \phi_j \rangle|^2 \le \|f\|_\mathbf{T}^2.$$

Our goal is to show that this is an equality.

If not, then Fatou's lemma implies that $\|\liminf_{n\to\infty} \mathscr{P}_n^\perp f\|_\mathbf{T} > 0$, whence $g := \liminf_{n\to\infty} \mathscr{P}_n^\perp f \ne 0$ on a set of positive Lebesgue measure. But note that $g \in \mathscr{P}_n^\perp$ for all n. Fix $\epsilon > 0$ and find a trigonometric polynomial $T \in \mathbf{S}_n$ for some large n such that $\|g - T\|_\mathbf{T} \le \epsilon$. Now expand:

$$(A.23) \qquad \begin{aligned} \epsilon^2 &\ge \|g - T\|_\mathbf{T}^2 \\ &= \|g\|_\mathbf{T}^2 + \|T\|_\mathbf{T}^2 - 2\langle g, T \rangle \\ &= \|g\|_\mathbf{T}^2 + \|T\|_\mathbf{T}^2 \\ &\ge \|g\|_\mathbf{T}^2. \end{aligned}$$

Thus, $g = 0$ almost everywhere, whence follows a contradiction. In fact, this argument shows that any subsequential limit of $\mathscr{P}_n^\perp f$ must be zero almost everywhere, and hence $\mathscr{P}_n^\perp f \to 0$ in $L^2(\mathbf{T})$. It follows that

$$(A.24) \qquad f = \lim_{n\to\infty} \mathscr{P}_n f = \sum_{j=-\infty}^\infty \langle f, \phi_j \rangle \phi_j \quad \text{in } L^2(\mathbf{T}),$$

as desired. $\square$

Bibliography

Adams, W. J. (1974). *The Life and Times of the Central Limit Theorem*. New York: Kaedmon Publishing Co.

Aldous, D. J. (1985). Exchangeability and Related Topics. In *École d'Été de Probabilités de Saint-Flour, XIII—1983*, Volume 1117 of *Lecture Notes in Math.*, pp. 1–198. Berlin: Springer.

Alon, N. and J. H. Spencer (1991). *The Probabilistic Method* (First ed.). New York: Wiley.

André, D. (1887). Solution directe du problème résolu par M. Bertrand. *C. R. Acad. Sci. Paris 105*, 436–437.

Azuma, K. (1967). Weighted sums of certain dependent random variables. *Tôhoku Math. J. (2) 19*, 357–367.

Bachelier, L. (1900). Théorie de la spéculation. *Ann. Sci. École Norm. Sup. 17*, 21–86. See also the 1995 reprint. Sceaux: Gauthier–Villars.

Bachelier, L. (1964). Theory of speculation. In P. H. Cootner (Ed.), *The Random Character of Stock Market Prices*, pp. 17–78. MIT Press. Translated from French by A. James Boness.

Banach, S. (1931). Über die Baire'sche kategorie gewisser Funkionenmengen. *Studia. Math. III*, 174–179.

Bass, R. F. (1998). *Diffusions and Elliptic Operators*. New York: Springer-Verlag.

Baxter, M. and A. Rennie (1996). *Financial Calculus: An Introduction to Derivative Pricing*. Cambridge: Cambridge University Press. Second (1998) reprint.

Berkés, I. (1998). Results and problems related to the pointwise central limit theorem. In *Asymptotic Methods in Probability and Statistics (Ottawa, ON, 1997)*, pp. 59–96. Amsterdam: North-Holland.

Bernoulli, J. (1713). *Ars Conjectandi (The Art of Conjecture)*. Basel: Basileæ: Impensis Thurnisiorum Frætrum.

Bernstein, S. N. (1912/1913). Démonstration du théorème de Weierstrass fondée sur le calcul des probabilités. *Comm. Soc. Math. Kharkow 13*, 1–2.

Bernšteĭn, S. N. (1964). On the property characteristic of the normal law. In *Sobranie sochinenii. Tom IV: Teoriya veroyatnostei. Matematicheskaya Statistika. 1911-1946*. "Nauka", Moscow.

Billingsley, P. (1995). *Probability and Measure* (Third ed.). New York: John Wiley & Sons Inc.

Bingham, N. H. (1986). Variants on the law of the iterated logarithm. *Bull. London Math. Soc. 18*(5), 433–467.

Birkhoff, G. D. (1931). Proof of the ergodic theorem. *Proc. Nat. Acad. Sci. 17*, 656–660.

Black, F. and M. Scholes (1973). Pricing of options and corporate liabilities. *J. Political Econ. 81*, 637–654.

Blumenthal, R. M. (1957). An extended Markov property. *Trans. Amer. Math. Soc. 85*, 52–72.

Borel, E. (1909). Les probabilités dénombrables et leurs applications arithmétique. *Rend. Circ. Mat. Palermo 27*, 247–271.

Borel, E. (1925). *Mécanique Statistique Classique* (Third ed.). Paris: Gauthier–Villars.

Bourke, C., J. M. Hitchcock, and N. V. Vinodchandran (2005). Entropy rates and finite-state dimension. *Theoret. Comput. Sci.* 349(3), 392–406.

Bovier, A. and P. Picco (1996). Limit theorems for Bernoulli convolutions. In *Disordered Systems (Temuco, 1991/1992)*, Volume 53, pp. 135–158. Paris: Hermann.

Breiman, L. (1992). *Probability*. Philadelphia, PA: Society for Industrial and Applied Mathematics (SIAM). See also the corrected reprint of the original (1968).

Bretagnolle, J. and D. Dacunha-Castelle (1969). Applications radonifiantes dans les espaces de type p. *C. R. Acad. Sci. Paris Sér. A-B* 269, A1132–A1134.

Broadbent, S. R. and J. M. Hammersley (1957). Percolation processes. I. Crystals and mazes. *Proc. Cambridge Philos. Soc.* 53, 629–641.

Buczolich, Z. and R. D. Mauldin (1999). On the convergence of $\sum_{n=1}^{\infty} f(nx)$ for measurable functions. *Mathematika* 46(2), 337–341.

Burkholder, D. L. (1962). Successive conditional expectations of an integrable function. *Ann. Math. Statist.* 33, 887–893.

Burkholder, D. L., B. J. Davis, and R. F. Gundy (1972). Integral inequalities for convex functions of operators on martingales. In *Proc. Sixth Berkeley Symp. Math. Statist. Probab., Vol. II*, Berkeley, Calif., pp. 223–240. Univ. California Press.

Burkholder, D. L. and R. F. Gundy (1970). Extrapolation and interpolation of quasi-linear operators on martingales. *Acta Math.* 124, 249–304.

Cantelli, F. P. (1917a). Su due applicazioni d'un teorema di G. Boole alla statistica matematica. *Atti della Reale Accademia Nazionale dei Lincei, Serie V, Rendicotti* 26, 295–302.

Cantelli, F. P. (1917b). Sulla probabilità come limite della frequenze. *Atti della Reale Accademia Nazionale dei Lincei, Serie V, Rendicotti* 26, 39–45.

Cantelli, F. P. (1933a). Considerazioni sulla legge uniforme dei grandi numeri e sulla generalizzazione di un fondamentale teorema del signor Lévy. *Giornale d. Istituto Italiano Attuari* 4, 327–350.

Cantelli, F. P. (1933b). Sulla determinazione empirica delle leggi di probabilita. *Giornale d. Istituto Italiano Attuari* 4, 421–424.

Carathéodory, C. (1948). *Vorlesungen über reelle Funktionen*. New York: Chelsea Publishing Company.

Champernowne, D. G. (1933). The construction of decimals normal in the scale of ten. *J. London Math. Soc.* 8, 254–260.

Chatterji, S. D. (1968). Martingale convergence and the Radon–Nikodym theorem in Banach spaces. *Math. Scand.* 22, 21–41.

Chebyshev, P. L. (1846). Démonstration élementaire d'une proposition génerale de la théorie des probabilités. *Crelle J. Math.* 33(2), 259–267.

Chebyshev, P. L. (1867). Des valeurs moyennes. *J. Math. Pures Appl.* 12(2), 177–184.

Chernoff, H. (1952). A measure of asymptotic efficiency for tests of a hypothesis based on the sum of observations. *Ann. Math. Statist.* 23, 493–507.

Chow, Y. S. and H. Teicher (1997). *Probability Theory: Independence, Interchangeability, Martingales* (Third ed.). New York: Springer-Verlag.

Chung, K. L. (1947). On the maximum partial sum of independent random variables. *Proc. Nat. Acad. Sci. U.S.A.* 33, 132–136.

Chung, K. L. (1974). *A Course in Probability Theory* (Second ed.). New York-London: Academic Press.

Chung, K.-L. and P. Erdős (1947). On the lower limit of sums of independent random variables. *Ann. of Math. (2)* 48, 1003–1013.

Chung, K. L. and P. Erdős (1952). On the application of the Borel–Cantelli lemma. *Trans. Amer. Math. Soc.* 72, 179–186.

Ciesielski, Z. and S. J. Taylor (1962). First passage times and sojourn times for Brownian motion in space and the exact Hausdorff measure of the sample path. *Trans. Amer. Math. Soc.* 103, 434–450.

Cifarelli, D. M. and E. Regazzini (1996). De Finetti's contribution to probability and statistics. *Statist. Sci.* 11(4), 253–282.

Coifman, R. R. (1972). Distribution function inequalities for singular integrals. *Proc. Nat. Acad. Sci. U.S.A.* 69, 2838–2839.

Copeland, A. H. and P. Erdős (1946). Note on normal numbers. *Bull. Amer. Math. Soc.* 52, 857–860.

Courtault, J.-M., Y. Kabanov, B. Bru, P. Crépel, I. Lebon, and A. Le Marchand (2000). Louis Bachelier. On the Centenary of *Théorie de la Spéculation*. *Math. Finance* 10(3), 341–353.

Cover, T. M. and J. A. Thomas (1991). *Elements of Information Theory*. New York: John Wiley & Sons Inc.

Cox, J. C., S. A. Ross, and M. Rubenstein (1979). Option pricing: a simplified approach. *J. Financial Econ. 7*, 229–263.

Cramér, H. (1936). Über eine Eigenschaft der normalen Verteilungsfunktion. *Math. Z. 41*, 405–415.

Csörgő, M. and P. Révész (1981). *Strong Approximations in Probability and Statistics*. New York: Academic Press Inc. [Harcourt Brace Jovanovich Publishers].

de Acosta, A. (1983). A new proof of the Hartman-Wintner law of the iterated logarithm. *Ann. Probab. 11*(2), 270–276.

de Finetti, B. (1937). La prévision: ses lois logiques, ses sources subjectives. *Ann. Inst. H. Poincaré 7*, 1–68.

de Moivre, A. (1718). *The Doctrine of Chances; or a Method of Calculating the Probabilities of Events in Play* (First ed.). London: W. Pearson.

de Moivre, A. (1733). Approximatio ad Summam terminorum Binomii $(a+b)^n$ in Serium expansi. Privately Printed.

de Moivre, A. (1738). *The Doctrine of Chances; or a Method of Calculating the Probabilities of Events in Play* (Second ed.). London: H. Woodfall.

Dellacherie, C. and P.-A. Meyer (1982). *Probabilities and Potential. B*. Amsterdam: North-Holland Publishing Co. Theory of martingales, Translated from French by J. P. Wilson.

Devaney, R. L. (2003). *An Introduction to Chaotic Dynamical Systems*. Boulder, CO: Westview Press. Reprint of the second (1989) edition.

Diaconis, P. and D. Freedman (1987). A dozen de Finetti-style results in search of a theory. *Ann. Inst. H. Poincaré Probab. Statist. 23*(2, suppl.), 397–423.

Diaconis, P. and J. B. Keller (1989). Fair dice. *Amer. Math. Monthly 96*(4), 337–339.

Donoho, D. L. and P. B. Stark (1989). Uncertainty principles and signal recovery. *SIAM J. Appl. Math. 49*(3), 906–931.

Doob, J. L. (1940). Regularity properties of certain families of chance variables. *Trans. Amer. Math. Soc. 47*, 455–486.

Doob, J. L. (1949). Application of the theory of martingales. In *Le Calcul des Probabilités et ses Applications*, pp. 23–27. Paris: Centre National de la Recherche Scientifique.

Doob, J. L. (1953). *Stochastic Processes*. New York: John Wiley & Sons Inc.

Doob, J. L. (1971). What is a martingale? *Amer. Math. Monthly 78*, 451–463.

Dubins, L. E. and D. A. Freedman (1965). A sharper form of the Borel-Cantelli lemma and the strong law. *Ann. Math. Statist. 36*, 800–807.

Dubins, L. E. and D. A. Freedman (1966). On the expected value of a stopped martingale. *Ann. Math. Statist 37*, 1505–1509.

Dudley, R. M. (1967). On prediction theory for nonstationary sequences. In *Proc. Fifth Berkeley Symp. Math. Statist. Probab., Vol. II*, pp. 223–234. Berkeley, Calif.: Univ. California Press.

Dudley, R. M. (2002). *Real Analysis and Probability*. Cambridge: Cambridge University Press. Revised reprint of the 1989 original.

Durrett, R. (1996). *Probability: Theory and Examples* (Second ed.). Belmont, CA: Duxbury Press.

Dynkin, E. and A. Jushkevich (1956). Strong Markov processes. *Teor. Veroyatnost. i Primenen. 1*, 149–155.

Erdős, P. (1948). Some remarks on the theory of graphs. *Bull. Amer. Math. Soc. 53*, 292–294.

Erdős, P. (1949). On the strong law of large numbers. *Trans. Amer. Math. Soc. 67*, 51–56.

Erdős, P. and G. Szekeres (1935). A combinatorial problem in geometry. *Composito. Math. 2*, 463–470.

Erdős, P. and A. Rényi (1959). On Cantor's series with convergent $\sum 1/q_n$. *Ann. Univ. Sci. Budapest. Eötvös. Sect. Math. 2*, 93–109.

Erdős, P. and A. Rényi (1970). On a new law of large numbers. *J. Analyse Math. 23*, 103–111.

Etemadi, N. (1981). An elementary proof of the strong law of large numbers. *Z. Wahrsch. Verw. Geb. 55*, 119–122.

Falconer, K. J. (1986). *The Geometry of Fractal Sets*, Volume 85. Cambridge: Cambridge University Press.

Fatou, P. J. L. (1906). Séries trigonométriques et séries de Taylor. *Acta Math. 69*, 372–433.

Feller, W. (1945). The fundamental limit theorems in probability. *Bulletin A.M.S. 51*, 800–832.

Feller, W. (1955a). On differential operators and boundary conditions. *Comm. Pure Appl. Math. 8*, 203–216.

Feller, W. (1955b). On second order differential operators. *Ann. of Math. (2) 61*, 90–105.

Feller, W. (1956). On generalized Sturm-Liouville operators. In *Proceedings of the Conference on Differential Equations (dedicated to A. Weinstein)*, pp. 251–270. University of Maryland Book Store, College Park, Md.

Feller, W. (1957). *An Introduction to Probability Theory and Its Applications. Vol. I* (Second ed.). New York: John Wiley & Sons Inc.

Feller, W. (1966). *An Introduction to Probability Theory and Its Applications. Vol. II.* New York: John Wiley & Sons Inc.

Fortuin, C. M., P. W. Kasteleyn, and J. Ginibre (1971). Correlation inequalities on some partially ordered sets. *Comm. Math. Phys. 22*, 536–564.

Fréchet, M. R. (1930). Sur la convergence en probabilité. *Metron 8*, 1–48.

Freiling, C. (1986). Axioms of symmetry: Throwing darts at the real number line. *J. Symbolic Logic 51*(1), 190–200.

Fristedt, B. and L. Gray (1997). *A Modern Approach to Probability Theory.* Boston, MA: Birkhäuser Boston Inc.

Garsia, A. M. (1965). A simple proof of E. Hopf's maximal ergodic theorem. *J. Math. Mech. 14*, 381–382.

Georgii, H.-O. (1988). *Gibbs Measures and Phase Transitions.* Berlin: Walter de Gruyter & Co.

Gerber, H. U. and S.-Y. R. Li (1981). The occurrence of sequence patterns in repeated experiments and hitting times in a Markov chain. *Stochastic Process. Appl. 11*(1), 101–108.

Glivenko, V. (1933). Sulla determinazione empirica delle leggi di probabilità. *Giornale d. Istituto Italiano Attuari 4*, 92–99.

Glivenko, V. (1936). Sul teorema limite della teoria delle funzioni caratteristiche. *Giornale d. Instituto Italiano attuari 7*, 160–167.

Gnedenko, B. V. (1967). *The Theory of Probability.* New York: Chelsea Publishing Co. Translated from the fourth Russian edition by B. D. Seckler.

Gnedenko, B. V. (1969). On Hilbert's Sixth Problem (Russian). In *Hilbert's Problems* (Russian), pp. 116–120. "Nauka", Moscow.

Gnedenko, B. V. and A. N. Kolmogorov (1968). *Limit Distributions for Sums of Independent Random Variables.* Reading, Mass.: Addison-Wesley Publishing Co. Translated from the original Russian, annotated, and revised by Kai Lai Chung. With appendices by J. L. Doob and P. L. Hsu. Revised edition.

Grimmett, G. (1999). *Percolation* (Second ed.). Berlin: Springer-Verlag.

Hamedani, G. G. and G. G. Walter (1984). A fixed point theorem and its application to the central limit theorem. *Arch. Math. (Basel) 43*(3), 258–264.

Hammersley, J. M. (1963). A Monte Carlo solution of percolation in a cubic lattice. In S. F. B. Alder and M. Rotenberg (Eds.), *Methods in Computational Physics*, Volume I. London: Academic Press.

Hardy, G. H. and J. E. Littlewood (1914). Some problems of diophantine approximation. *Acta Math. 37*, 155–239.

Hardy, G. H. and J. E. Littlewood (1930). A maximal theorem with function-theoretic applications. *Acta Math. 54*, 81–166.

Harris, T. E. (1960). A lower bound for the critical probability in a certain percolation process. *Proc. Cambridge Philos. Soc. 56*, 13–20.

Harrison, J. M. and D. Kreps (1979). Martingales and arbitrage in multiperiod securities markets. *J. Econ. Theory 20*, 381–408.

Harrison, J. M. and S. R. Pliska (1981). Martingales and stochastic integrals in the theory of continuous trading. *Stoch. Proc. Their Appl. 11*, 215–260.

Hartman, P. and A. Wintner (1941). On the law of the iterated logarithm. *Amer. J. Math. 63*, 169–176.

Hausdorff, F. (1927). *Mengenlehre.* Berlin: Walter De Gruyter & Co.

Hausdorff, F. (1949). *Grundzüge der Mengenlehre.* New York: Chelsea publishing Co.

Helms, L. L. and P. A. Loeb (1982). A nonstandard proof of the martingale convergence theorem. *Rocky Mountain J. Math. 12*(1), 165–170.

Hoeffding, W. (1963). Probability inequalities for sums of bounded random variables. *Amer. Statist. Assoc. 58*, 13–30.

Hoeffding, W. (1971). The L_1 norm of the approximation error for Bernstein-type polynomials. *J. Approximation Theory 4*, 347–356.

Houdré, C., V. Pérez-Abreu, and D. Surgailis (1998). Interpolation, correlation identities, and inequalities for infinitely divisible variables. *J. Fourier Anal. Appl. 4*(6), 651–668.

Hunt, G. A. (1956). Some theorems concerning Brownian motion. *Trans. Amer. Math. Soc. 81*, 294–319.

Hunt, G. A. (1957). Markoff processes and potentials. I, II. *Illinois J. Math. 1*, 44–93, 316–369.

Hunt, G. A. (1966). *Martingales et processus de Markov*. Paris: Dunod.

Ionescu Tulcea, A. and C. Ionescu Tulcea (1963). Abstract ergodic theorems. *Trans. Amer. Math. Soc. 107*, 107–124.

Isaac, R. (1965). A proof of the martingale convergence theorem. *Proc. Amer. Math. Soc. 16*, 842–844.

Itô, K. (1944). Stochastic integral. *Proc. Imp. Acad. Tokyo 20*, 519–524.

Jones, R. L. (1997/1998). Ergodic theory and connections with analysis and probability. *New York J. Math. 3A*(Proceedings of the New York Journal of Mathematics Conference, June 9–13, 1997), 31–67 (electronic).

Kac, M. (1937). Une remarque sur les polynômes de M.S. Bernstein. *Studia Math. 7*, 49–51.

Kac, M. (1939). On a characterization of the normal distribution. *Amer. J. Math. 61*, 726–728.

Kac, M. (1949). On deviations between theoretical and empirical distributions. *Proc. Nat. Acad. Sci. U.S.A. 35*, 252–257.

Kac, M. (1956). Foundations of kinetic thoery. In J. Neyman (Ed.), *Proc. Third Berkeley Symp. on Math. Statist. Probab.*, Volume 3, pp. 171–197. Univ. of Calif.

Kahane, J.-P. (1997). A few generic properties of Fourier and Taylor series. In *Trends in Probability and Related Analysis (Taipei, 1996)*, pp. 187–196. River Edge, NJ: World Sci. Publishing.

Kahane, J.-P. (2000). Baire's category theorem and trigonometric series. *J. Anal. Math. 80*, 143–182.

Kahane, J.-P. (2001). Probabilities and Baire's theory in harmonic analysis. In *Twentieth Century Harmonic Analysis—A Celebration (Il Ciocco, 2000)*, Volume 33 of *NATO Sci. Ser. II Math. Phys. Chem.*, pp. 57–72. Dordrecht: Kluwer Acad. Publ.

Karlin, S. and H. M. Taylor (1975). *A First Course in Stochastic Processes* (Second ed.). Academic Press [Harcourt Brace Jovanovich Publishers], New York-London.

Karlin, S. and H. M. Taylor (1981). *A Second Course in Stochastic Processes*. New York: Academic Press Inc. [Harcourt Brace Jovanovich Publishers].

Keller, J. B. (1986). The probability of heads. *Amer. Math. Monthly 93*(3), 191–197.

Kesten, H. (1980). The critical probability of bond percolation on the square lattice equals $\frac{1}{2}$. *Comm. Math. Phys. 74*(1), 41–59.

Khintchine, A. and A. Kolmogorov (1925). Über Konvergenz von Reihen deren Glieder durch den Zufall bestimmt werden. *Rec. Math. Moscow 32*, 668–677.

Khintchine, A. Y. (1923). Über dyadische Brüche. *Math. Z. 18*, 109–116.

Khintchine, A. Y. (1924). Ein Satz der Wahrscheinlichkeitsrechnung. *Fund. Math. 6*, 9–10.

Khintchine, A. Y. (1929). Sur la loi des grands nombres. *C. R. Acad. Sci. Paris 188*, 477–479.

Khintchine, A. Y. (1933). *Asymptotische Gesetz der Wahrscheinlichkeitsrechnung*. Springer.

Kinney, J. R. (1953). Continuity properties of sample functions of Markov processes. *Trans. Amer. Math. Soc. 74*, 280–302.

Knight, F. B. (1962). On the random walk and Brownian motion. *Trans. Amer. Math. Soc. 103*, 218–228.

Knight, F. B. (1981). *Essentials of Brownian Motion and Diffusion*. Providence, R.I.: American Mathematical Society.

Knuth, D. E. (1981). *The Art of Computer Programming. Vol. 2* (Second ed.). Reading, Mass.: Addison-Wesley Publishing Co. Seminumerical Algorithms.

Kochen, S. and C. J. Stone (1964). A note on the Borel–Cantelli lemma. *Illinois J. Math. 8*, 248–251.

Kolmogorov, A. (1930). Sur la loi forte des grandes nombres. *C. R. Acad. Sci. Paris 191*, 910–911.

Kolmogorov, A. N. (1929). Über das Gesetz des iterierten Logarithmus. *Math. Ann. 101*, 126–135.

Kolmogorov, A. N. (1933). *Grundbegriffe der Wahrscheinlichkeitsrechnung*. Berlin: Springer.

Kolmogorov, A. N. (1950). *Foundations of Probability*. New York: Chelsea Publishig Company. Translation edited by Nathan Morrison.

Krickeberg, K. (1963). *Wahrscheinlichkeitstheorie*. Stuttgart: Teubner.

Krickeberg, K. (1965). *Probability Theory*. Reading, Massachusetts: Addison–Wesley.

Kyburg, Jr., H. E. and H. E. Smokler (1980). *Studies in Subjective Probability* (Second ed.). Huntington, N.Y.: Robert E. Krieger Publishing Co. Inc.

Lacey, M. T. and W. Philipp (1990). A note on the almost sure central limit theorem. *Statist. Probab. Lett. 9*(3), 201–205.

Lamb, C. W. (1973). A short proof of the martingale convergence theorem. *Proc. Amer. Math. Soc. 38*, 215–217.

Lange, K. (2003). *Applied Probability*. New York: Springer-Verlag.

Laplace, P. S. (1782). Mémoire sur les approximations des formules qui sont fonctions de trés-grands nombres. Technical report, Histoire de l'Académie Royale des Sciences de Paris.

Laplace, P.-S. (1805). *Traité de Mécanique Céleste*, Volume 4. Chez J. B. M. Duprat, an 7 (Crapelet). Reprinted by the Chelsea Publishing Co. (1967). Translated by N. Bowditch.

Laplace, P.-S. (1812). *Théorie Analytique des Probabilités, Vol. I and II*. V[iéme] Courcier. Reprinted in *Oeuvres complétes de Laplace*, Volume VII (1886), Paris: Gauthier-Villars.

Lebesgue, H. (1910). Sur l'intégration des fonctions discontinues. *Ann. Ecole Norm. Sup. 27*(3), 361–450.

Levi, B. (1906). Sopra l'integrazione delle serie. *Rend. Instituto Lombardino di Sci. e Lett. 39*(2), 775–780.

Lévy, P. (1925). *Calcul des Probabilités*. Paris: Gauthier–Villars.

Lévy, P. (1937). *Théorie de l'Addition des Variables Aléatoires*. Paris: Gauthier–Villars.

Lévy, P. (1951). La mesure de Hausdorff de la courbe du mouvement brownien à n dimensions. *C. R. Acad. Sci. Paris 233*, 600–602.

Li, S.-Y. R. (1980). A martingale approach to the study of occurrence of sequence patterns in repeated experiments. *Ann. Probab. 8*(6), 1171–1176.

Liapounov, A. M. (1900). Sur une proposition de la théorie des probabilités. *Bulletin de l'Académie Impériale des Sciences de St. Petérsbourg 13*(4), 359–386.

Liapounov, A. M. (1922). Nouvelle forme du théorème sur la limite de probabilité. *Mémoires de l'Académie Impériale des Sciences de St. Petérsbourg 12*(5), 1–24.

Lindeberg, J. W. (1922). Eine neue Herleitung des Exponentialgesetzes in der Wahrscheinlichkeitsrechnung. *Math. Z. 15*, 211–225.

Lindvall, T. (1982). Bernstein polynomials and the law of large numbers. *Math. Sci. 7*(2), 127–139.

Lipschitz, R. (1876). Sur la possibilité d'intégrer complèment un système donné d'équations différentielles. *Bull. Sci. Math. 10*, 149–159.

Mahmoud, H. M. (2000). *Sorting: A distribution theory*. Wiley-Interscience, New York.

Markov, A. A. (1910). Recherches sur un cas remarquable d'épreuves dépendantes. *Acta Math. 33*, 87–104.

Mattner, L. (1999). Product measurability, parameter integrals, and a Fubini–Tonelli counterexample. *Enseign. Math. (2) 45*(3-4), 271–279.

Mazurkiewicz, S. (1931). Sur les fonctions non dérivables. *Studia. Math. III*, 92–94.

McShane, E. J. (1969). *A Riemann-type integral that includes Lebesgue-Stieltjes, Bochner and stochastic integrals*. Memoirs of the American Mathematical Society, No. 88. Providence, R.I.: American Mathematical Society.

Merton, R. C. (1973). Theory of rational option pricing. *Bell J. of Econ. and Management Sci. 4*(1), 141–183.

Mukherjea, A. (1972). A remark on Tonelli's theorem on integration in product spaces. *Pacific J. Math. 42*, 177–185.

Mukherjea, A. (1973/1974). Remark on Tonelli's theorem on integration in product spaces. II. *Indiana Univ. Math. J. 23*, 679–684.

Nash, J. (1958). Continuity of solutions of parabolic and elliptic equations. *Amer. J. Math. 80*, 931–954.

Nelson, E. (1967). *Dynamical Theories of Brownian Motion*. Princeton, N.J.: Princeton University Press.

Norris, J. R. (1998). *Markov Chains*. Cambridge: Cambridge University Press. Reprint of 1997 original.

Okamoto, M. (1958). Some inequalities relating to the partial sum of binomial probabilities. *Ann. Inst. Statist. Math. 10*, 29–35.

Paley, R. E. A. C., N. Wiener, and A. Zygmund (1933). Notes on random functions. *Math. Z. 37*, 647–668.

Paley, R. E. A. C. and A. Zygmund (1932). A note on analytic functions in the unit circle. *Proc. Camb. Phil. Soc. 28*, 366–372.

Perrin, J. B. (1913). *Les Atomes*. Paris: Librairie F. Alcan. See also *Atoms*. The revised reprint of the second (1923) English edition. Van Nostrand, New York. Translated by Dalziel Llewellyn Hammick.

Pitman, J. (1981). A note on L_2 maximal inequalities. In *Seminar on Probability, XV (Univ. Strasbourg, Strasbourg, 1979/1980) (French)*, Volume 850 of *Lecture Notes in Math.*, pp. 251–258. Berlin: Springer.

Plancherel, M. (1910). Contribution à l'étude de la représentation d'une fonction arbitraire par des intégrales définies. *Rend. Circ. Mat. Palermo 30*, 289–335.

Plancherel, M. (1933). Sur les formules de réciprocité du type de Fourier. *J. London Math. Soc. 8*, 220–226.

Plancherel, M. and G. Pólya (1931). Sur les valeurs moyennes des fonctions réelles définies pour toutes les valuers de la variables. *Comment. Math. Helv. 3*, 114–121.

Poincaré, H. (1912). *Calcul des Probabilités*. Paris: Gauthier-Villars.

Pollard, D. (2002). *A User's Guide to Measure Theoretic Probability*. Cambridge: Cambridge University Press.

Pólya, G. (1920). Über den zentralen Grenzwertsatz der Wahrscheinlichkeitstheorie und das Momentenproblem. *Math. Zeit. 8*, 171–181.

Rademacher, H. (1919). Über partielle und totale Differenzierbarkeit I. *Math. Ann. 89*, 340–359.

Rademacher, H. (1922). Einige Sätze über Reihen von allgemeinen Orthogonalfunktionen. *Math. Ann. 87*, 112–138.

Raikov, D. (1936). On some arithmetical properties of summable functions. *Math. Sb. 1*(43:3), 377–383.

Ramsey, F. P. (1930). On a problem of formal logic. *Proc. London Math. Soc. 30*(2), 264–286.

Rényi, A. (1962). *Wahrscheinlichkeitsrechnung. Mit einem Anhang über Informationstheorie*. Berlin: VEB Deutscher Verlag der Wissenschaften.

Resnick, S. I. (1999). *A Probability Path*. Boston, MA: Birkhäuser Boston Inc.

Revuz, D. and M. Yor (1999). *Continuous Martingales and Brownian Motion* (Third ed.). Berlin: Springer-Verlag.

Rosenlicht, M. (1972). Integration in finite terms. *Amer. Math. Monthly 79*, 963–972.

Schnorr, C.-P. and H. Stimm (1971/1972). Endliche Automaten und Zufallsfolgen. *Acta Informat. 1*(4), 345–359.

Schrödinger, E. (1946). *Statistical Thermodynamics*. Cambridge: Cambridge University Press.

Shannon, C. E. (1948). A mathematical theory of communication. *Bell System Tech. J. 27*, 379–423, 623–656.

Shannon, C. E. and W. Weaver (1949). *The Mathematical Theory of Communication*. Urbana, Ill.: Univ. of Illinois Press.

Shultz, H. S. and B. Leonard (1989). Unexpected occurences of the number e. *Math. Magazine 62*(4), 269–271.

Sierpiński, W. (1920). Sur les rapport entre l'existence des integrales $\int_0^1 f(x,y)\,dx$, $\int_0^1 f(x,y)\,dy$ et $\int_0^1 dx \int_0^1 f(x,y)\,dy$. *Fund. Math. 1*, 142–147.

Skolem, T. (1933). Ein kombinatorischer Satz mit anwendung auf ein logisches Entscheidungsproblem. *Fund. Math. 20*, 254–261.

Skorohod, A. V. (1961). *Issledovaniya po teorii sluchainykh protsessov*. Kiev. Univ., Kiev. [*Studies in the Theory of Random Processes*. Translated from Russian by Scripta Technica, Inc., Addison-Wesley Publishing Co., Inc., Reading, Mass. (1965). See also the second edition (1985), Dover, New York.].

Skorohod, A. V. (1965). *Studies in the theory of random processes*. Addison-Wesley Publishing Co., Inc., Reading, Mass.

Slutsky, E. (1925). Über stochastische Asymptoten und Grenzwerte. *Metron 5*, 3–89.

Solovay, R. M. (1970). A model of set theory in which every set of reals is Lebesgue measurable. *Ann. Math. 92*, 1–56.

Steinhaus, H. (1922). Les probabilités dénombrables et leur rapport à la théorie de mesure. *Fund. Math. 4*, 286–310.

Steinhaus, H. (1930). Sur la probabilité de la convergence de série. *Studia Math. 2*, 21–39.

Stigler, S. M. (1986). *The History of Statistics*. Cambridge, MA: The Belknap Press of Harvard University Press.

Stirling, J. (1730). *Methodus Differentialis*. London: Whiston & White.

Strassen, V. (1967). Almost sure behavior of sums of independent random variables and martingales. In *Proc. Fifth Berkeley Sympos. Math. Statist. and Probability (Berkeley, Calif., 1965/66)*, pp. Vol. II, Part 1, pp. 315–343. Berkeley, Calif.: Univ. California Press.

Stroock, D. W. (1993). *Probability Theory, An Analytic View*. Cambridge: Cambridge University Press.

Stroock, D. W. and O. Zeitouni (1991). Microcanonical distributions, Gibbs states, and the equivalence of ensembles. In *Random Walks, Brownian Motion, and Interacting Particle Systems*, pp. 399–424. Boston, MA: Birkhäuser Boston.

Tandori, K. (1983). The Life and Works of Lipót Fejér. In *Functions, series, operators, Vol. I, II (Budapest, 1980)*, Volume 35 of *Colloq. Math. Soc. János Bolyai*, pp. 77–85. Amsterdam: North-Holland.

Trotter, H. F. (1959). An elementary proof of the central limit theorem. *Arch. Math. 10*, 226–234.

Turing, A. M. (1934). On the Gaussian error function. Technical report, Unpublished Fellowship Dissertation, King's College Library, Cambridge.

Varadhan, S. R. S. (2001). *Probability Theory*. New York: New York University Courant Institute of Mathematical Sciences.

Varberg, D. E. (1966). Convergence of quadratic forms in independent random variables. *Ann. Math. Statist. 37*, 567–576.

Veech, W. A. (1967). *A Second Course in Complex Analysis*. New York, Amsterdam: W. A. Benjamin, Inc.

Ville, J. (1939). *Etude Critique de la Notion de Collectif*. Paris: Gauthier–Villars.

Ville, J. (1943). Sur l'application, à un critère d'indépendance, du dénombrement des inversions présentées par une permutation. *C. R. Acad. Sci. Paris 217*, 41–42.

von Neumann, J. (1940). On rings of operators, III. *Ann. Math. 41*, 94–161.

von Smoluchowski, M. (1918). Über den Begriff des Zufalls und den Ursprung der Wahrscheinlichkeit. *Die Naturwissenschaften 6*(17), 253–263.

Wagon, S. (1985). Is π normal? *Math. Intelligencer 7*(3), 65–67.

Wiener, N. (1923a). Differential space. *J. Math. Phys. 2*, 131–174.

Wiener, N. (1923b). The homogeneous chaos. *Amer. J. Math. 60*, 879–036.

Williams, D. (1991). *Probability with Martingales*. Cambridge: Cambridge University Press.

Wong, C. S. (1977). Classroom Notes: A Note on the Central Limit Theorem. *Amer. Math. Monthly 84*(6), 472.

Woodroofe, M. (1975). *Probability with Applications*. New York: McGraw-Hill Book Co.

Young, L. C. (1970). Some new stochastic integrals and Stieltjes integrals. I. Analogues of Hardy-Littlewood classes. In *Advances in Probability and Related Topics, Vol. 2*, pp. 161–240. New York: Dekker.

Zabell, S. L. (1995). Alan Turing and the central limit theorem. *Amer. Math. Monthly 102*(6), 483–494.

Index

Symbols

$\mathscr{B}(\Omega)$, Borel sigma-algebra on Ω 24
$\mathrm{Bin}(n,p)$, binomial distribution 8
$C(\mathbf{X})$, continuous functions from $\mathbf{X}$ to $\mathbf{R}$ 49
$\mathbf{C}$, the complex numbers xv
$C_b(\mathbf{X})$, bounded continuous functions from $\mathbf{X}$ to $\mathbf{R}$ 94
$C_c(\mathbf{X})$, compactly supported continuous functions from $\mathbf{X}$ to $\mathbf{R}$ 94
$C_c^\infty(\mathbf{R}^k)$, infinitely differentiable functions of compact support from $\mathbf{R}^k$ to $\mathbf{R}$ 111
$\mathrm{Cov}(X,Y)$, covariance between X and Y 67
$\mathrm{E}X$, expectation of X 39
$\mathrm{E}[X;A] = \mathrm{E}[X\mathbf{1}_A] = \int_A X\,d\mathrm{P}$ 39
$\mathrm{Geom}(p)$, geometric distribution 8
L^p, random variables with p finite absolute moments 39
$\mathscr{L}^p$, completion of L^p 43
$N(\mu, \sigma^2)$, normal distribution 11
$\mathbf{N}$, the natural numbers xv
$\mathscr{P}(\Omega)$, the power set of Ω 24
$\mathrm{Poiss}(\lambda)$, Poisson distribution 9
$\mathbf{Q}$, the rationals xv
$\mathbf{R}$, the real numbers xv
$\mathbf{S}^{n-1}$, the unit sphere in $\mathbf{R}^n$ 102
$\mathrm{SD}(X)$, standard deviation of X 67
$\mathrm{Unif}(a,b)$, uniform distribution 11
$\mathrm{Var}\,X$, variance of X 67
$\mathbf{X}_+$, the positive elements of $\mathbf{X}$ xv
$\mathbf{Z}$, the integers xv
a.e., almost everywhere 43
a.s., almost surely 43
$d\nu/d\mu$, Radon–Nikodým derivative 47
$f^+ = \max(f,0)$ 38
$f^- = \max(-f,0)$ 38
i.i.d., independent identically distributed 68
$\mathrm{mc}(\mathscr{A})$, monotone class generated by $\mathscr{A}$ 30
$\|f\|_{L^p(\mu)} = (\int |f|^p\,d\mu)^{1/p}$ 39
$\|X\|_p = (\mathrm{E}[|X|^p])^{1/p}$ 39
$\nu \ll \mu$, absolute continuity 47
$\sigma(X), \sigma(\mathscr{A}),\ldots$ sigma-algebra generated by X, $\mathscr{A}$, etc. **24**, 49, 120
$\wedge, \vee$, the min and max operators xvi

A

Absolute continuity .. 11, *see also* Measure
Adams, William J. 22
Adapted process 126, 182
Aldous, David J. 157
Almost everywhere convergence 43
Almost sure
  central limit theorem 90
  convergence 43
Alon, Noga 89
André, Desiré 179, 180
Approximation to the identity 205
Avogadro, Lorenzo Romano Amedeo Carlo 159
Azuma, Kazuoki 157
Azuma–Hoeffding inequality ..155, *see also* Hoeffding's inequality

B

Bachelier, Louis Jean Baptiste Alphonse 159, 177, 180
Backward [or reversed] martingale 155
Banach, Stefan 157
Bass, Richard Franklin 201
Baxter, Martin 157
Berkés, Istvan 90

217

Bernoulli trials 8
Bernoulli, Jacob 18, 22, 71
Bernstein
   polynomials 77
   proof of the Weierstrass theorem 77
Bernstein [Bernshtein], Sergei Natanovich 77, 117
Bessel's inequality 208
Bessel, Friedrich Wilhelm 208
Bingham, Nicholas H. 157
Binomial distribution .... 8, *see also* Distribution
Birkhoff, George David 90
Black, Fischer 144, 145
Black–Scholes Formula 145
Blumenthal's zero-one law 177
Blumenthal, Robert M. 177, 180
Borel set and/or sigma-algebra 24
Borel, Émile .. 52, 73, 86, 89, 109, 117, 137
Borel–Cantelli lemma ..... 63, **73**, *see also* Paley–Zygmund inequality
   Dubins–Freedman 156
   Lévy's 136
Borel–Carathéodory lemma 109
Bounded convergence theorem 45
   conditional 121
Bourke, Chris 90
Bovier Anton 113
Bretagnolle, Jean 158
Broadbent, S. R. 83
Brown, Robert 159
Brownian bridge 177
Brownian motion
   and the heat equation 177, 178, 196, 198
   as a Gaussian process 163
   Einstein's predicates 160
   exit distribution .. 196, 197, 200, *see also* Chung's formula
   filtration .... 171, *see also* Filtration
   gambler's ruin formula 200
   nowhere differentiability of 168
   quadratic variation 163
   Wiener's construction 166–168
   with drift 201
Bru, Bernard 180
Buczolich, Zoltán 90
Bunyakovsky, Viktor Yakovlevich 40
Burkholder, Donald L. 52, 156

C

Call options 144
Cantelli, Francesco Paolo 73, 80, 85, 89, 138
Cantor set 88
Cantor, Georg 34
Cantor–Lebesgue function 88
Carathéodory Extension Theorem 27

Carathéodory, Constantine 27, 109
Cauchy
   sequence 46
   summability test 183
Cauchy, Augustin Louis .. 40, 108, 113, 183
Cauchy–Bunyakovsky–Schwarz inequality 40
Central limit theorem . 19, 22, 89, **100**, 102
   Bovier-Picco 113
   de Moivre–Laplace 19
   Liapounov's 114
   Lindeberg's 115
   projective 102
   via Liapounov's method 105
   Ville's 116
   with error estimates 105, 116
Champernowne, David Gawen 90
Characteristic function **96**, 96–117
   convergence theorem **99**, 102
   inversion theorem 112
   uniqueness theorem 99
Chatterji, Srishti Dhav 157
Chebyshev's inequality 18, **43**
   conditional 152
   for sums 63
Chebyshev, Pafnutii Lvovich 18, 43, 63
Chernoff's inequality 51
Chernoff, Herman 51, 52, 87
Chung's formula **197**, 200
Chung, Kai Lai .... xii, 89, 197
Ciesielski, Zbigniew 201
Cifarelli, Donato Michele 158
Coifman, Ronald R. 52
Compact support [or compactly supported]
   function 94
Compact-support
   process 182
Complete
   measure space 28
   topological space 42
Completion 29
Conditional expectation 120–125
   and prediction 122
   classical 124
   properties 120–121, 123
   towering property 123
Conditional probability 4, 125
Consistent measures 59
Convergence
   almost everywhere 43
   almost sure 43
   in L^p 43
   in measure 43
   in probability 43
   weak 91
Convergence theorem .... 102, *see also* Characteristic function
Convex function **40**, 50, 156

Convolution 15, **98**, 113
Cootner, Paul H. 159
Copeland, Arthur H. 90
Correlation 67
Countable (sub-) additivity 24
Courtault, Jean-Michel 180
Covariance 67
    matrix 161
Cover, Thomas M. 89
Cox, John C. 144
Crépel, Pierre 180
Cramér's theorem 107
Cramér, Harald 102, 107, 117
Cramér–Wold device 102
Csörgő, Miklós 157
Cumulative distribution function ... 66, *see also* Distribution function
Cylinder set 58

D

Dacunha-Castelle, Didier 158
Davis, Burgess J. 52
de Acosta, Alejandro 156, 158
de Finetti's theorem 156
de Finetti, Bruno 156–158
de Moivre, Abraham 15, 19, 22
de Moivre–Laplace central limit theorem 19, *see also* Central limit theorem
de Moivre's formula . 19, *see also* Stirling's formula
Degenerate normal 11, *see also* Distribution
Dellacherie, Claude 201
Density function 10, 14, 48
Devaney, Robert L. 90
Diaconis, Persi 22, 117
Dimension 59
Dini, Ulisse 182
Dini-continuous process
    in $L^2(\mathrm{P})$ 182
Distribution 49, **65**
    binomial 8
        characteristic function 97
        connection to Poisson 10, 19
        mean and variance 12
    Cauchy 14, 113
        non-existence of the mean 14
    discrete 7
    discrete uniform 15
    exponential 13
        characteristic function 97
        connection to uniform 15
        mean and variance 13
    function 13, 33, **66**
    gamma 13
        characteristic function 112
        mean and variance 13
    geometric 8

        mean and variance 12
    hypergeometric 13
        mean and variance 13
    infinitely divisible 113
    negative binomial 13
        mean and variance 13
    normal 11, 28
        characteristic function 97, 161
        degenerate 11
        mean and variance 12
        multi-dimensional 28, 161
        standard 11
    Poisson 9, 10
        characteristic function 97
        connection to binomial 10, 19
        mean and variance 13
    uniform 11, 15, 28
        characteristic function 97
        connection to discrete uniform 15
        connection to exponential 15
        mean and variance 12
Dominated convergence theorem 46
    conditional 121
Donoho, David L. 115
Doob's
    decomposition **128**, 152
    martingale convergence theorem 134
    martingales 127, 200
    maximal inequality 134
        continuous-time 189
    optional stopping theorem 130
    strong (p,p)-inequality
        continuous-time 189
    strong L^1-inequality 156
    strong L^p-inequality 153
        Pitman's improvement 153
Doob, Joseph Leo .. xii, 127, 130, 134, 152, 153, 156, 157, 189, 200
Dubins, Lester E. 154, 158
Dudley, Richard Mansfield 22, 51
Durrett, Richard 90, 158
Dyadic filtration **142**, 148
Dyadic interval **142**, 148
Dynkin, Eugene B. 180

E

Einstein, Albert 159
Elementary function 37
Entropy 79
Erdős, Paul [Pál] 81, 88–90
Etemadi, Nasrollah 89
European options 144
Event .... 35, *see also* Measurable set
Exchangeable 155, 156
Expectation 12, **39**

F

Falconer, Kenneth K. 34
Fatou's lemma **45**, 50
    conditional 121
Fatou, Pierre Joseph Louis 45, 52
Fejér, Léopold 117
Feller, William K. 16, 117, 157, 201
Fermi, Enrico 84
Filtration 126
    Brownian 171
    right-continuous 171
Fischer, Ernst Sigismund 168
FKG inequality 63
Fortuin, C. M. 64
Fourier series 205–208
Fourier transform .... 96, see also
    Characteristic function
Fourier, Jean Baptiste Joseph 96
Fréchet, Maurice René 52
Freedman, David A. 117, 154, 158
Freiling, Chris 23
Fubini, Guido 55
Fubini–Tonelli theorem 55
    inapplicability of 56–58, 62, 63

G

Gambler's ruin formula .... 133, *see also* Random walk, *see also* Brownian motion
Garsia, Adriano M. 90
Gaussian process 163
Geometric distribution .... 8, *see also* Distribution
Georgii, Hans-Otto 158
Gerber, Hans U. 157
Ginibre, Jean 64
Glivenko, Valerii Ivanovich 80, 117
Glivenko–Cantelli theorem 80
Gnedenko, Boris Vladimirovich 52, 117
Grimmett, Geoffrey R. 83
Gundy, Richard F. 52

H

Hadamard's Inequality 51
Hadamard, Jacques Salomon 51
Hamedani, Gholamhossein Gharagoz .. 117
Hammersley, John M. 83, 89
Hardy, Godfrey Harold . 138, 140, 143, 187
Hardy–Littlewood maximal function .. 143, 187
Harris, Theodore E. 83
Harrison, J. Michael 144
Hartman, Philip 138
Hausdorff measure 34
Hausdorff, Felix 34, 138
Helms, Lester L. 157
Hilbert space 203

Hilbert, David 52
Hitchcock, John M. 90
Hoeffding's inequality ..... 51, **87**, *see also* Azuma–Hoeffding
Hoeffding, Wassily 51, 52, 87, 89, 157
Hölder
    continuous function 78, 199
    inequality 39
        conditional 121
        generalized 51
Hölder, Otto Ludwig 39, 51, 199
Houdré, Christian 117
Hunt, Gilbert A. 130, 171, 180

I

Independence 14, 62, **68**
Indicator function 36
Infinitely divisible 113
Information inequality 87
Inner product 203
Integrable function 39
Integral 39
Inverse image xvi
Inversion theorem .... 112, *see also* Characteristic function
Ionescu Tulcea, Alexandra 157
Ionescu Tulcea, Cassius 157
Isaac, Richard 157
Itô
    formula 194, 195
    integral 181–185
        indefinite 185
        under Dini-continuity 184
    isometry 184
    lemma .... *see also* Itô formula
Itô, Kiyosi 181, 184, 185, 194, 195

J

Jensen's inequality 40
    conditional 121
Jensen, Johann Ludwig Wilhelm Waldemar 40
Jones, Roger L. 52
Jushkevich [Yushkevich], Alexander A. 180

K

Kabanov, Yuri 180
Kac, Mark 15, 78, 115–117
Kahane, Jean-Pierre 157
Kasteleyn, Pieter Willem 64
Keller, Joseph B. 22
Kesten, Harry 83
Khintchine [Khinchin], Aleksandr Yakovlevich ...... xii, 71, 72, 88, 89, 138, 153, 177
Khintchine's
    inequality 88

Index

weak law of large numbers .. 72, *see also* Law of large numbers
Kinney, John R. 180
Knight, Frank B. 178, 201
Knuth, Donald E. 84
Kochen, Simon 89
Kolmogorov
    consistency theorem .... 60, *see also* Kolmogorov extension theorem
    extension theorem 60
    maximal inequality 74
    one-series theorem 85
    strong law of large numbers . 73, *see also* Law of large numbers
    zero-one law **69**, 136
Kolmogorov, Andrei Nikolaevich xii, 52, 60, 69, 73, 74, 85, 89, 103, 117, 138, 153
Kreps, David M. 144
Krickeberg's decomposition 128
Krickeberg, Klaus 128
Kyburg, Henry E., Jr. 158

L

Lacey, Michael T. 90
Lamb, Charles W. 157
Laplace, Pierre-Simon . 16, 19, 21, 157, 206
Law of large numbers 71–88
    and Monte Carlo simulation 83
    and Shannon's theorem 79
    and the Glivenko–Cantelli theorem ... 80
    Erdős–Rényi 88
    strong **73**, 85
    weak 72
Law of rare events 19
Law of the iterated logarithm 138, 154, 156
    for Brownian motion 177
Law of total probability 5
Lebesgue
    differentiation theorem **140**, 155
    measurable ..... 30, *see also* Measurable
    measure .... 25, *see also* Measure
Lebesgue, Henri Léon 52, 113, 140
Lebon, Isabelle 180
Leonard, Bill 86
Levi's monotone convergence theorem .. 46
Levi, Beppo 52
Lévy's
    Borel–Cantelli lemma 136
    concentration inequality 112
    equivalence theorem 155
    forgery theorem 178
Lévy, Paul .. 90, 92, 99, 112, 117, 136, 155, 160, 178, 196, 200
Li, Robert Shuo-Yen 149, 157
Liapounov [Lyapunov], Aleksandr Mikhailovich 104, 114
Lindeberg, Jarl Waldemar 104, 115

Lindvall, Torgny 89
Liouville, Joseph 11, 16, 108
Lipschitz continuous 147, 183
Lipschitz, Rudolf Otto Sigismund 157
Littlewood, John Edensor ... 138, 140, 143, 187
Loeb, Peter A. 157

M

Mahmoud, Hosam M. 157
Le Marchand, Arnaud 180
Markov
    inequality 43
    property ... 174, *see also* Strong Markov
    of Brownian motion **160**, 163
Markov, Andrei Andreyevich 43, 89
Martingale 126
    and likelihood ratios 152
    continuous-time 187
    convergence theorem 134, *see also* Doob's
    representations 145
    reversed 155
    transforms 127
Mass function **7**, 14
Mattner, Lutz 64
Mauldin, R. Daniel 90
Maximal inequality 74–76, 87, 134, 189
Maxwell, James Clerk 117
Mazurkiewicz, Stefan 157
McShane, Edward James 199
Mean .... 39, *see also* Expectation
Measurable
    function 35
    Lebesgue 30
    set 24
    space .... 25, *see also* Measure space
Measure 24
    absolutely continuous 47
    counting 33
    finite product 54
    Lebesgue 25
        invariance properties 33
    probability 25
    space 25
    support of .... 33, *see also* Support
Merton, Robert C. 144
Meyer, Paul-André 201
Minkowski's inequality 40
    conditional 121
Minkowski, Hermann 40
Mixture 51
Modulus of continuity 77, 182
Monotone class 30
    theorem 30
Monotone convergence theorem 46, *see also* Levi
    conditional 121

Monte Carlo
  integration 84
  simulation ..... 83, see also Law of large numbers, 84
Mukherjea, Arunava 64

N

Nash's Poincaré inequality 116
Nash, John 116, 117
Negative part of a function 38
Nelson, Edward 180
Newton's method 87
Newton, Isaac 87
Nikodým, Otton Marcin 47
Normal distribution .... 11, see also Distribution
Normal numbers 86, 90
Norris, James R. 90

O

Okamoto, Masashi 87
Optional stopping
  continuous-time 187
Optional stopping theorem 130
Options 144
Orthogonal 203
Orthogonal decomposition theorem 203

P

Paley, Raymond Edward Alan Christopher 86, 157, 168
Paley–Zygmund inequality 86
Parseval des Chênes, Marc-Antoine 117
Parseval's identity 117
Percolation 82
Pérez-Abreu, Victor 117
Perrin, Jean Baptiste 180
Philipp, Walter 90
Picco, Pierre 113
Piecewise continuous function 11
Pitman's L^2 inequality .... 153, see also Doob's strong L^p inequality
Pitman, James W. 153
Plancherel theorem 99, 115
Plancherel, Michel 34, 97, 115
Pliska, Stanley R. 144
Poincaré inequality 116
Poincaré, Jules Henri 16, 22, 159
Point-mass 25
Poisson distribution 9, see also Distribution
Poisson, Siméon Denis 9, 19
Poissonization 10
Pólya's urns 152
Pólya, George 34, 117, 152
Positive part of a function 38
Positive-type function 113
Power set 24

Previsible process 127
Probability space ..... 25, see also Measure space
Product
  measure .... 54, see also Measure
  topology 58

Q

Quadratic variation 163, see also Brownian motion

R

Rademacher's theorem 147
Rademacher, Hans 89, 147
Radon, Johann 47
Radon–Nikodým theorem 47
Raikov's ergodic theorem 88, 90
Raikov, Dmitrii Abramovich 88, 90
Ramsey number 81
Ramsey, Frank Plumpton 81
Random
  permutation ..3, 9, 22, 116, 155, see also Central limit theorem, Ville's
  set 64
  variable 35
    absolutely continuous 10, 14
    discrete 7, 14
    vector 14
Random walk 131
  gambler's ruin formula 133, 152
  nearest neighborhood 132
  simple 132
Reflection principle 175, 179
Regazzini, Eugenio 158
Rennie, Andrew 157
Rényi, Alfred 88, 89
Reversed martingale 155
Révész, Pál 157
Revuz, Daniel 201
Riemann, Georg Friedrich Bernhard ... 113
Riemann–Lebesgue lemma 113
Riesz reresentation theorem 49
Riesz, Frigyes 49, 168
Rosenlicht, Maxwell 16
Ross, Stephen A. 144
Rubenstein, Mark 144

S

Schnorr, Claus-Peter 90
Scholes, Myron 144, 145
Schrödinger, Erwin 86
Schwarz's lemma 108
Schwarz, Hermann Amandus 40, 108
Semimartingale 126
Shannon's theorem 79
Shannon, Claude Elwood 78, 79
Shultz, Harris S. 86

Sierpiński, Wacław 57, 64
Sigma-algebra 23
  Borel 24
  generated by a random variable . 49, **120**
Simple function 37
Simple walk ...114, see also Random walk, 132
Simpson, Thomas 22
Simulation ....83, see also Law of large numbers
Skolem, Thoralf Albert 81
Skorohod [Skorokhod], Anatoli Vladimirovich 116, 178
Skorohod embedding 178
Skorohod's theorem 116
Slutsky, Evgeny 50, 52
Slutsky's theorem 50
Smokler, Howard E. 158
Solovay, Robert M. 23, 34
Spencer, Joel 89
Stable distribution 178
Standard deviation 12, **67**
Standard normal . 11, see also Distribution
Stark, Philip B. 115
Steinhaus probability space 44
Steinhaus, Hugo 44, 89, 138, 154
Stigler, Steven Mack 89
Stimm, H. 90
Stirling's formula ..21, 22, 82, 156, see also de Moivre's formula
Stirling, James 21
Stochastic integral 179, see also Itô integral
Stochastic process 126
  continuous-time 181
Stone, Charles J. 89
Stopping time **129**, **170**
  simple 170
Strassen, Volker 157
Strong law of large numbers ...73, see also Law of large numbers
  Cantelli's 85
  for dependent variables 85
  for exchangeable variables 155
Strong Markov property 174
Stroock, Daniel W. 117
Submartingale 126
  continuous-time 200
Supermartingale 126
  continuous-time 200
Support 33
Surgailis, Donatas 117
Szekeres, Gábor [György] 81

T
Tandori, Károly 117
Taylor, Samuel James 201
Thomas, Joy A. 89

Tonelli, Leonida 55
towering property of conditional expectations 123
Triangle inequality 38, 41
Trigonometric polynomial 205
Trotter, Hale Freeman 117
Turing, Alan Mathison 117
Turner, James A. 152

U
U-statistics 151, 155
Ulam, Stanisław 84
Uncertainty principle 51, 52, 115
Uncorrelated random variables 67
Uniform distribution ....11, see also Distribution
  on S^{n-1} 103
Uniform integrability **51**, 154
  and weak convergence 115
Uniqueness theorem ....99, see also Characteristic function

V
Varberg, Dale E. 151
Variance 12, **67**
  conditional 152
Veech, William A. 117
Viéte, François 117
Ville, Jean 117, 157
Vinodchandran, N. V. 90
von Neumann, John 48, 84
von Smoluchowski, Marian 159

W
Wagon, Stanley 90
Wald's identity 131, 153
Wald, Abraham 131, 153
Wallis, John 114
Walter, Gilbert G. 117
Weak convergence 91
Weak law of large numbers ....72, see also Law of large numbers
Weaver, Warren 78
Weierstrass approximation theorem 77, see also Bernstein
  Hoeffding's refinement 89
  Kac's refinement 78
Weierstrass, Karl Theodor Wilhelm 77
Weyl, Hermann 52
White noise 178, 179
Wiener process ....180, see also Brownian motion
Wiener, Norbert ...157, 159, 166, 168, 179
Williams, David 157
Wintner, Aurel 138
Wold, Herman O. A. 102
Wong, Chi Song 22

Woodroofe, Michael 154

Y

Yor, Marc 201
Young integral 199, 200
Young's inequality 51
Young, Laurence Chisholm 199
Young, William Henry 51

Z

Zabell, Sandy L. 117
Zeitouni, Ofer 117
Zero-one law
    Blumenthal's 177
    Kolmogorov's **69**, 136
Zygmund, Antoni 86, 157, 168